Big Data Analytics Using Splunk

Peter Zadrozny

Raghu Kodali

Apress·

Big Data Analytics Using Splunk

ISBN-13 (pbk): 978-1-4302-5761-5

ISBN-13 (electronic): 978-1-4302-5762-2

Trademarked names, logos, and images may appear in this book. Rather than use a trademark symbol with every occurrence of a trademarked name, logo, or image we use the names, logos, and images only in an editorial fashion and to the benefit of the trademark owner, with no intention of infringement of the trademark.

The use in this publication of trade names, trademarks, service marks, and similar terms, even if they are not identified as such, is not to be taken as an expression of opinion as to whether or not they are subject to proprietary rights.

While the advice and information in this book are believed to be true and accurate at the date of publication, neither the authors nor the editors nor the publisher can accept any legal responsibility for any errors or omissions that may be made. The publisher makes no warranty, express or implied, with respect to the material contained herein.

President and Publisher: Paul Manning
Lead Editors: Jonathan Gennick, Tom Welsh
Technical Reviewer: Paul Stout
Editorial Board: Steve Anglin, Mark Beckner, Ewan Buckingham, Gary Cornell, Louise Corrigan, Morgan Ertel, Jonathan Gennick, Jonathan Hassell, Robert Hutchinson, Michelle Lowman, James Markham, Matthew Moodie, Jeff Olson, Jeffrey Pepper, Douglas Pundick, Ben Renow-Clarke, Dominic Shakeshaft, Gwenan Spearing, Matt Wade, Tom Welsh
Coordinating Editor: Katie Sullivan
Copy Editor: Laura Lawrie
Compositor: SPi Global
Indexer: SPi Global
Artist: SPi Global
Cover Designer: Anna Ishchenko

Distributed to the book trade worldwide by Springer Science+Business Media New York, 233 Spring Street, 6th Floor, New York, NY 10013. Phone 1-800-SPRINGER, fax (201) 348-4505, e-mail orders-ny@springer-sbm.com, or visit www.springeronline.com. Apress Media, LLC is a California LLC and the sole member (owner) is Springer Science + Business Media Finance Inc (SSBM Finance Inc). SSBM Finance Inc is a Delaware corporation.

For information on translations, please e-mail rights@apress.com, or visit www.apress.com.

Apress and friends of ED books may be purchased in bulk for academic, corporate, or promotional use. eBook versions and licenses are also available for most titles. For more information, reference our Special Bulk Sales–eBook Licensing web page at www.apress.com/bulk-sales.

Any source code or other supplementary materials referenced by the author in this text is available to readers at www.apress.com/9781430257615. For detailed information about how to locate your book's source code, go to www.apress.com/source-code/.

*To my wife Graciela and my children Johannes and Gracie for their infinite patience
and incredible support*

—Peter Zadrozny

*To my wife Lakshmi, our kids Yash and Nikhil, and my parents Laila and Chandra Sekhara Rao
for their perpetual support and unconditional love*

—Raghu Kodali

Contents at a Glance

Contents

About the Authors

**Peter Zadrozny** is a veteran of the software industry. He is the Founder and Chief Technology Officer of Opallios, a company focused on harnessing the power of big data and cloud technologies while providing value to its customers through outsourced product development. Peter is also a lecturer on big data topics at San José State University. He has held various executive and technical roles in many countries around the world with startups as well as Fortune 500 companies. Highlights of his career include starting the operations of WebLogic in Europe and Sun Microsystems in México. He is an accomplished author with various successful books on Java Enterprise Edition and performance engineering, and was an original contributor to The Grinder, a popular open source project.

**Raghu Kodali** is vice president of Product Management & Strategy at Solix Technologies, where he is responsible for product vision, management, strategy, user design, and interaction. His work includes the next-generation data optimization platform, industry-leading data discovery platform, enterprise data management-as-a-service, application development using Big Data platforms and cloud. Before Solix he was with Oracle for 12 years, holding senior management positions responsible for Product Management & Strategy for Oracle Fusion Middleware products. In addition, Raghu was Oracle's SOA Evangelist leading next-generation Java EE initiatives. Raghu authored a leading technical reference on Java computing, "Beginning EJB 3 Application Development: From Novice to Professional" (Apress, 2006), and has published numerous articles on enterprise technologies. He also was a contributing author to "Oracle Information Integration, Migration and Consolidation" (PACKT Publishing, 2011).

About the Technical Reviewer

Paul Stout manages the use of Splunk for operational intelligence and IT security for Splunk Inc. Paul deployed the first enterprise-scale Splunk instance at the company and frequently speaks on how Splunk gains operational and security intelligence from its own big data set. Before his role at Splunk, Paul held various positions in IT management, operations management, online marketing, and web application development.

Acknowledgments

We start by thanking the engineering team of Splunk for building a superb product and continuously improving it. In no particular order, we thank Stephen Sorkin, for all his guidance, feedback, and comments throughout the book; Siegfried Puchbauer, for his invaluable help with everything DB Connect; Alice Cotti, for her help massaging the Foursquare data; Sunny Choi, for her support and all the performance information; Dhivya Srinivasan, for getting the globe app to work correctly; David Carasso, for his support with the sentiment analysis app; and David Foster, for his help with the Twitter app. We also acknowledge Rob Das, for creating the original content of Chapter 7, "Using Log Files to Create Advanced Analytics"; Omcar Paradkar, for being the patient sounding board and shooting down the so-so and bad ideas; and, last but not least, Rob Reed, for his encouragement and editing skills.

Special thanks to GoGrid (www.gogrid.com) for providing free cloud services to Peter's course, "Introduction to Big Data Analytics," at San José State University, which we also used for the development of all the projects in this book. We also thank Peter's Fall 2012 and Spring 2013 semester students for being subject to the experimentation of lectures that ended up defining the content of this book.

---0---

It would be unfair to leave out my family, all of whom had to bear the brunt of my absence; although physically present, they only saw the back of my head. I thank them for their support and understanding.

—Peter Zadrozny

I have to say that I have been very lucky to get rock-solid support from several incredible people who have made it possible to write this book. Thanks to Peter who has constantly provided guidance on the storyboard, helping me to focus on the right things.

Thanks to Paul Stout for his comments during technical review and making sure that the chapters describe the Splunk features as accurately as possible.

Thanks to Tom Welsh for his honest feedback during the editorial process.

Thanks to my parents Laila and Chandra Sekhara Rao Kodali, who are my biggest fans, for their infinite patience. They always believed in me, and they made a trip to California all the way from India to provide encouragement. They took care of the boys while I was hiding away to write this book.

Thanks to my wife Lakshmi, who encouraged me to jump into this project and supported me; to my eight-year-old son Yash, who had a keen interest in following this project to see how I was doing and kept checking to see if he could be of some help; and to my five-year-old son Nikhil, who asked me what big data was and constantly reminded me that I needed to finish writing so that he could see his name in the book.

—Raghu Kodali

CHAPTER 1

■ ■ ■

Big Data and Splunk

In this introductory chapter we will discuss what big data is and different ways (including Splunk) to process big data.

What Is Big Data?

Big data is, admittedly, an overhyped buzzword used by software and hardware companies to boost their sales. Behind the hype, however, there is a real and extremely important technology trend with impressive business potential. Although big data is often associated with social media, we will show that it is about much more than that. Before we venture into definitions, however, let's have a look at some facts about big data.

Back in 2001, Doug Laney from Meta Group (an IT research company acquired by Gartner in 2005) wrote a research paper in which he stated that e-commerce had exploded data management along three dimensions: volumes, velocity, and variety. These are called the three Vs of big data and, as you would expect, a number of vendors have added more Vs to their own definitions.

Volume is the first thought that comes with big data: the *big* part. Some experts consider Petabytes the starting point of big data. As we generate more and more data, we are sure this starting point will keep growing. However, volume in itself is not a perfect criterion of big data, as we feel that the other two Vs have a more direct impact.

Velocity refers to the speed at which the data is being generated or the frequency with which it is delivered. Think of the stream of data coming from the sensors in the highways in the Los Angeles area, or the video cameras in some airports that scan and process faces in a crowd. There is also the click stream data of popular e-commerce web sites.

Variety is about all the different data and file types that are available. Just think about the music files in the iTunes store (about 28 million songs and over 30 billion downloads), or the movies in Netflix (over 75,000), the articles in the New York Times web site (more than 13 million starting in 1851), tweets (over 500 million every day), foursquare check-ins with geolocation data (over five million every day), and then you have all the different log files produced by any system that has a computer embedded. When you combine these three Vs, you will start to get a more complete picture of what big data is all about.

Another characteristic usually associated with big data is that the data is unstructured. We are of the opinion that there is no such thing as unstructured data. We think the confusion stems from a common belief that if data cannot conform to a predefined format, model, or schema, then it is considered unstructured.

An e-mail message is typically used as an example of unstructured data; whereas the body of the e-mail could be considered unstructured, it is part of a well-defined structure that follows the specifications of RFC-2822, and contains a set of fields that include *From, To, Subject,* and *Date.* This is the same for Twitter messages, in which the body of the message, or tweet, can be considered unstructured as well as part of a well-defined structure.

In general, free text can be considered unstructured, because, as we mentioned earlier, it does not necessarily conform to a predefined model. Depending on what is to be done with the text, there are many techniques to process it, most of which do not require predefined formats.

Relational databases impose the need for predefined data models with clearly defined fields that live in tables, which can have relations between them. We call this Early Structure Binding, in which you have to know in advance what questions are to be asked of the data, so that you can design the schema or structure and then work with the data to answer them.

As big data tends to be associated with social media feeds that are seen as text-heavy, it is easy to understand why people associate the term *unstructured* with big data. From our perspective, multistructured is probably a more accurate description, as big data can contain a variety of formats (the third V of the three Vs).

It would be unfair to insist that big data is limited to so-called unstructured data. Structured data can also be considered big data, especially the data that languishes in secondary storage hoping to make it some day to the data warehouse to be analyzed and expose all the golden nuggets it contains. The main reason this kind of data is usually ignored is because of its sheer volume, which typically exceeds the capacity of data warehouses based on relational databases.

At this point, we can introduce the definition that Gartner, an Information Technology (IT) consultancy, proposed in 2012: "Big data are high volume, high velocity, and/or high variety information assets that require new forms of processing to enable enhanced decision making, insight discovery and processes optimization." We like this definition, because it focuses not only on the actual data but also on the way that big data is processed. Later in this book, we will get into more detail on this.

We also like to categorize big data, as we feel that this enhances understanding. From our perspective, big data can be broken down into two broad categories: human-generated digital footprints and machine data. As our interactions on the Internet keep growing, our digital footprint keeps increasing. Even though we interact on a daily basis with digital systems, most people do not realize how much information even trivial clicks or interactions leave behind. We must confess that before we started to read Internet statistics, the only large numbers we were familiar with were the McDonald's slogan "Billions and Billions Served" and the occasional exposure to U.S. politicians talking about budgets or deficits in the order of trillions. Just to give you an idea, we present a few Internet statistics that show the size of our digital footprint. We are well aware that they are obsolete as we write them, but here they are anyway:

- By February 2013, Facebook had more than one billion users, of which 618 million were active on a daily basis. They shared 2.5 billion items and "liked" other 2.7 billion every day, generating more than 500 terabytes of new data on a daily basis.

- In March 2013, LinkedIn, which is a business-oriented social networking site, had more than 200 million members, growing at the rate of two new members every second, which generated 5.7 billion professionally oriented searches in 2012.

- Photos are a hot subject, as most people have a mobile phone that includes a camera. The numbers are mind-boggling. Instagram users upload 40 million photos a day, like 8,500 of them every second, and create about 1,000 comments per second. On Facebook, photos are uploaded at the rate of 300 million per day, which is about seven petabytes worth of data a month. By January 2013, Facebook was storing 240 billion photos.

- Twitter has 500 million users, growing at the rate of 150,000 every day, with over 200 million of the users being active. In October 2012, they had 500 million tweets a day.

- Foursquare celebrated three billion check-ins in January 2013, with about five million check-ins a day from over 25 million users that had created 30 million tips.

- On the blog front, WordPress, a popular blogging platform reported in March 2013 almost 40 million new posts and 42 million comments per month, with more than 388 million people viewing more than 3.6 billion pages per month. Tumblr, another popular blogging platform, also reported, in March 2013, a total of almost 100 million blogs that contain more than 44 billion posts. A typical day at Tumblr at the time had 74 million blog posts.

- Pandora, a personalized Internet radio, reported that in 2012 their users listened to 13 billion hours of music, that is, about 13,700 years worth of music.

- In similar fashion, Netflix announced their users had viewed one billion hours of videos in July 2012, which translated to about 30 percent of the Internet traffic in the United States. As if that is not enough, in March 2013, YouTube reported more than four billion hours watched per month and 72 hours of video uploaded every minute.

- In March 2013, there were almost 145 million Internet domains, of which about 108 million used the famous ".com" top level domain. This is a very active space; on March 21, there were 167,698 new and 128,866 deleted domains, for a net growth of 38,832 new domains.

- In the more mundane e-mail world, Bob Al-Greene at Mashable reported in November 2012 that there are over 144 billion e-mail messages sent every day, with about 61 percent of them from businesses. The lead e-mail provider is Gmail, with 425 million active users.

Reviewing these statistics, there is no doubt that the human-generated digital footprint is huge. You can quickly identify the three Vs; to give you an idea of how big data can have an impact on the economy, we share the announcement Yelp, a user-based review site, made in January 2013, when they had 100 million unique visitors and over one million reviews: "A survey of business owners on Yelp reported that, on average, customers across all categories surveyed spend $101.59 in their first visit. That's everything from hiring a roofer to buying a new mattress and even your morning cup of joe. If each of those 100 million unique visitors spent $100 at a local business in January, Yelp would have influenced over $10 billion in local commerce."

We will not bore you by sharing statistics based on every minute or every second of the day in the life of the Internet. However, a couple of examples of big data in action that you might relate with can consolidate the notion; the recommendations you get when you are visiting the Amazon web site or considering a movie in Netflix, are based on big data analytics the same way that Walmart uses it to identify customer preferences on a regional basis and stock their stores accordingly. By now you must have a pretty good idea of the amount of data our digital footprint creates and the impact that it has in the economy and society in general. Social media is just one component of big data.

The second category of big data is machine data. There is a very large number of firewalls, load balancers, routers, switches, and computers that support our digital footprint. All of these systems generate log files, ranging from security and audit log files to web site log files that describe what a visitor has done, including the infamous abandoned shopping carts.

It is almost impossible to find out how many servers are needed to support our digital footprint, as all companies are extremely secretive on the subject. Many experts have tried to calculate this number for the most visible companies, such as Google, Facebook, and Amazon, based on power usage, which (according to a Power Usage Effectiveness indicator that some of these companies are willing to share) can provide some insight as to the number of servers they have in their data centers. Based on this, James Hamilton in a blog post of August 2012 published server estimates conjecturing that Facebook had 180,900 servers and Google had over one million servers. Other experts state that Amazon had about 500 million servers in March 2012. In September 2012, the *New York Times* ran a provocative article that claimed that there are tens of thousands of data centers in the United States, which consume roughly 2 percent of all electricity used in the country, of which 90 percent or more goes to waste, as the servers are not really being used.

We can only guess that the number of active servers around the world is in the millions. When you add to this all the other typical data center infrastructure components, such as firewalls, load balancers, routers, switches, and many others, which also generate log files, you can see that there is a lot of machine data generated in the form of log files by the infrastructure that supports our digital footprint.

What is interesting is that not long ago most of these log files that contain machine data were largely ignored. These log files are a gold mine of useful data, as they contain important insights for IT and the business because they are a definitive record of customer activity and behavior as well as product and service usage. This gives companies end-to-end transaction visibility, which can be used to improve customer service and ensure system security, and also helps to meet compliance mandates. What's more, the log files help you find problems that have occurred and can assist you in predicting when similar problems can happen in the future.

In addition to the machine data that we have described so far, there are also sensors that capture data on a real-time basis. Most industrial equipment has built-in sensors that produce a large amount of data. For example, a blade in a gas turbine used to generate electricity creates 520 Gigabytes a day, and there are 20 blades in one of these turbines. An airplane on a transatlantic flight produces several Terabytes of data, which can be used to streamline maintenance operations, improve safety, and (most important to an airline's bottom line) decrease fuel consumption.

Another interesting example comes from the Nissan Leaf, an all-electric car. It has a system called CARWINGS, which not only offers the traditional telematics service and a smartphone app to control all aspects of the car but wirelessly transmits vehicle statistics to a central server. Each Leaf owner can track their driving efficiency and compare their energy economy with that of other Leaf drivers. We don't know the details of the information that Nissan is collecting from the Leaf models and what they do with it, but we can definitely see the three Vs in action in this example.

In general, sensor-based data falls into the industrial big data category, although lately the "Internet of Things" has become a more popular term to describe a hyperconnected world of *things* with sensors, where there are over 300 million connected devices that range from electrical meters to vending machines. We will not be covering this category of big data in this book, but the methodology and techniques described here can easily be applied to industrial big data analytics.

Alternate Data Processing Techniques

Big data is not only about the data, it is also about alternative data processing techniques that can better handle the three Vs as they increase their values. The traditional relational database is well known for the following characteristics:

- Transactional support for the ACID properties:

    - Atomicity: Where all changes are done as if they are a single operation.

    - Consistency: At the end of any transaction, the system is in a valid state.

    - Isolation: The actions to create the results appear to have been done sequentially, one at a time.

    - Durability: All the changes made to the system are permanent.

- The response times are usually in the subsecond range, while handling thousands of interactive users.

- The data size is in the order of Terabytes.

- Typically uses the SQL-92 standard as the main programming language.

In general, relational databases cannot handle the three Vs well. Because of this, many different approaches have been created to tackle the inherent problems that the three Vs present. These approaches sacrifice one or more of the ACID properties, and sometimes all of them, in exchange for ways to handle scalability for big volumes, velocity, or variety. Some of these alternate approaches will also forgo fast response times or the ability to handle a high number of simultaneous users in favor of addressing one or more of the three Vs.

Some people group these alternate data processing approaches under the name NoSQL and categorize them according to the way they store the data, such as key-value stores and document stores, where the definition of a document varies according to the product. Depending on who you talk to, there may be more categories.

The open source Hadoop software framework is probably the one that has the biggest name recognition in the big data world, but it is by no means alone. As a framework it includes a number of components designed to solve the issues associated with distributed data storage, retrieval and analysis of big data. It does this by offering two basic functionalities designed to work on a cluster of commodity servers:

- A distributed file system called HDFS that not only stores data but also replicates it so that it is always available.

- A distributed processing system for parallelizable problems called MapReduce, which is a two-step approach. In the first step or Map, a problem is broken down into many small ones and sent to servers for processing. In the second step or Reduce, the results of the Map step are combined to create the final results of the original problem.

Some of the other components of Hadoop, generally referred to as the Hadoop ecosystem, include Hive, which is a higher level of abstraction of the basic functionalities offered by Hadoop. Hive is a data warehouse system in which the user can specify instructions using the SQL-92 standard and these get converted to MapReduce tasks. Pig is another high-level abstraction of Hadoop that has a similar functionality to Hive, but it uses a programming language called Pig Latin, which is more oriented to data flows.

HBase is another component of the Hadoop ecosystem, which implements Google's Bigtable data store. Bigtable is a distributed, persistent multidimensional sorted map. Elements in the map are an uninterpreted array of bytes, which are indexed by a row key, a column key, and a timestamp.

There are other components in the Hadoop ecosystem, but we will not delve into them. We must tell you that in addition to the official Apache project, Hadoop solutions are offered by companies such as Cloudera and Hortonworks, which offer open source implementations with commercial add-ons mainly focused on cluster management. MapR is a company that offers a commercial implementation of Hadoop, for which it claims higher performance.

Other popular products in the big data world include:

- Cassandra, an Apache open source project, is a key-value store that offers linear scalability and fault tolerance on commodity hardware.

- DynamoDB, an Amazon Web Services offering, is very similar to Cassandra.

- MongoDB, an open source project, is a document database that provides high performance, fault tolerance, and easy scalability.

- CouchDB, another open source document database that is distributed and fault tolerant.

In addition to these products, there are many companies offering their own solutions that deal in different ways with the three Vs.

What Is Splunk?

Technically speaking, Splunk is a time-series indexer, but to simplify things we will just say that it is a product that takes care of the three Vs very well. Whereas most of the products that we described earlier had their origins in processing human-generated digital footprints, Splunk started as a product designed to process machine data. Because of these humble beginnings, Splunk is not always considered a player in big data. But that should not prevent you from using it to analyze big data belonging in the digital footprint category, because, as this book shows, Splunk does a great job of it. Splunk has three main functionalities:

- Data collection, which can be done for static data or by monitoring changes and additions to files or complete directories on a real time basis. Data can also be collected from network ports or directly from programs or scripts. Additionally, Splunk can connect with relational databases to collect, insert or update data.

- Data indexing, in which the collected data is broken down into events, roughly equivalent to database records, or simply lines of data. Then the data is processed and a high performance index is updated, which points to the stored data.

- Search and analysis. Using the Splunk Processing Language, you are able to search for data and manipulate it to obtain the desired results, whether in the form of reports or alerts. The results can be presented as individual events, tables, or charts.

Each one of these functionalities can scale independently; for example, the data collection component can scale to handle hundreds of thousands of servers. The data indexing functionality can scale to a large number of servers, which can be configured as distributed peers, and, if necessary, with a high availability option to transparently handle fault tolerance. The search heads, as the servers dedicated to the search and analysis functionality are known, can also scale to as many as needed. Additionally, each of these functionalities can be arranged in such a way that they can be optimized to accommodate geographical locations, time zones, data centers, or any other requirements. Splunk is so flexible regarding scalability that you can start with a single instance of the product running on your laptop and grow from there.

You can interact with Splunk by using SplunkWeb, the browser-based user interface, or directly using the command line interface (CLI). Splunk is flexible in that it can run on Windows or just about any variation of Unix.

Splunk is also a platform that can be used to develop applications to handle big data analytics. It has a powerful set of APIs that can be used with Python, Java, JavaScript, Ruby, PHP, and C#. The development of apps on top of Splunk is beyond the scope of this book; however, we do describe how to use some of the popular apps that are freely available. We will leave it at that, as all the rest of the book is about Splunk.

About This Book

We have a couple of objectives with this book. The first one is to provide you with enough knowledge to become a *data wrangler* so that you can extract wisdom from data. The second objective is that you learn how to use Splunk, a simple yet extremely powerful tool that will allow you to "click for gold" in the data you analyze.

The book has been designed so that you become exposed to big data from digital footprints and machine data. It starts by presenting simple concepts and progressively introducing slightly more difficult approaches. It is meant to be a hands-on guide for big data analytic projects that involve machine data, social media, and mining existing data warehouses. We do this through real projects, which review in detail how to collect data, load it into Splunk, process and analyze it, and visualize the results so that they can be easily consumed by the intended audience. We have broken the book into four parts:

- Splunk's Basic Operation, in which we introduce basic data collection, processing, analysis, and visualization of results. We use machine data in this part of the book to introduce you to the basic commands of the Splunk Processing Language. The last chapter in this part presents a way to create advanced analytics using log files.

- The airline on-time performance project. Once you are familiar with the basic concepts and commands of Splunk, we take you through the motions of a typical big data analytics project. We present you with a simple methodology, which we then apply to the project at hand, the analysis of airline performance data over the last 26 years. The data of this project falls under the category of mining an existing data warehouse. Using this project, we go over collecting data that is available in CSV format, as well as picking it up directly from a relational database. In both cases, there are some special considerations regarding the timestamp that is available in this data set, and we go in detail on how to handle them. This interesting project allows us to introduce some new Splunk commands and other features of commands that were presented in the first part of the book.

- The third part of the book is dedicated to social media. We go in detail into how to collect, process, and analyze tweets and Foursquare check-ins, as well as providing a full chapter dedicated to sentiment analysis. These chapters provide you with the necessary knowledge to wrangle any big data project that involves a social media stream.

- The fourth part of the book goes into detail on the architecture and topology of Splunk: how to scale Splunk to cover your needs, and the basic concepts of distributed processing and high availability.

- We also included a couple of appendices that cover the performance of Splunk as well as a quick overview of the various apps that are available.

The book is not meant to describe in detail each of the commands of Splunk, as the company's online documentation is very good and it does not make sense to repeat it. Our focus is on hands-on big data projects through which you can learn how to use Splunk and also become versed on handling big data projects. The book has been designed so that you can go directly to any chapter and be able to work with it without having to refer to previous chapters. Having said that, if you are new to Splunk, you will benefit from reading the book from the beginning. If you do read the book that way, you might find some of the information related to collecting the data and installing apps repetitive, as we have targeted the material to those who wish to jump directly into specific chapters.

■ **Note**   The searches presented in this book have been formatted to make them more readable. SplunkWeb, the user interface of Splunk, expects the searches as a single continuous line.

All of the data used in the book is available in the download package, either as raw data, as programs that create it or collect it, or as links where you can download it. This way you are able to participate in the projects as you read the book.

We have worked to make this book as practical and hands-on as possible so that you can get the most out of your learning experience. We hope that you enjoy it and learn enough to be able to become a proficient data wrangler; after all, there is so much data out there and so few people that can tame it.

CHAPTER 2

■ ■ ■

Getting Data into Splunk

In this chapter, you will learn how to get the data into Splunk. We will look at different sources of data and different ways of getting them into Splunk. We will make use of a data generator to create user activity for a fictitious online retail store MyGizmoStore.com, and we will load sample data into Splunk. You will also learn how Splunk Technology Add-ons provide value with some specific sources of data from operating systems such as Windows and Unix. Before wrapping up the chapter, you will get an overview of the Splunk forwarders concept to understand how to load remote data into Splunk.

Variety of Data

A typical enterprise information technology (IT) infrastructure today consists of network and server components that could range from mainframes to distributed servers. On top of that hardware infrastructure you will find databases that store information about transactions related to customers, vendors, orders, shipping, supply chain, and so on. These are captured, processed, and analyzed by several types of business applications. Traditionally, enterprises have used all this structured data to make their business decisions. The challenge has been mainly in integrating and making sense of all the data that comes from so many different sources. Whereas this has been the focus of the traditional IT organizations, we are seeing the definition of data and usage of data going beyond that traditional model. Most enterprises these days want to process and analyze data, which could fall in broad categories such as:

- Traditional structured data that is residing in databases or data warehouses

- Unstructured data or documents stored in content repositories

- Multistructured data available in different types of logs

- Clickstream data

- Network data

- Data originated by social media applications, and so on

You can see these newer categories of data such as logs, network, clickstream, and social media becoming part of the mainstream data analysis done by enterprises to make better business decisions. These types of data are sometimes also known as machine data or operational data. Some of the typical examples of enterprises wanting to make use of these types of data sources include:

- Web log files, which are created by web servers such as Apache and IIS. These log files provide information about the different types of activity happening on the web sites and the associated applications.

- Clickstream data files provide information down to the detail of what visitors have done while visiting a web site. This can be used to analyze shopping patterns and special behaviors such as abandoned shopping carts.

9

- Application log data, which typically has have plenty of information about the execution of applications, that can be used for operational purposes, such as optimizing the use of servers.

- Operating system level logs that could be used for performance and system monitoring.

- Firewall logs to better analyze security issues.

- Data from social media sources such as Twitter, Foursquare, and so on, which can be used for a myriad of marketing and sales purposes.

Gone are the days when machine data or log data was considered to be something for system administrators, who are sitting in dark data centers to debug and analyze why the systems went down or why the performance is not meeting the Service Level Agreements (SLAs). Although that use case is still valid, there is a complete paradigm shift on what data enterprises want to look at, process, and analyze for real-time, near real-time, or traditional business intelligence and reporting. The question now is, can Splunk handle all these sources of machine data or operational data and work with traditional data sources such as databases and data warehouses? The short answer is yes, and we will learn how we can get the data into Splunk in the following sections of this chapter.

How Splunk deals with a variety of data

For any practical purpose, Splunk can deal with pretty much any type of data coming from a wide variety of different sources including web logs, application logs, network feeds, system metrics, structured data from databases, social data, and so on. Splunk needs to be configured with individual sources of data and that each source can become a specific data input. The data coming into Splunk can be local, meaning that the data is sitting or available on the same computer where Splunk is running, or the data can be coming from any remote device connected to the server(s) running Splunk. You will see how remote data can be loaded into Splunk later in this chapter. Splunk broadly categorizes the sources of data that can be loaded as:

- Files & Directories

- Network sources

- Windows data

- Other sources

You will look into each one of these sources in detail. Splunk provides different options to define and configure the above sources as data inputs:

- Splunk Web—This is the standard user interface, which is the easiest way to interact with Splunk.

- Splunk CLI—The command line interface (CLI) can also be used to interact with Splunk, but it is used mainly by scripted programs, which could handle batch processes.

- Apps or Add-ons—These are specialized applications that sit on top of the Splunk framework and make it easy to work with one or more types of data sources. We will discuss the differences between Apps and Add-ons and how they can be used with an example later in this chapter.

- Configuration files—Splunk provides various configuration files that can be edited to configure and point to different sources of data. Irrespective of the option that is used to configure the sources of data inputs.conf file always gets updated either by the Splunk Web, Splunk CLI, Apps and Add-ons, or manually.

Independently of which option you chose to work with Splunk, the definition and configuration of data inputs is ultimately stored in the configuration files. For the examples in this book, we will be using Splunk Web, the user interface. One of the most popular forms of machine or log data, widely analyzed by enterprises, comprises web logs, or access logs as they are also known. We will use web logs as a starting point to explore and get familiar with what can be done with Splunk. In order to simulate to what would happen in a real-world online web application, we have created a fictitious ecommerce web site called MyGizmoStore.com, which sells widgets. The data for MyGizmoStore.com is created by a generator, which is described later in this chapter. This generator simulates the log files created by typical user activity, which includes browsing the catalog of widgets, adding to the shopping cart and potentially making the final purchase.

Files & Directories

Splunk makes it very easy to get data from files or files stored within a directory structure. You can load data from a static file as a one-time operation, also known as a oneshot, or you can ask Splunk to monitor a set of directories for certain types of files. We start by loading a single file. In order to make this easy we have generated an access log for MyGizmoStore.com that has approximately 250 log entries, which represent user activity over a period of two days in the life of the store. The file access.log is part of the download package of the book. Once you have the download package, copy the access.log file to the directory /opt in case of Linux, or C:\opt in case of Windows.

Splunk will give you the option of adding data based on the type or the source of the data. For this initial example, we will work with a source, the access log file. Once you have logged into the Splunk instance, go to the Splunk home page and click on "Add data" button in the "Do more with Splunk" section. In the Add Data to Splunk page you will see different options are available under two categories.

- Choose a Data Type—allows to select a pre-determined type of logs such as access logs, sys logs etc.

- Choose a Data Source—allows to bind determined type of sources such as windows registry data or get an output from a script which will be a data input into Splunk.

Click on the "From files and directories" link under "Choose a Data Source" section. The difference between this option and a similar option in the "Choose a Data Type" section is that we get an additional option to make use of a forwarder to send the data to a Splunk Server in the data types category. As you have not yet been introduced to the concept of forwarders, we will choose the simple one to get started, as seen in Figure 2-1.

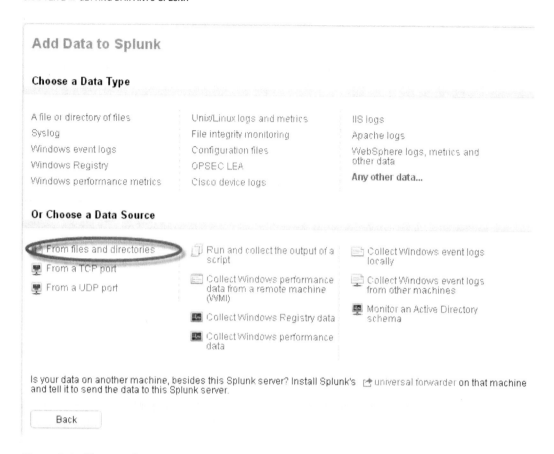

Figure 2-1. *Choose a data source*

This will take you into the Data preview page, as seen in Figure 2-2. Because this is the first time that we are loading the data into Splunk, it makes sense to get a preview of the data. The preview option provides an opportunity to see how the log entries are going to get processed before they are committed into the Splunk data store, which is called index. Next you select the "Preview data before indexing" radio button and choose the access.log file under \opt directory if you are on Linux or C:\opt if you are in Windows environment. Then click on the Continue button.

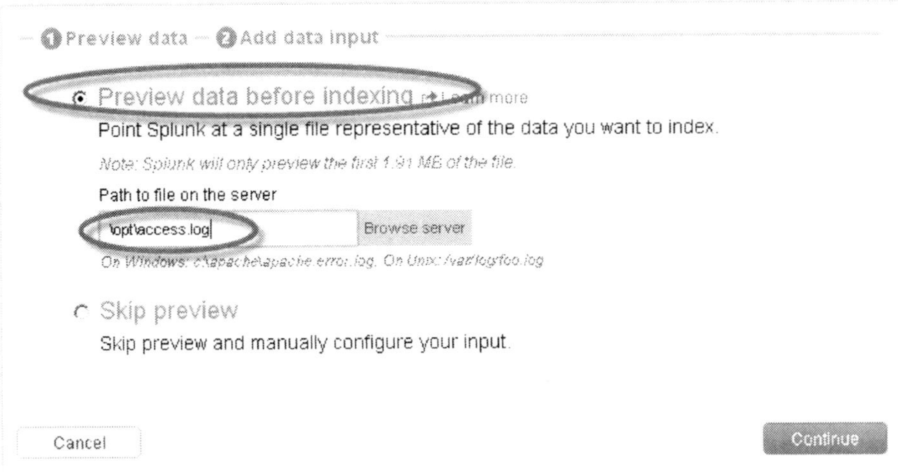

Figure 2-2. Preview data

The dialog box that comes up gives you options to set the source type. By default Splunk parses the data files provided as input and determines what the source type for the file is. It then loads or indexes the file according to the source type characteristics. In this case, because we are loading an access log file, which has been identified as a combined access log file, we accept the default option of "Use auto-detected source type" as seen in Figure 2-3 and click the Continue button. In Chapter 3, we will explain in detail the format of an access combined log file and what information is stored in the log entries. In Chapter 9, we will review other options such as defining the input of custom data files, which needs to be processed differently.

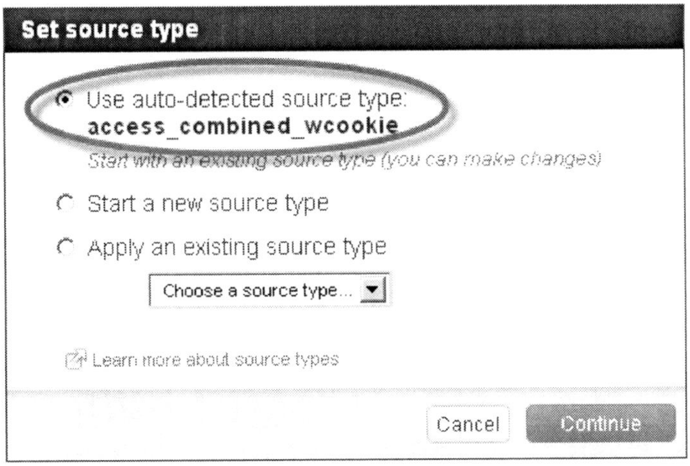

Figure 2-3. Set source type

What are typically known as records, or just plain lines of data, are referred to in Splunk as events, and every event has a timestamp. Throughout the book we will be using, analyzing, and manipulating timestamps as they are a key element of Splunk and big data analysis. The next screen in the data loading process presents the way Splunk has broken down the log entries into different events along with associated timestamps. The default behavior is to break an event on the timestamp, but if Splunk cannot find the timestamp it will present one single event that contains all the lines of the file. We can then customize where to break the entries into different events, which we will learn in Chapter 9. In this case, Splunk breaks down very nicely the entries of the access log into events as can be seen in Figure 2-4. Here you can see that each event has the information about a user activity that happened on MyGizmoStore.com. The preview option also shows the number of events extracted from the log file, which in this case contains 243 events.

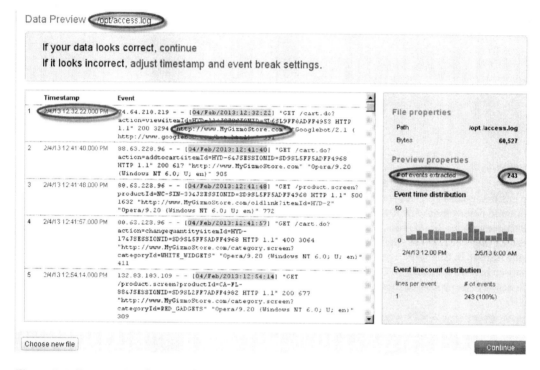

Figure 2-4. *Data preview for access.log*

■ **Note**   If you are familiar with relational database concepts, it may help to realize that each event broken down by Splunk is conceptually equal to a single row in a relational database table.

Now that we have confirmed that the data looks good and has been processed correctly by Splunk, you can click on the Continue button, which will take us to the "Add new" page, where we will customize some of the settings before we get the data finally indexed into Splunk. Because this is a one-time file processing operation, we will select the "Index a file once from this Splunk server" radio button under Source. Because our computer has a rather complex name, we also chose to change the Host file value to BigDBook-Test.

In the previous step, we asked Splunk to automatically determine the sourcetype, so we will see the sourcetype is set to automatic. By default, when data is loaded into Splunk it goes into the main index, which is the mechanism used to store, process, and analyze data. Splunk also offers the ability to define and use other indexes, which can help you better organize and manage your data, especially regarding data governance, such as access, protection, and retention policies. For this example, we will use the main, but we will see how to create a new one and make use of it in the next section. Click on the Save button. The next page shows a Success message if Splunk is able to process the data completely.

■ **Note** If you are familiar with schemas in Oracle, the Splunk index is very similar to that concept. An index in Splunk is a collection of data, and a schema has a collection of tables with the data.

Once Splunk indexes the data successfully, you can start to review it to make sure it was done correctly. To list all the log entries that were loaded into Splunk we type host=BigDBook-Test in the search bar and hit enter. This search gives complete listing of all events along with the default fields, which are the timestamp, the host from where the data comes, the source type and the source of the data. Splunk will always have those fields available. In Figure 2-5, you can see that the total count is 243 events, which is the same number of log entries we had in the file we just indexed. You can also see the timestamp to the left side of each event and the other three default fields presented below each event.

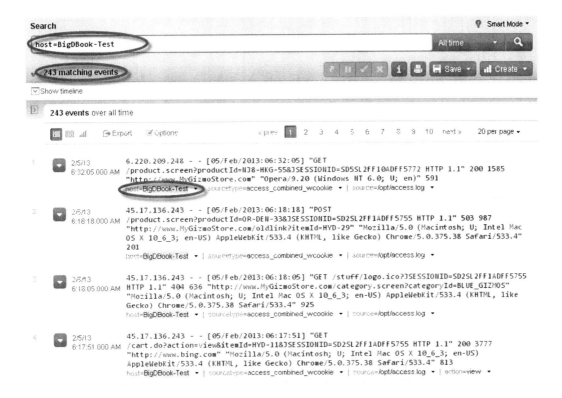

Figure 2-5. Search results

When individual sources of data in large volumes are ingested into Splunk they can be better managed with separate indexes, which could be placed on different tiers of storage. We will go ahead and create a separate index for the MyGizmoStore.com log files. To create an index click on the "Manager" menu item on the upper right corner of the user interface and in the Data section click on the "Indexes" link as shown in Figure 2-6.

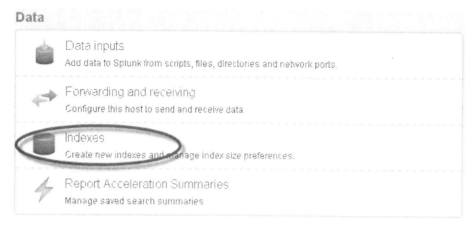

Figure 2-6. *Index creation*

In the Indexes page, click on the New button and name the index mygizmostoreindex. You can ignore the cold and thawed path options of the form. They are advanced options that can be used to move older or unused data in a Splunk index to, for example, lower cost storage. It also provides the capabilities to archive data outside of Splunk using the concept of a frozen archive path. In our case, we will leave these options at their defaults as the simulated data set is not going to be massive. Click on the Save button. Once the new index has been created successfully, it will show up in the indexes page as shown in Figure 2-7.

Indexes   New

Showing 1-10 of 10 items   Results per page 25 ▾

Index name ⬍	Max size (MB) of entire index ⬍	Frozen archive path ⬍	Current size (in MB) ⬍	Event count ⬍	Earliest event ⬍	Latest event ⬍	Home path ⬍
_audit	500,000	None	13	120,384	Jan 12, 2013 4:33:52 AM	Feb 9, 2013 6:15:32 AM	/splunk/splunk/var/lib/splunk/audit/db
_blocksignature	0	None	1	0	N/A	N/A	/splunk/splunk/var/lib/splunk/blockSignature/db
_internal	500,000	None	222	3,875,358	Nov 12, 2012 7:11:02 PM	Feb 9, 2013 6:15:32 AM	/splunk/splunk/var/lib/splunk/_internaldb/db
_thefishbucket	500,000	None	1	0	N/A	N/A	/splunk/splunk/var/lib/splunk/fishbucket/db
history	500,000	None	1	0	N/A	N/A	/splunk/splunk/var/lib/splunk/historydb/db
main	500,000	None	4	8,658	Dec 13, 2012 6:39:16 AM	Feb 5, 2013 6:32:05 AM	/splunk/splunk/var/lib/splunk/defaultdb/db
mygizmostoreindex	500,000	None	1	0	N/A	N/A	/splunk/splunk/var/lib/splunk/mygizmostoreindex/db
os	500,000	None	44	471,978	Aug 29, 2007 12:19:26	Feb 9, 2013 6:15:37	/splunk/splunk/var/lib/splunk/os/db

Figure 2-7. *List of indexes*

What we have seen so far is how to get data into Splunk using one-time file indexing. In real-world use cases, we would be seeing that log files are being continuously updated with new entries, and these expanded log files need to be processed and analyzed on a continuous basis. This is where Splunk provides the ability to monitor certain directories where files are being continuously updated. Splunk's directory monitoring capability lets us specify the directory that needs to be monitored and the files in that directory will be used as data input. Before we explore the monitoring option, let us go back to MyGizmoStore.com one more time. We have used a static log file from the online store to see how we can get that kind of data into Splunk. In the real-world MyGizmoStore.com would probably be running on multiple web servers on different hosts and writing out log files which need to be monitored, processed, and analyzed on a continuous basis.

To simulate this behavior we have come up with a test data generator to generate log files for MyGizmoStore.com. We will review how the test data generator operates and how we can start generating the data.

Data Generator

The sample data generator is written using the Python scripting language. It is designed to run on both Microsoft Windows and Linux operating systems. The sample data generator uses a random generator to create random IP addresses, which point to the visitor of the web site; a predetermined set of product identification codes that look like "CA-NY-99" and item identification codes in the form of HYD-19. The gizmos sold on the store can be categorized as follows:

- BLUE_GIZMOS

- RED_GADGETS

- WHITE_WIDGETS

- ORANGE_WATCHMACALLITS

- PURPLE_DOOHICKEYS

- BLACK_DOODADS

The visitors to the web site can perform the following actions:

- Purchase

- Add an item to the shopping cart (Addtocart)

- Remove an item from the shopping cart (Remove)

- View the catalog of gizmos (View)

- Change the quantity of an item in the shopping cart (Changequantity)

The HTTP protocol used by web sites includes a return code that either signifies success in the operation or describes a problem. The code 200 indicates a successful operation, whereas codes in the 400s and 500s indicate some sort of a problem. The data generator creates a realistic percentage of error codes. Additionally, the generator includes a random selection of user agents. These are a fancy name for the description of the combination of the browser and operating system used by the visitors, which also include the version number of both.

- Mozilla/4.0 (compatible; MSIE 6.0; Windows NT 5.1; SV1)

- Opera/9.01 (Windows NT 5.1; U; en)

- Mozilla/5.0 (Macintosh; U; Intel Mac OS X 10_6_3; en-US) AppleWebKit/533.4 (KHTML, like Gecko) Chrome/5.0.375.38 Safari/533.4

- Mozilla/4.0 (compatible; MSIE 6.0; Windows NT 5.1)

- Googlebot/2.1 (http://www.googlebot.com/bot.html)

- Mozilla/4.0 (compatible; MSIE 6.0; Windows NT 5.1; SV1; .NET CLR 1.1.4322)

- Mozilla/5.0 (Windows; U; Windows NT 5.1; en-GB; rv:1.8.1.6) Gecko/20070725 Firefox/2.0.0.6

- Opera/9.20 (Windows NT 6.0; U; en)

The data generator script includes the ability to customize the percentage of user agents and HTTP error codes, the default settings are 25 percent and 30 percent, respectively. The generator has a default value for maximum number of events, which is 50,000 for a period of 30 days. The generator creates the sample data starting from the current date and goes backward 30 days. For this example, we have taken the defaults as described.

To run the data generator script, you will need to have Python 2.7 or above installed on your system. You can make use of the Python that is bundled with Splunk. You will be able to find Python installed in the $SPLUNK_HOME/bin directory (where $SPLUNK_HOME is the directory where Splunk is installed). The data generator script has been tested with Python 2.7 that comes with Splunk and also with Python 3.1.5. For more information on Python you can visit http://www.python.org. The sample data generation script included in the download package of the book is called Generate_Apache_Logs.py. We have installed that script into /datagen/BigDBook directory on our Linux machine.

Generate Sample Data

To generate the sample data, you will execute the python script by typing command shown in Figure 2-8. The same command applies to both Windows and Unix.

Figure 2-8. *Run test data generator*

On successful execution of the Python script, you will be able find the generated log files in the /opt/log directory if it is a Unix operating system and in case of a Windows operating system the files are placed in the directory c:\opt\log. As we mentioned earlier, a typical ecommerce web site runs on various servers, thus our data generator simulates log entries for three different hosts that are named BigDBook-www1, BigDBook-www2, and BigDBook-www3. The location of generated files can be customized in the script file. For this chapter the log files were generated in the following directories:

- /opt/log/BigDBook-www1/access.log

- /opt/log/BigDBook-www2/access.log

- /opt/log/BigDBook-www3/access.log

If you type the following Unix command ls -Rla in the /opt/log directory it will list the files as seen in Figure 2-9 and we can see that the access log files were created for the three different hosts in separate directories.

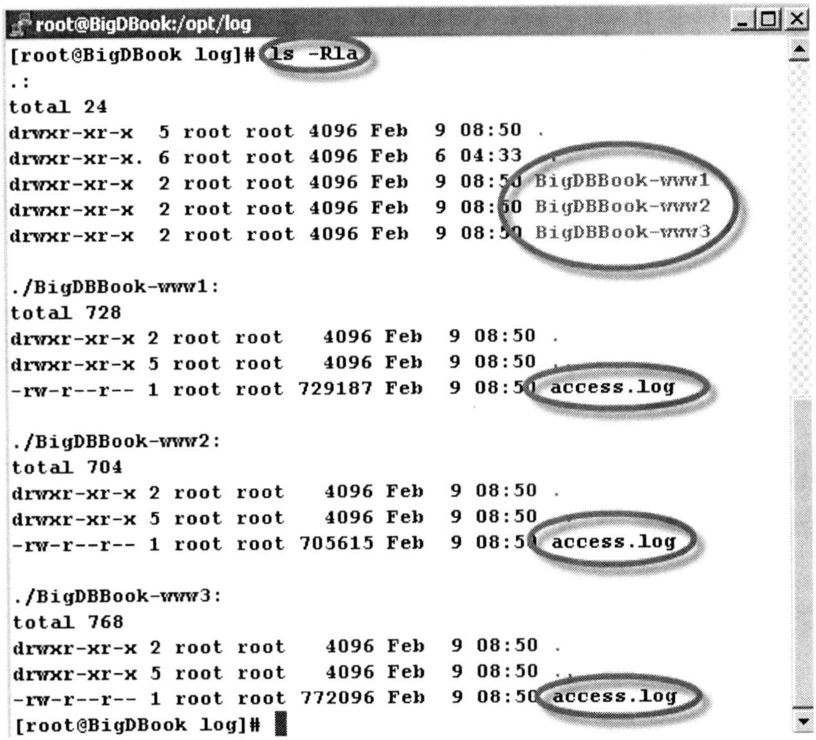

```
root@BigDBook:/opt/log                                    _ □ X
[root@BigDBook log]# ls -Rla
.:
total 24
drwxr-xr-x  5 root root 4096 Feb   9 08:50 .
drwxr-xr-x. 6 root root 4096 Feb   6 04:33 ..
drwxr-xr-x  2 root root 4096 Feb   9 08:50 BigDBBook-www1
drwxr-xr-x  2 root root 4096 Feb   9 08:50 BigDBBook-www2
drwxr-xr-x  2 root root 4096 Feb   9 08:50 BigDBBook-www3

./BigDBBook-www1:
total 728
drwxr-xr-x 2 root root    4096 Feb   9 08:50 .
drwxr-xr-x 5 root root    4096 Feb   9 08:50 ..
-rw-r--r-- 1 root root 729187 Feb   9 08:50 access.log

./BigDBBook-www2:
total 704
drwxr-xr-x 2 root root    4096 Feb   9 08:50 .
drwxr-xr-x 5 root root    4096 Feb   9 08:50 ..
-rw-r--r-- 1 root root 705615 Feb   9 08:50 access.log

./BigDBBook-www3:
total 768
drwxr-xr-x 2 root root    4096 Feb   9 08:50 .
drwxr-xr-x 5 root root    4096 Feb   9 08:50 ..
-rw-r--r-- 1 root root 772096 Feb   9 08:50 access.log
[root@BigDBook log]# █
```

Figure 2-9. *Generated access log files*

Now that the MyGizmoStore access logs are created and ready, you can configure Splunk to monitor the directory where these log files are being placed. To do this go to the manager screen, as was done earlier, and click on the "Data inputs" link (as seen in Figure 2-1). In the Data inputs page, click on the "Add new" link for the Files & directories option. Now you can select to skip the data preview, as we already did this earlier with the same data, and click on the continue button. In the Add new page, under the source option, select the radio button for "continuously index data from a file or directory this Splunk instance can access." Because the sample generator is writing out logs to the /opt/log directory (in the case of Linux), we will use that as an input for the "Full path to your data" option. Because we created a separate index for MyGizmoStore.com, called mygizmostoreindex, we will use it to illustrate how to load data into an index other than the main one. Select the check box for "More" settings. One of the options we have here is to set the host name. This is very useful, as you can do specific searches based on host name. The Set host option provides the following choices:

- You can define a constant value for the host name, which is useful when you want to have a single host name for all the log files.

- The RegEx option, which allows you to extract the hostname from a string using a regular expression.

- The segment option, which allows you to make use of a particular segment in the full file pathname.

In our case for MyGizmoStore.com, we want to use multiple host names as the test generator is creating files under /opt/log/BigDBook-www1, /opt/log/BigDBook-www2, /opt/log/BigDBook-www3. Because the hostname is the third value in all pathnames, we will make use of the segment option and specifying a value as 3. As we have did when we loaded the single file, we will let Splunk set the source type automatically, and in the index we will specify the newly created `mygizmostoreindex` index. We do this by selecting that index from drop-down box. The advanced options include a whitelist and a blacklist, which help specifying which files in the directory should be monitored and which should be ignored. In our case the sample data generator is only generating one log file under each subdirectory, so we leave them blank, as there is nothing to black or white list. Figure 2-10 shows the settings that we have configured. After all is defined, click on the Save button.

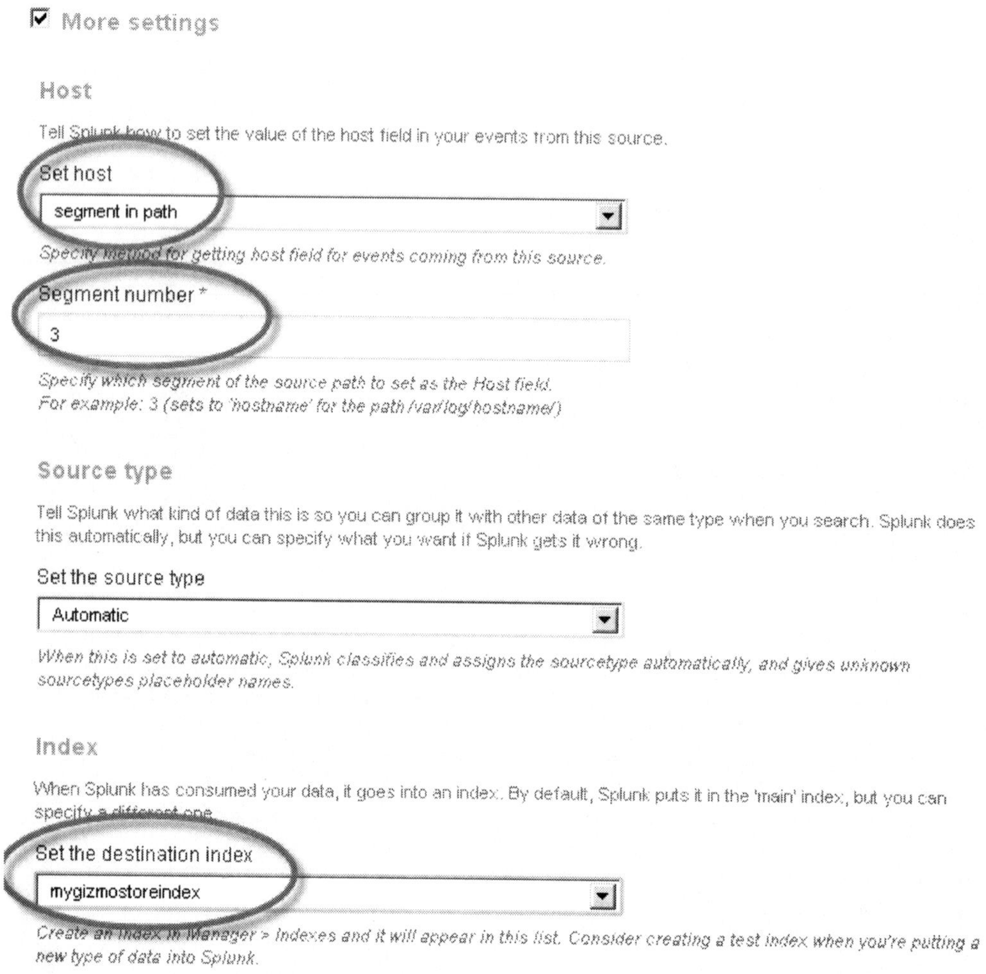

☑ More settings

Host

Tell Splunk how to set the value of the host field in your events from this source.

Set host

    segment in path ▾

Specify method for getting host field for events coming from this source.

Segment number *

    3

Specify which segment of the source path to set as the Host field.
For example: 3 (sets to 'hostname' for the path /var/log/hostname/)

Source type

Tell Splunk what kind of data this is so you can group it with other data of the same type when you search. Splunk does this automatically, but you can specify what you want if Splunk gets it wrong.

Set the source type

    Automatic ▾

When this is set to automatic, Splunk classifies and assigns the sourcetype automatically, and gives unknown sourcetypes placeholder names.

Index

When Splunk has consumed your data, it goes into an index. By default, Splunk puts it in the 'main' index, but you can specify a different one.

Set the destination index

    mygizmostoreindex ▾

Create an index in Manager > Indexes and it will appear in this list. Consider creating a test index when you're putting a new type of data into Splunk.

Figure 2-10. *Monitoring a directory*

As with the previous example, you will be able to see the newly saved configuration in the Data inputs page, as shown in Figure 2-11.

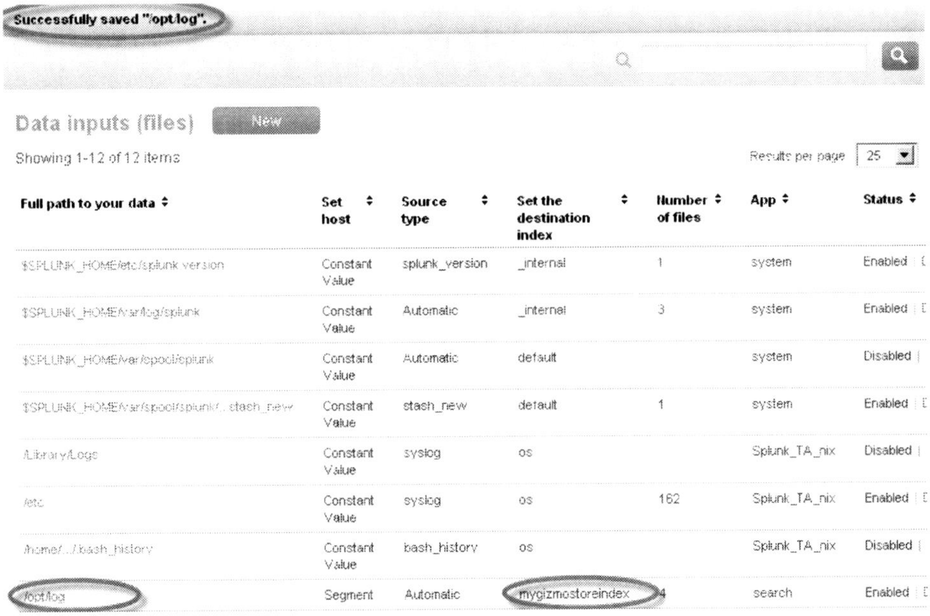

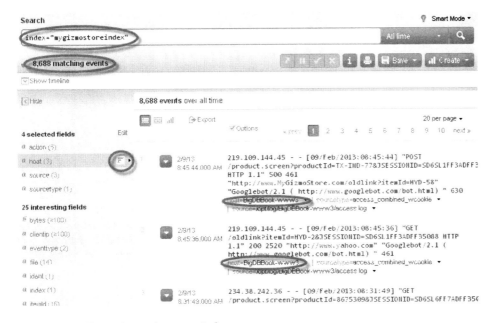

Figure 2-11. *Saved configuration*

Once Splunk indexes the files successfully, you can review the data. To do this, type `index=mygizmostoreindex` in the search bar and hit enter, as shown in Figure 2-12. This search lists all the events in the specified index, and as before it includes the default fields. You can see that the host name is set correctly according to our specifications, which were to use the third segment of the file pathname. The source type and source fields also appear to be correct. Because the data seems to be correctly indexed based on our simplistic review, we can feel comfortable that Splunk is monitoring those directories on a regular basis and appending the new information into the `mygizmostoreindex`.

Figure 2-12. *Events in mygizmostoreindex*

We can now introduce the left side bar, also known as the field bar. This side bar always presents the default fields and additional fields, which Splunk calls fields of interest. These are defined as fields that show up in 50 percent or more of the events of that particular index. It is a very useful tool to quickly gain a better understanding of the data you are working with. For example, we can quickly see that the host field has three values. If we want to know which are those values you can click on the bar graph icon against the host field. As seen in Figure 2-13, this will bring up a dialog box that shows all the values of the host field. Thus we avoid having to review a number of events to verify that the three expected values are present. Not only that, the dialog box also presents some summary statistics about the field, such as the total count of events that contain each value and the percentage. Additionally, it presents a bar chart with that information, making it a very compelling and easy way to gain a good understanding of a specific field.

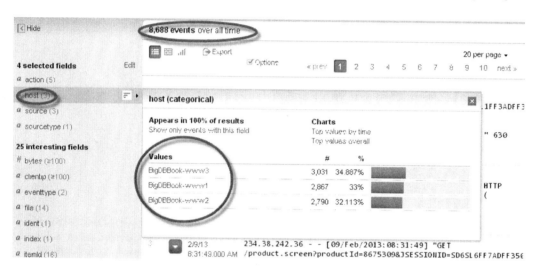

Figure 2-13. *Events in each host*

Most of the time the generated log files are very big; they get compressed using zip technology. Additionally, most of the servers have log rotation capability, where log files are moved to another place after a certain time or size has been reached. The Splunk directory monitoring facility is intelligent enough to address these real-world use cases, as it can unzip the compressed files in the directory before indexing them. It can also automatically detect the log rotation and keep track of where the last log entry that was indexed is located; this way it can start from that point. This is useful in case of Splunk restart or when maintenance tasks are executed.

Network Sources

A number of system applications and network devices such as routers switches relay events over network ports using the TCP or UDP protocols. Some applications make use of the SNMP standard to send events over UDP. Syslog, which is a standard for computer data logging is another set of sources where there is a wealth of information that could be captured at a network port level. Splunk can be enabled to accept input from a TCP or UDP port. To do this, you can use the Splunk Web user interface and configure a network input source where all you have to specify is the host, port, and sourcetype. Once you save the configuration, Splunk will start indexing the data coming out of the specified network port. This kind of network input can be used to capture syslog information that gets generated on remote machines and the data does not reside locally to a Splunk instance. Splunk forwarders can also be used to gather data on remote hosts. We will discuss forwarders in the last part of this chapter.

Windows data

The Windows operating system churns out a number of log files that have information about Windows events, registry, Active Directory, WMI, performance, and other data. Splunk recognizes Windows log streams as a source type and allows adding one more of these log streams to be indexed as input for further processing and analysis. Although Windows sources such as Active Directory or others can be individually configured, Splunk provides a better and easy way of dealing with these Windows logs or events by using the Splunk App for Windows or the Splunk Technology Add-on for Windows. We will explore this later in this chapter.

Other Sources

Splunk supports scripting as a mechanism to get data from other sources that are not provided with a specific default configuration. There is no limit on types of data that scripting can touch, as long as the scripting output can be provided as an input that Splunk can understand or can be tweaked to make it understand. Examples include a script that could be getting data from a database or a script that could be getting data from Twitter—we will look into this example in Chapter 12. Technology Add-ons could also be using scripting as the mechanism to get the data, but they provide an abstraction to bring the added value of making it easy to get the data into Splunk for further processing and analysis. This gives is a good opportunity for a quick overview of Splunk Apps and Add-ons.

Apps and Add-ons

The Splunk user interface defaults to the Search app. However, Splunk is designed as a platform that serves as an infrastructure where third-party developers or ISVs can build specialized applications that provide extensions to Splunk. There are two ways of building these extensions: Apps and Technology Add-ons.

Splunk Apps package the extended functionality together with standard features such as saved searches, dashboards, and defined inputs. Additionally, they bundle their own user interface layered on top of Splunk's user interface. By contrast, Add-ons, or simply TAs, are smaller components as compared to Apps, which include the additional functionality without their own user interface. We will have to use the standard Splunk Search application against the indexed data configured through add-ons. Apps and Add-ons can be written by anybody with a decent programming knowledge. Splunk has a vibrant community that constantly creates and shares Apps and Add-ons. It is hosted at http://splunk-base.splunk.com/apps/. In Appendix B, you will find a list of useful Apps and Add-ons that you can use in your enterprise.

To explore Windows sources, we will make use of the Splunk Technology Add-on for Windows. This will allow us to get the windows data into Splunk. You can find the Add-on by going to directly to splunkbase or you can search and find apps directly from Splunk instance. To do that click on "Find more apps … " under the "App" menu as shown in Figure 2-14.

Figure 2-14. *Find Splunk apps*

In the search bar of the user interface type the word `windows` and hit enter. At the time of this writing the fourth result came up as Splunk for Windows technology add-on, which is the one we want use. Click on the Install free button, as shown in Figure 2-15. You will require a Splunk login to do this. It is free to register.

Splunk for Windows technology add-on

The Splunk for Windows technology add-on includes predefined inputs to collect data from Windows systems and maps to normalize the data to the Common Information Model... ☐ Read more

Author: Splunk   **Version:** 4.6.1   **Last updated:** 11/02/12   **Downloads:** 6013   **License:** Splunk Software License Agreement

Figure 2-15. *Splunk TA for Windows*

You will be prompted to enter the credentials for your Splunk Web site account. Once you enter your credentials, the add-on will be installed onto the Splunk instance. You should see a message stating that a restart is required. Click on Restart Splunk button. On successful restart and login into the Splunk instance, you will be able to find the newly installed Splunk TA for Windows under the "App" menu, as shown in Figure 2-16 below.

Figure 2-16. *Splunk TA for Windows*

Specific Windows event logs or files and directories can be enabled by clicking on the add-on link in the "App" menu. Once the specific sources are enabled the add-on will automatically configure the data to be loaded into the Splunk instance. We will enable Application, System, and Security event logs as seen in Figure 2-17 to do some basic testing and show how the Windows TA helps to simplify the process of loading Windows sources data into Splunk.

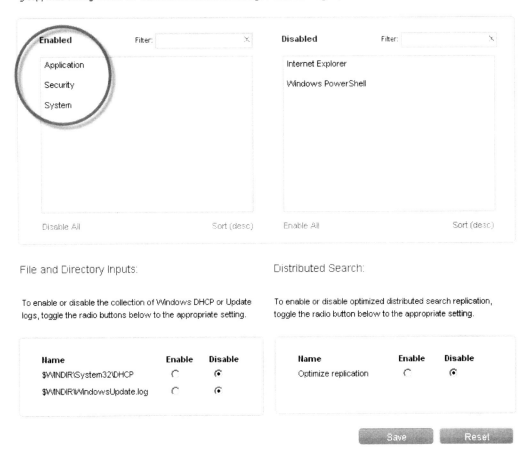

Figure 2-17. Enabling Windows event logs

■ **Note**   Windows Technology add-on can be installed on Splunk running on Windows. If you are running Splunk on Linux then the Windows TA can be installed on a forwarder running on a Windows machine. Forwarders are explained later in this chapter and in Chapter 15. We also recommend reviewing the Splunk administrative manuals when installing a combination of Apps and Add-ons to see what configuration(s) are supported.

Now that we have installed Windows TA and enabled some of the available sources, we can go to Splunk Search App and search for indexed data to test whether the TA is able to get the data into Splunk. We can pick Application, which is one of the sources that we have enabled, and see if events related to Windows applications are indexed into Splunk. For this type in the search bar the following command:

```
sourcetype="WinEventLog:Application"
```

As seen in Figure 2-18, the search came back with 12897 events from the Application events log. You can pick other event logs that have been enabled and search for events related to those logs. As this chapter is focused on understanding on how to get data into Splunk, we are not going to explore all of the available events that have been indexed into Splunk. Chapter 3 will take you into details of how to process and analyze the data once it is in Splunk and Chapter 4 will go into details of visualizing the data indexed into Splunk.

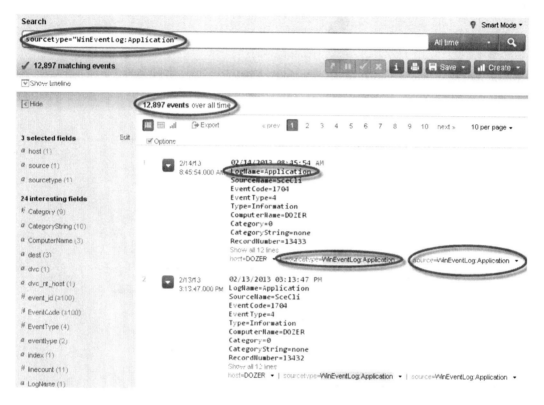

Figure 2-18. *Events from Windows Application indexed into Splunk*

Splunk also provides similar Technology Add-ons for Linux and Unix known as *Nix. This Add-on makes use of both log files and scripting to get different sets of event and log data available in Linux or Unix into Splunk. You can install *Nix technology add-on using the same process we used to install the Windows Technology Add-on:

- Go to "App" menu and click on "Find more apps ... " as shown in Figure 2-14.

- Search for *nix and the second entry at the time of this writing comes up as "Splunk for Unix and Linux technology add-on."

- Click on the Install free button; after a successful install you will be prompted to Restart Splunk.

- On a successful restart, go to the "App" menu as seen in Figure 2-19; here you can see "Splunk *Nix (TA)" as a new menu item.

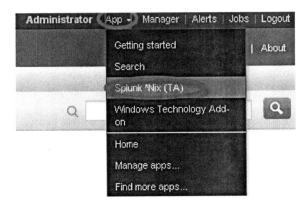

Figure 2-19. Splunk *Nix (TA)

Click on the link for "Splunk *Nix(TA)" so that you can setup the sources of data for your flavor of Unix . This brings up the setup screen as seen in Figure 2-20. Here you can see that the Add-on makes use of the Files & Directory monitoring approach that we used earlier with the MyGizmoStore.com to monitor different log files in Unix. It also makes use of scripting approach that we just discussed to load specific data inputs, such as CPU usage into Splunk. This really shows the power of a TA, which abstracts the layer of annoying details and makes it very simple to get data loaded into Splunk. To do a quick test, we enable the /var/log directory under the "file directory inputs" as shown in Figure 2-20. Additionally, we enable the cpu.sh script under "scripted inputs" title as seen in Figure 2-21. Clicking on the Save button saves the settings to the corresponding configuration file. If you go to the "Indexes" page, you will see that all the *Nix data is being loaded into a separate index named os.

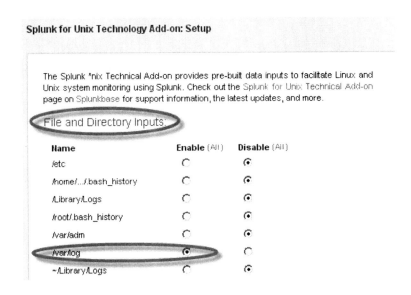

Figure 2-20. Configure *Nix TA

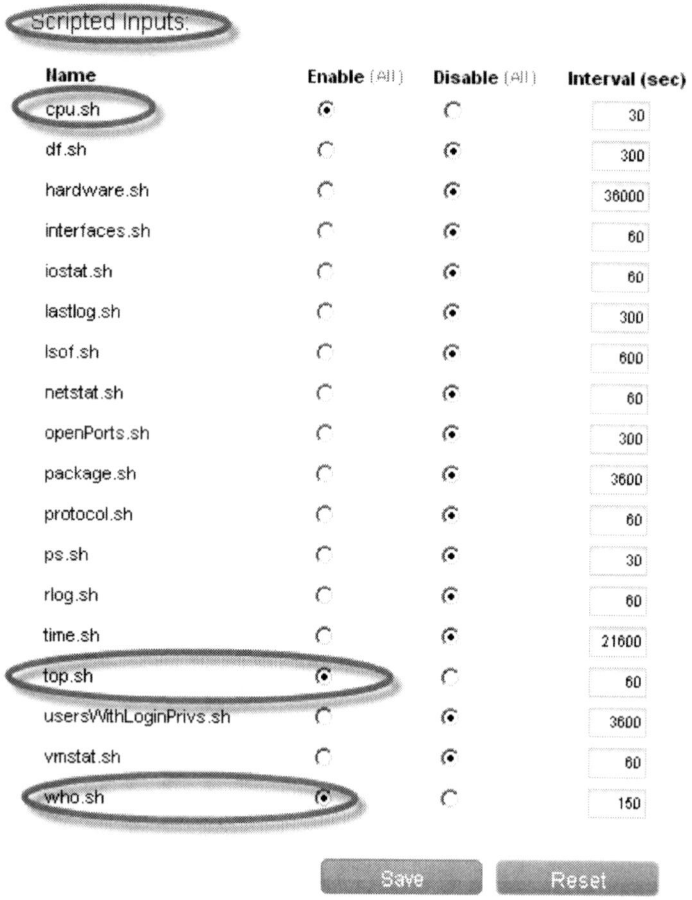

Scripted Inputs:

Name	Enable [All]	Disable [All]	Interval (sec)
cpu.sh	●	○	30
df.sh	○	●	300
hardware.sh	○	●	36000
interfaces.sh	○	●	60
iostat.sh	○	●	60
lastlog.sh	○	●	300
lsof.sh	○	●	600
netstat.sh	○	●	60
openPorts.sh	○	●	300
package.sh	○	●	3600
protocol.sh	○	●	60
ps.sh	○	●	30
rlog.sh	○	●	60
time.sh	○	●	21600
top.sh	●	○	60
usersWithLoginPrivs.sh	○	●	3600
vmstat.sh	○	●	60
who.sh	●	○	150

[Save] [Reset]

Figure 2-21. Scripted inputs in *Nix TA

Now that you have installed the *Nix TA and enabled some of the available sources, you can go to the Splunk Search App and search for indexed data to test out whether the *Nix TA was able to load the data. As we mentioned earlier, the *Nix TA loads the data into separate index called os, so you can go ahead and search for all the events by entering index=os in the search bar, and this will retrieve all the events that have been indexed into Splunk for the enabled sources. You can see in Figure 2-22, that the search shows the events captured from top, who, and various log file under /var/log directory.

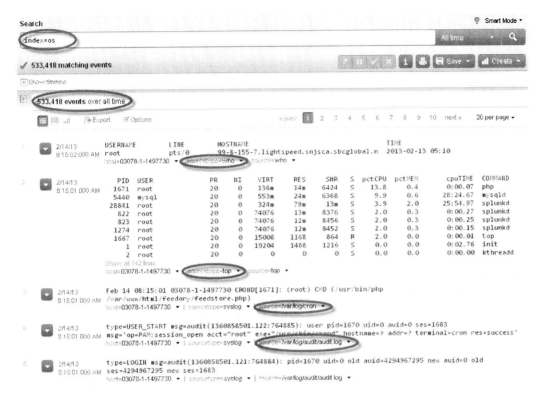

Figure 2-22. *Events captured using *Nix*

Forwarders

In the real world, enterprises have numerous applications and most of them will be running on a heterogeneous infrastructure, which includes all sorts of hardware, databases, middleware, and application programs. It will not be possible to have Splunk running locally or near to each of the applications or infrastructure, meaning the data will not be local to Splunk. What we have seen in this chapter is how we can get data into Splunk which is local to it. The use cases assumed that Splunk will be able to access files or directories, which could be on local or file systems that have remote data, but they are attached to the machine where Splunk is running.

To address the use case of getting remote data into Splunk, we will explain Splunk forwarders. A Splunk forwarder is the same as a standard Splunk instance but with only the essential components that are required to forward data to receivers, which could be the main Splunk instance or indexer. The forwarder's primary job is of gathering data that is remote to Splunk, forwarding it to a main Splunk instance or indexer that will load the remote data. Splunk universal forwarder is a downloadable component from splunk.com that can be set up to forward the required data. Technology add-ons such as the Windows TA can work with a universal forwarder to send Windows source data to a Splunk instance running on Unix environment. Using Splunk forwarders is the recommended deployment or best practice to work with remote data. Forwarders provide many benefits, including:

- They automatically buffer data at remote locations, which can be useful when the main instance of Splunk goes down for whatever reason.

- They support making use of Technology Add-ons to gather different sources of data available remotely.

- They can be administered remotely.

- They support securely sending the remote data, with compression and data acknowledgment.

- They provide support for load balancing, and are better suited for scalability and performance.

They can work with any available network port on the remote location, whose data need to be gathered. We will go in more details about Splunk deployments in Chapters 14 and 15, which discuss distributed topologies and high availability.

Summary

In this chapter, we have seen the different types of sources from which data can be loaded into Splunk. We discussed in detail how to get data using the Files & Directories option and how to make use of Splunk's monitoring capability to get MyGizmoStore.com access logs. We touched on how technology add-ons like Windows and *Nix can be configured to gather data from Windows and Linux boxes. Finally, you learned about Splunk forwarders and how they can help get hold of data that is remote to Splunk.

CHAPTER 3

Processing and Analyzing the Data

In this chapter you will learn how to process and analyze the data using Splunk's Search Processing Language (SPL). We will continue with the data-loading work that we did in Chapter 2; this will help you to get a good understanding of the combined access log format and the information that the log entries provide. You will then learn how to process the data of MyGizmoStore.com using SPL's reporting, sorting, filtering, modifying, and grouping commands.

Getting to Know Combined Access Log Data

One of the points that we stress in this book is the need to understand the data set that you want to process and analyze; that is, getting intimately acquainted with the data you will work with first. In this chapter we are going to take the first step of explaining the data of the combined access log format that we used to generate the sample data in Chapter 2.

Log files are generated by almost all kinds of applications and servers—whether they are end-user applications, web servers, or complex middleware platforms that serve as an infrastructure to run the applications used by consumers or business users. Operating systems and firmware also generate huge amounts of raw data into log files. The challenge lies in understanding, analyzing, and mining the raw data in the log files and making sense out of it.

Combined access logs generated by web servers such as Apache or Microsoft IIS provide information about activity, performance, and problems that are happening, whether intermittently or continuously. These logs contain information about all of the requests processed by the server. Both Apache and IIS allow customization of the combined access log format, which is commonly used and well understood by many log analysis applications that interpret and process the entries in these log files. The log entries produced in the combined access log format look like this:

```
127.0.0.1 - JohnDoe [10/Oct/2000:13:55:36 -0700] "GET /apache_pb.gif HTTP/1.0" 200 2326
"http://www.google.com" "Opera/9.20 (Windows NT 6.0; U; en)"
```

The meaning of each individual field in this example is described in Table 3-1.

Table 3-1. *Description of fields in combined access log*

Field	Description
127.0.0.1	This is the IP address of the client (the machine, host, or proxy server) that was making an HTTP request to access either a web application or an individual web page. The value in the field could be represented as hostname
-	This field is used to identify the client making the HTTP request. Because the contents of this field are highly unreliable, a hyphen is typically used, which indicates the information is not available
JohnDoe	This is the user id of the user who is requesting the web page or an application
10/Jan/2013:10:32:55 -0800	The timestamp of when the server finished processing the request. The format can be controlled using web server settings
"GET /apache_pb.gif HTTP/1.0"	This is the request line that is received from the client. It shows the method information, in this example GET, the resource that the client was requesting, in this case /apache_pb.gif, and the protocol used, in this case HTTP/1.0
200	This is the status code that the server sends back to the client. Status codes are very important information as they tell whether the request from the client was successfully fulfilled or failed, in which case some action needs to be taken. 200 in this case indicates that the request has been successful
2326	This number indicates the size of the data returned to the client. In this case 2326 bytes were sent back to the client. If no content was returned to the client, this value will be a hyphen "-"
"http://www.google.com"	This field is known as a referrer field and shows from where the request has been referred. You could be seeing web site URLs like http://www.google.com, http://www.yahoo.com, or http://www.bing.com as the values in the referrer field. Referrer information helps web sites or online applications to see how the users are coming in to the web site and this information could be used to determine where the online advertisement dollars should be spent. As you may notice that referrer has an extra "r". That is intentional and originated from the original proposal submitted in the HTTP specification. In browsers like Chrome where users can use incognito mode, or have referrers disabled, the values in the field will not be accurate. In HTML5 the user agent that is reporting this information can be instructed not to send the referrer information
"Opera/9.20 (Windows NT 6.0; U; en)"	This is the user-agent field, and it has the information that the client browser reports about itself. You will see values like "Opera/9.20 (Windows NT 6.0; U; en)", which means that the request is coming from an Opera browser running on a Windows NT (actually Windows Vista or Windows Server 2008) operating system. User-agent information helps to optimize web sites and web applications and cater for requests coming from smaller form factor devices such as the iPad and mobile phones

Now let us look at some of the sample log entries that we generated in Chapter 2 for `MyGizmoStore.com`. Here are sample entries from the */opt/log/BigDBBook-www1/access.log* file. You can see that there are different status codes as well as user agents or browsers.

```
196.65.184.6 - - [28/Dec/2012:06:54:46] "GET /product.screen?productId=CA-NY-
99&JSESSIONID=SD5SL8FF8ADFF4974 HTTP 1.1" 200 992 "http://www.bing.com" "Opera/9.20 (Windows NT 6.0;
U; en)" 597

92.189.220.86 - - [29/Dec/2012:02:58:28] "GET /cart.do?action=purchase&itemId=HYD-
2&JSESSIONID=SD2SL1FF4ADFF5176 HTTP 1.1" 500 1058 "http://www.MyGizmoStore.com/oldlink?itemId=HYD-2"
"Mozilla/4.0 (compatible; MSIE 6.0; Windows NT 5.1; SV1; .NET CLR 1.1.4322)" 604

189.228.151.119 - - [30/Dec/2012:18:18:50] "GET /product.screen?productId=8675309&JSESSIONID=SD6S
L9FF2ADFF6808 HTTP 1.1" 404 3577 "http://www.MyGizmoStore.com/product.screen?productId=CA-NY-99"
"Opera/9.01 (Windows NT 5.1; U; en)" 916

218.123.191.148 - - [31/Dec/2012:04:28:45] "GET /category.screen?categoryId=BLUE
GIZMOS&JSESSIONID=SD0SL1FF1ADFF7226 HTTP 1.1" 500 2992 "http://www.MyGizmoStore.com/category.
screen?categoryId=BLUE GIZMOS" "Mozilla/5.0 (Macintosh; U; Intel Mac OS X 10_6_3; en-US)
AppleWebKit/533.4 (KHTML, like Gecko) Chrome/5.0.375.38 Safari/533.4" 928

78.65.68.244 - - [31/Dec/2012:02:22:40] "GET /category.screen?categoryId=ORANGE
WATCHMACALLITS&JSESSIONID=SD1SL5FF9ADFF7146 HTTP 1.1" 200 2120 "http://www.bing.com" "Mozilla/4.0
(compatible; MSIE 6.0; Windows NT 5.1; SV1)" 338
```

Searching and Analyzing Indexed Data

Searching and analyzing machine or log data can provide tremendously useful intelligence on how the applications, systems, web servers, load balancers, and firewalls are working. This information can also be used for debugging, root cause analysis, and, in general, getting a deeper understanding of external or internal customer behavior in terms of usage or buying patterns. By analyzing machine data, enterprises can start asking questions that haven't been thought of before to find out what is happening with the IT infrastructure.

We will use the sample data of `MyGizmoStore.com` and see if we can find answers to a number of questions that typical IT organizations would like to ask about their web sites and applications. We can start by trying a few simple search commands to see if the field names are properly aligned with the information in Table 3-1. In the Splunk search bar, type `sourcetype=access_combined_wcookie` and you can see in Figure 3-1 that 8,688 matching events are retrieved. This number will vary based on the customizations you have made to the sample data generator used in Chapter 2.

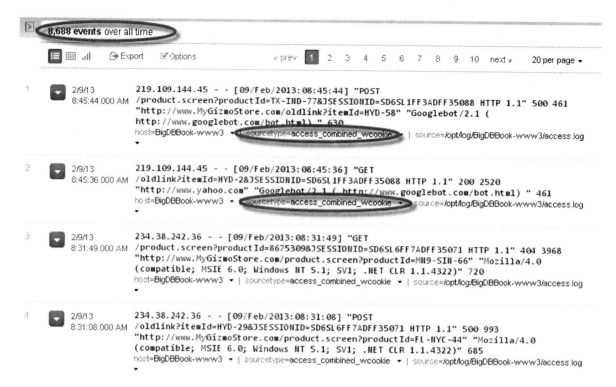

Figure 3-1. Total number of events for MyGizmoStore.com

To further validate the total number of events, we can find how many events are related to each of the hosts BigDBook-www1, BigDBook-www2, and BigDBook-www3. The total number of events for all of the hosts should match up to the earlier result we got using the sourcetype field search. Type host=BigDBook* into the search bar and you will see in Figure 3-2 that 8,688 events are retrieved and the value for host field is highlighted in all events. The number of events is equal to the number that we got from the previous search.

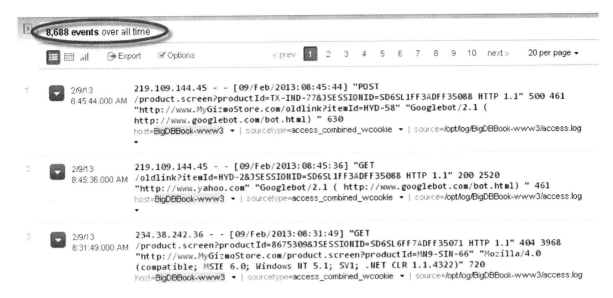

Figure 3-2. *Total number of events for all hosts*

We can also check on individual fields to further validate the data. The sample data generator by default had the following categories:

- BLUE_GIZMOS
- RED_GADGETS
- WHITE_WIDGETS
- ORANGE_WATCHMACALLITS
- PURPLE_DOOHICKEYS
- BLACK_DOODADS

You can go the left bar in the Splunk search app as shown in Figure 3-3; this side bar always presents the default fields and additional fields, which Splunk calls fields of interest. These are defined as fields that show up in 50 percent or more of the events of that particular index.

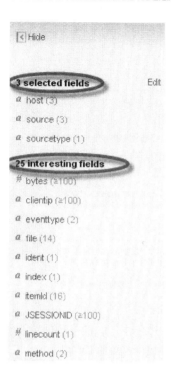

Figure 3-3. *Side bar with selected and interesting fields*

You can see if we have got a field where we can find the list of categories. Because we cannot find a categories field in either of the fields lists, click on the "View all 44 fields" link, which is at the bottom. This will bring up a dialog box that shows the complete list of fields. You can see `categoryId` listed as the third field from the top, as shown in Figure 3-4.

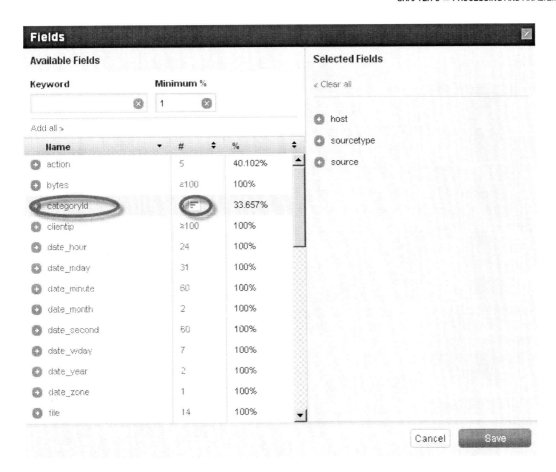

Figure 3-4. *List of fields*

To see the list of categories, click on the small bar graph icon shown in column 2 of the categoryId field. You will see the list of categories grouped across all the events in the popup as shown in Figure 3-5, and the list in the popup matches the list of default categories that we used in the sample data generator. In addition, the dialog box presents the category field values; it also shows some summary statistics about the field, such as the total count of events for each category and the percentage. Additionally, it presents a bar chart, making this a very compelling and easy way to gain a good understanding of a specific field.

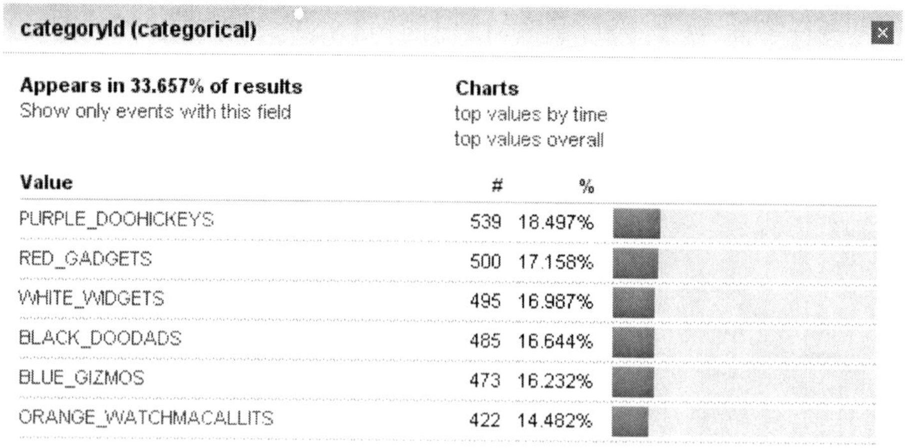

categoryId (categorical) ☒

Appears in 33.657% of results
Show only events with this field

Charts
top values by time
top values overall

Value	#	%	
PURPLE_DOOHICKEYS	539	18.497%	▓
RED_GADGETS	500	17.158%	▓
WHITE_WIDGETS	495	16.987%	▓
BLACK_DOODADS	485	16.644%	▓
BLUE_GIZMOS	473	16.232%	▓
ORANGE_WATCHMACALLITS	422	14.482%	▓

Figure 3-5. *Categories list*

You can always add additional fields to the side bar. We can do this while we are in the "Field" dialog box; click on the arrow sign against the field action, which is first in the list of fields, and click the Save button. You can do a quick validation on the action field. In our sample data generator, we have used the following methods for the actions that MyGizmoStore.com visitors can perform:

- Purchase

- Add an item to the shopping cart (Addtocart)

- Remove an item from the shopping cart (Remove)

- View the catalog of gizmos (View)

- Change the quantity of an item in the shopping cart (Changequantity)

In the left bar, you can now see that the action field is listed in the selected fields list. Click on the small bar graph icon for that field. Figure 3-6 shows the popup with the list of methods for the action field; these match with the list we have. We can see that the view method has the highest number of events. You will analyze this as you learn how to process the events using SPL.

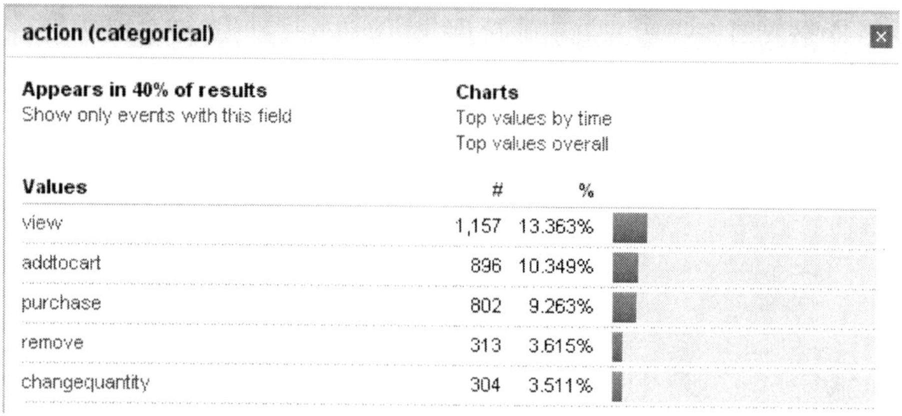

action (categorical) ☒

Appears in 40% of results
Show only events with this field

Charts
Top values by time
Top values overall

Values	#	%	
view	1,157	13.363%	▓
addtocart	896	10.349%	▓
purchase	802	9.263%	▓
remove	313	3.615%	▌
changequantity	304	3.511%	▌

Figure 3-6. *List of action methods*

The basic checks and validations that we have just performed on the sample data helped us to establish that we got the right data; now we can proceed to work on processing and analyzing the data set in detail. The Splunk search application provides a very useful time picker that helps us zoom into a subset of the data based on time, which can be useful for familiarizing ourselves and validating the data. We will employ the same query we have used before, which is:

```
sourcetype=access_combined_wcookie
```

The results are shown in Figure 3-7. The graphical time line shows the events in a bar graph across the range of one month, in this case between Dec 28, 2012 and Jan 27, 2013. The time picker on the top right-hand side shows that the option we have used is "All time", which in our case maps to 30 days of sample data we have generated. You will also see the field sourcetype is highlighted against each event listed.

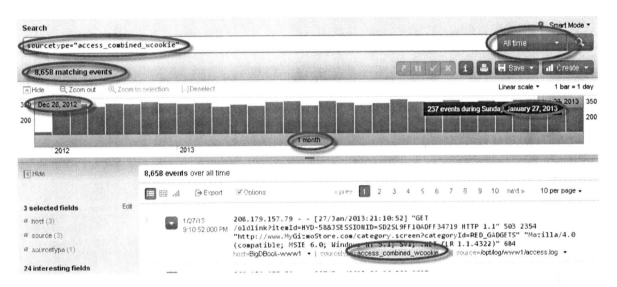

Figure 3-7. *Splunk Search App with time picker*

In real-world use cases, you may be working with data sets that could be spread over several months or years; or you could have millions of events over a very short span of time. In any case, you would need to drill down into a specific time window and find out what is happening. This is where the Splunk time picker comes in very handy. Clicking on the time picker shows us different time lines that can be used, as shown in Figure 3-8. The listed options allow us to view the events in short spans, such as minutes to hours to days, or in custom time periods that can be specified using the "Custom time..." option.

Figure 3-8. *Splunk time picker*

Select the "Last 7 days" option in the time picker and see how the timeline bar gets updated. In Figure 3-9, we can see the event distribution between Jan 21, 2013 and Jan 28, 2013, and you will see 1,923 events.

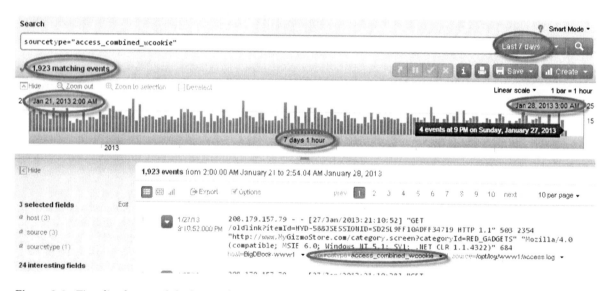

Figure 3-9. *Time line bar graph for last seven days*

Splunk provides time range commands that we can use in the search bar to control the behavior of time periods in which the events have to be processed. You will learn about these time range commands in Chapters 5 and 11.

To process and analyze real-world use cases, we would need to know more than the number of events that we saw against the fields in left bar and time line. This is where SPL comes in. It provides easy yet powerful ways to analyze the data loaded into Splunk. In the next sections of this chapter you will learn and explore different SPL commands in the following categories:

- Reporting
- Sorting

- Filtering
- Adding and evaluating fields
- Grouping

Reporting

Reports are essential for every IT organization and for business leaders in enterprises. They are used daily to get information about the current state of IT systems, or how the business is functioning in terms of sales, manufacturing, shipping, and so on. Typically, reports aggregate information to present summaries that are easily consumable for the intended audience. We will explore reporting commands in Splunk starting with top. This command comes in very handy to answer the following type of questions that are most commonly asked in IT organizations that have web sites or web applications:

- Which are the top browsers?
- Which are my top five IP addresses?
- Which are the top referral web sites?

The top command can be used with one or more fields that are available in a Splunk event. It works on the given dataset, finds the most frequently occurring field values, and returns the count and percentage for the fields specified after the command. By default, the top command returns the top ten, but you can control the number returned using the limit option. You may or may not need all of the columns returned by the top command in every search. Splunk lets you eliminate some of the fields from search results using the fields command. The columns that need to be eliminated have to be specified with a – in front.

Which Are the Top Browsers?

In Table 3-1, we showed different fields that are available in combined access logs. One of the fields is user agent, which has the information that the client browser reports about itself. Online retail stores such as MyGizmoStore.com would like to know which are the top user agents (browsers) that customers are using to view the products or make purchases. Knowing which browsers are being used will help MyGizmoStore.com to optimize the online store pages and make sure that the pages are rendered properly in the most popular browsers that the users are using to get to the web site. The top command makes it really easy for us to get that information. Before you start using top, make sure that you have selected the "All time" option in the time picker. The useragent field contains the browser information. Splunk automatically mapped the fields in Table 3-1 to its own field names while indexing, making it easy for us to find the information or field we are looking for. Type in the following into the search bar:

```
sourcetype=access_combined_wcookie
| top useragent
```

When the first command is a search, you don't have to type it, as it is assumed to be so. The first clause of the above search is equivalent to: search sourcetype=access_combined_wcookie. The pipe (|) sign in our search passes the results from each Splunk command as inputs to the next command. Sometimes it passes a specified subset of the data from previous search evaluations to the next set of evaluations. In this search, you will get the subset of events based on the sourcetype field and only those events piped into the top command to find the top browsers. As you can see in Figure 3-10, the search found 6,254 matching events and found the top browsers along with their count and percentage. You can see that Mozilla 5.0 on Mac OS was the most popular browser, closely followed by the same browser on Windows.

Figure 3-10. *Top browsers*

___

■ **Note**   In SQL terms, you could compare this to using a *TOP* keyword along with a select statement in MySQL or using *rownum* with a where clause in Oracle. Both cases produce the same result set.

___

Splunk allows us to save the search results so that we don't always have to type in the search query again to see the results. As seen in Figure 3-11, click on the "Save" link at the top right-hand side of the Splunk search application. In the drop-down list, you will see three different options:

- *Save results* option—allows us to save the current results from the search

- *Save search* option—used to save the search that can be retrieved back using the "Searches & Reports" link in the menu of the Splunk search application

- *Save & share results* option—allows us to save the search query and also share the results with other Splunk users who have been provisioned with the right security role to access the searches. This feature in Splunk makes it a collaborative environment in which one or more power users can create simple to complex searches which can then be shared with a set of business users

Figure 3-11. *Save search*

When you click on the "Save search" option, it brings up the dialog shown in Figure 3-12. This dialog box provides options to keep the search private or share it with other Splunk users. You can also see an option to accelerate the searches that helps to optimize the searches so that the historical sets of events are already summarized to process the searches faster. You will learn more about search and report acceleration in Chapter 10. For this example, we will select the radio button for the option "Share as read-only to all users of current app" and click on the Finish button.

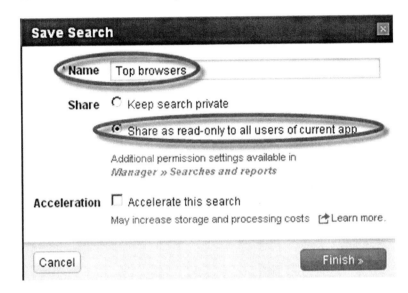

Figure 3-12. *Save search options*

On successfully saving the search, Splunk presents a unique URL that is similar to `http://<hostname>:<port#>/en-US/app/search/%40go?s=Top%20browsers`. This URL can be shared with other users who can then get direct access to the server where this search has been saved. As mentioned earlier, saved searches can be easily accessed via the "Searches & Reports" link in the main menu of Splunk search app. Clicking on that link will show you the list of accessible reports. Figure 3-13 shows the report that we have just saved.

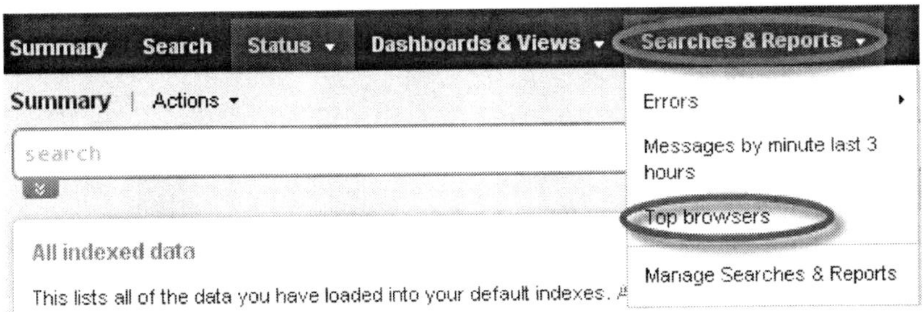

Figure 3-13. *Saved reports*

Saved searches can also be accessed using the "Manager" menu, by clicking on the "Searches and reports" link under the Knowledge section. This will show a detailed list of searches and reports as seen in Figure 3-14. The key difference in accessing the reports via this alternative method is that you will get access to additional actions for each of the reports, such as Delete, Move, and Disable. The available actions are shown on the right-hand side of the table based on the security roles that have been provisioned to the user. You will also be able to set alerts for the saved searches by clicking on the individual search and specifying the alert options. We will explore Splunk alerts in Chapter 5.

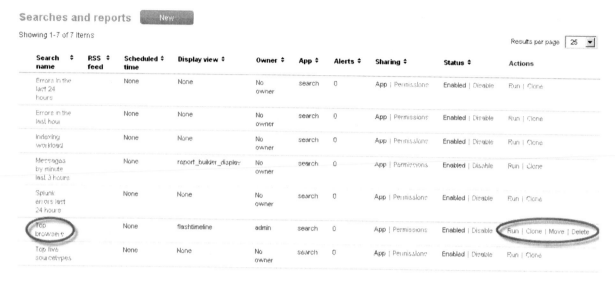

Figure 3-14. *Running saved searches*

Top Five IP Addresses

We will make use of the `top` command to find the top five IP addresses that have made requests to `MyGizmoStore.com`. In this search we will also use the `limit` option along with the `top` command to get only the top five values in the report. In the left side bar, we can see `clientip` as one of the fields; this field holds the IP address where the individual client requests have come from:

```
sourcetype=access_combined_wcookie
| top limit=5 clientip
```

Splunk processes 8,688 events and reports the top five (this is where the `limit` option can help) IP addresses that have been hitting `MyGizmoStore.com`, as shown in Figure 3-15. In Chapter 4, you will see how we can use Splunk Technology Add-Ons or Apps to make more sense of these IP addresses either by finding the geographical location or the domain name for the IP addresses.

Figure 3-15. *Top five users based on IP addresses*

Which Are the Top Referral Web Sites?

These days, users mostly reach web sites by clicking on third-party web sites or search engines such as Google, Yahoo, Bing, and so on. In some cases, users find online advertisements that pop up on the web sites they are looking at and they click on them to get into a web site and take some action. For online retail stores such as MyGizmoStore.com, it makes perfect sense to find where the users are coming from, what keywords are really working, and where to spend online advertising dollars.

For our MyGizmoStore.com, we want to find out which are the top referrer web sites other than MyGizmoStore.com itself. We will use the following search, in which we are going to find the events where the referrer does not have the MyGizmo in the value. We do this by using != which means "not equal". The pipe sends the results from the first clause into the top command to provide a report on the sites that are referring to MyGizmoStore.com and finally eliminate the percent field from the result set:

```
sourcetype=access_combined_wcookie  referer != *MyGizmo*
| top referer
| fields - percent
```

Note that the first clause of the search implies an AND between the sourcetype and the referer. By default the search command assumes an AND, so this clause is equivalent to:

```
sourcetype=access_combined_wcookie
AND referer != *MyGizmo*
```

Figure 3-16 shows that 2,256 events matched the search condition, and Bing, Yahoo, and Google are shown as the top three referrer sites other than MyGizmoStore.com itself. You will also notice that the percent column is eliminated from the results.

Figure 3-16. Referral sites to `MyGizmoStore.com`

Although it is useful to get reports on different sets of top activities or items, you also want to know what is not working. This is where we can take advantage of the reporting command `rare`. This command does exactly the opposite of what the `top` command does. We could just replace `top` with `rare` in all of the previous searches to find which browsers are not common, or which referrers that we were expecting to provide more hits haven't been doing so, and so on.

The next reporting command we will explore is the `stats` command. This command is the workhorse, as it is used to calculate aggregated statistics over a given set of data. The `Stats` command comes with several functions such as `count`, `average`, `min`, `median`, `mode`, `sum`, and so on, where each of these functions takes the values from multiple events and groups them on certain criteria to provide a single value or more meaningful data about a set of events. This is very similar to SQL aggregation.

Online web sites such as `MyGizmoStore.com` could be failing to process requests coming from clients for different reasons, and web servers such as Apache and IIS return HTTP status codes that are in the classes 4xx and 5xx. Table 3-2 provides descriptions of some of the common 4xx and 5xx status codes.

Table 3-2. Description of HTTP status 4xx and 5xx

HTTP status code	Description
400	Bad Request—The request sent by the client could not be understood by the server due to incorrect syntax
401	Unauthorized—The request sent by the client requires user authentication
403	Forbidden—The request sent by the client is understood by the server but the server refuses or declines to process the request. For example, most web sites prevent users from browsing the directory structure of the site
404	Not Found—A particular resource requested by the client, such as a web page or an image, could not be found on the server
405	Method Not Allowed—The request sent by the client has a method that is not allowed
407	Proxy Authentication Required—This is very similar to the 401 status code, but it means the request sent by the client must first authenticate itself with the proxy server
408	Request Timeout—This means that the server has timed out waiting for the client to produce a request

(continued)

Table 3-2. (*continued*)

HTTP status code	Description
500	Internal Server Error—The server encountered a condition that prevents it from fulfilling the request
501	Not Implemented—At present the server does not support the requested functionality
502	Bad Gateway—The server receives an invalid response for the request it has sent to an internal server to fulfill the incoming request
503	Service Unavailable—The server is overloaded and unable to process incoming requests
504	Gateway Timeout—The server does not receive a response from another server such as LDAP server to process the request further
505	HTTP Version Not Supported—The server does not support the HTTP protocol version that was used in the client request

To explore the `stats` command, we will look at a couple of use cases that are typical of online retail stores like `MyGizmoStore.com`:

- Finding all of the events where a client request failed producing a HTTP status code like a 404, and finding the count of them based on different hosts or web servers

- Finding all the events that have purchase actions and the count of the products that are part of those events

Web sites need to keep track of status codes such as 404, which means that the requested resource is not found or available. This resource could be an HTML page, an image, or similar items. Tracking the specific status codes or class of status codes such as 4xx messages helps web site maintainers fix the web site and make sure that users are finding what they are looking for. We'll work on these two use cases in a step-by-step fashion, in which we will first find the events that have 404 status.

How Many Events Have HTTP 404 Status?

The `status` field is the one that has information about the HTTP status code that is returned for each request coming from the client. Splunk provides a simple way of adding conditions as criteria to a search using the fields and values associated with it. To find all of the events that have a 404 code, we will use the following search:

```
sourcetype="access_combined_wcookie" status=404
```

Figure 3-17 shows the results for the above search. You can see that there are 696 events in total that had a 404 status and that each event that met the search condition has a 404 status code.

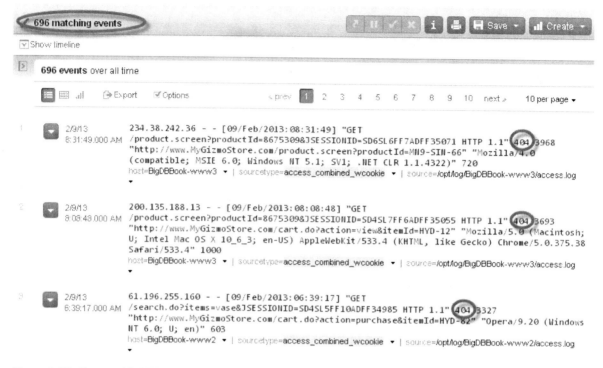

Figure 3-17. *Events with HTTP status 404*

We will take the search that we have just built and pipe (|) the results to a stats command that uses the count function. The count function will aggregate the count of each host. Enter the following in the search bar:

```
sourcetype=access_combined_wcookie status=404
| stats count by host
```

Figure 3-18 shows that 696 events have a 404 status code, and how they are distributed across the hosts BigDBook-www1, BigDBook-www2, and BigDBook-www3.

Figure 3-18. *Events with HTTP status 404 for each host*

The second use case involves using the stats command with products that are available on MyGizmoStore.com. A very useful report would be one showing which products are best-selling. Instead of finding out all of the products that users have looked at, it would be more interesting to know the products that were involved in the purchase process or the ones that were actually bought. What this means is that we would want to know how many times the product has been part of the purchase process. The stats command will help to count the products using the count function. First find the events that are part of a purchase request, and then use the action field that was added to the selected field list earlier.

How Many Events Have Purchase Action?

Use the following search in which the value of field action as purchase is specified:

```
sourcetype=access_combined_wcookie action=purchase
```

As seen in Figure 3-19, there are 767 events that match the condition, and each event shows the URL action as purchase.

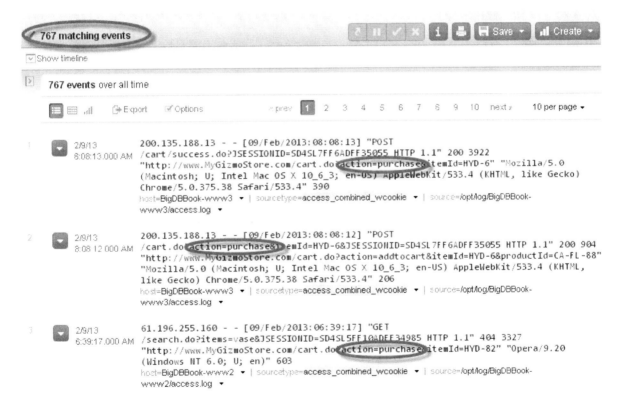

Figure 3-19. *Events with purchase action*

Now that we have got the events that have the purchase action, the next step would be to do a statistical aggregation and find the count for each of the products that are involved in these events.

List of Products That Are Part of a Purchase

We will take the query we just used and we will pipe (|) the results to a stats command with a count function. The count function will aggregate the count the each productId. Enter the following in the search bar:

```
sourcetype=access_combined_wcookie action=purchase
| stats count by productId
```

Figure 3-20 shows how the same 767 events are used to perform the statistical analysis, and productid along with the corresponding count is listed in a table. The sample data generated in Chapter 2 only contains the product identifier, but you don't have the actual description that corresponds to each product identifier. In Chapter 10 you will see how you can make use of the lookup command in Splunk to correlate data across different sources such as getting product description or price for product identifiers from a database or a CSV file and linking them so that the reports become more business-friendly and easier to understand.

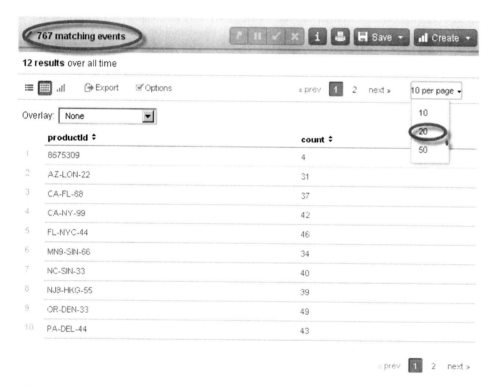

Figure 3-20. *Count of products that have been part of a purchase action*

The stats command has a function called distinct_count or dc that counts the distinct values for a particular field. The *sum* function can be used to calculate the totals of a column. For example, we can take the search that we built earlier to find the top referrers that don't have MyGizmoStore.com, and extend that to find the total referrals that are not from MyStoreGizmo.com. The updated search will look as follows:

```
sourcetype=access_combined_wcookie   referer != *MyGizmo*
| top referer
| fields - percent
| stats sum(count)
```

We can replace the sum function in this query with the avg function to find the average referrals from each source that is not MyGizmoStore.com. The search for that would look as follows:

```
sourcetype=access_combined_wcookie    referer != *MyGizmo*
| top referer
| fields - percent
| stats avg(count)
```

You will want to show the report with all of the product identifiers and a total count for each one of them, but you also will want a grand total column for the total number of products sold. This is where you can make use of the addcoltotals command, which adds up all the aggregated values and comes up with a new event that has the total for the field. The search for that is:

```
sourcetype=access_combined_wcookie   action=purchase
| stats count by productId
| addcoltotals labelfield=Total label=AllProducts
```

As you can see in Figure 3-21, all of the information you had in the previous report has been retained, but now you can add all of the products in the report. In addition to that, the addcolTotals command allows you to create a new event that will retain the summary in an event called Total; it is labeled AllProducts. Because we have more than 13 results, we can make the Splunk search App display 20 records per page by changing the number of records per page to 20 using the page scroller option highlighted on the right-hand side, as seen in Figure 3-21.

Figure 3-21. Total for all the products

So far we have looked at the SPL reporting commands top, rare, and stats, and we've explored typical reporting use cases that we can solve using these commands. Other SPL reporting commands are chart and timechart. We will explore them in Chapter 4 when we start visualizing the data.

Sorting

An important requirement for reports is being able to sort the results, as that will help to understand the information we see in the report. Sort is the main SPL command in the sorting category. As you saw in Figure 3-21 the results for the number of products that have been counted are not sorted in any particular order which makes it difficult to understand the highs and lows for the products. The sort command sorts resulting events using the list of fields specified after the command. The ascending and descending order is controlled using (+) or (-) sign before the field. We will take the same query that we used in our previous example and pipe the results into a sort command to get the results in descending order so that we know what the most popular products are:

```
sourcetype=access_combined_wcookie  action=purchase
| stats count by productId
| sort -count
```

Figure 3-22 shows the sorted products in descending order. You will see that the productid OR-DEN-33 was most popular, with a count of 49.

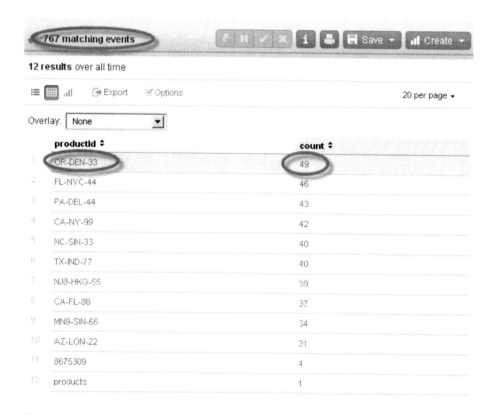

Figure 3-22. *Products sorted in descending order*

Filtering

The third category of SPL commands is filtering commands. Filtering commands take a set of events or results that are piped into them and filter them into a smaller set of results. Filtering helps to speed up analysis and also helps you to zoom into specific results where additional intelligence can be gathered.

If you want to limit the number of events that have to be processed and analyzed, Splunk provides the head command, which allows you to process only a subset of events. Take a sample use case of finding the top products for the first 100 qualifying events. To do this, take the previous search query where we found the products count and sorted them; now we will make use of the head command to find which products are most popular in the last 100 qualifying events.

In the following search, pipe the events that meet the condition of having a purchase action and use the head command to take the first 100 events and then aggregate the result set with count function to sort the final results:

```
sourcetype=access_combined_wcookie   action=purchase
| head 100
| stats count by productId
| sort -count
```

Figure 3-23 shows the popular products using the first 100 events that have qualified for purchase action. You can see the product OR-DEN-33 is the most popular one.

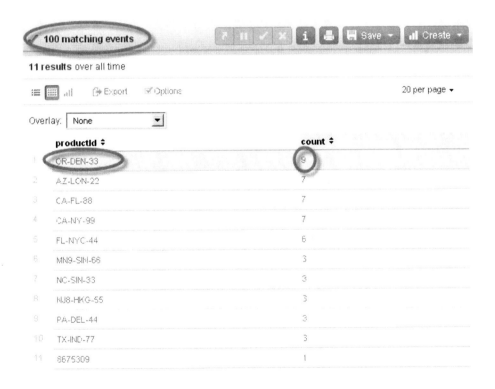

Figure 3-23. *Most popular product for the first 100 events*

Splunk also provides a `tail` command, which takes the last n number of events and processes them for further analysis. We can modify the last search query and replace the head command with `tail`, as follows:

```
sourcetype=access_combined_wcookie  action=purchase
| tail 100
| stats count by productId
| sort -count
```

Figure 3-24 shows the updated results using the last 100 events instead of the first 100. The most popular product is PA-DEL-44 instead of OR-DEN-33. Although the dataset is small, these techniques can be used to analyze the first or last n events and identify new trends or patterns to identify which product might be getting more popular and if there is a particular reason for it, such as seasonal change or holiday.

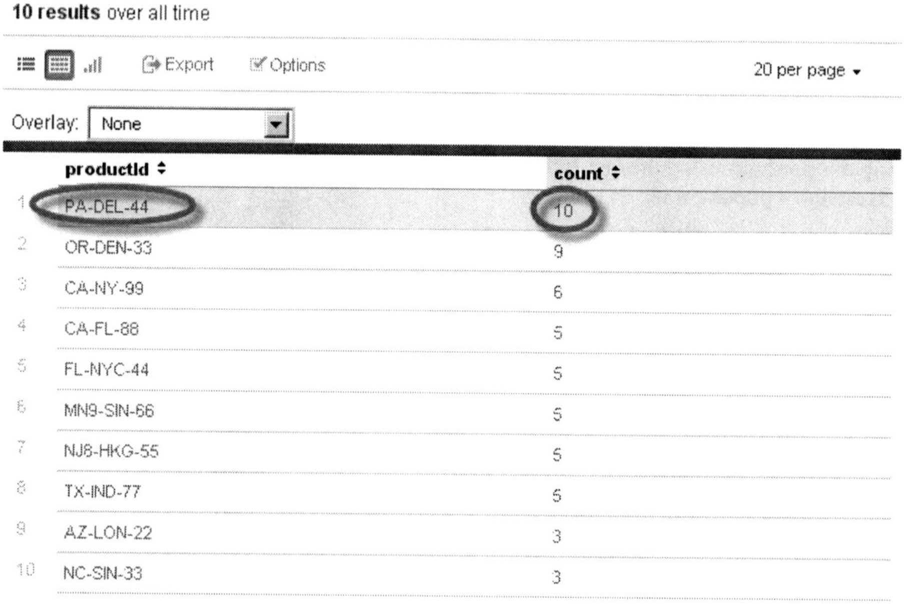

10 results over all time

| | | | Export | Options | | 20 per page ▾ |

Overlay: None ▾

	productId ⬍	count ⬍
1	PA-DEL-44	10
2	OR-DEN-33	9
3	CA-NY-99	6
4	CA-FL-88	5
5	FL-NYC-44	5
6	MN9-SIN-66	5
7	NJ8-HKG-55	5
8	TX-IND-77	5
9	AZ-LON-22	3
10	NC-SIN-33	3

Figure 3-24. Most popular product based on last 100 events

Two other useful filtering commands are dedup and `where`. The Dedup command is used to remove redundant data, meaning that the command keeps only the first count of results for each combination of the values of the specified fields and discards the remainder. We will make use of the dedup command extensively in Chapter 10.

Adding and Evaluating Fields

The next category of commands is used for filtering, modifying, and adding fields. We will look at one of the most powerful commands, which is eval. An eval command can be used to calculate an expression and store the value in a new field. Eval can work with arbitrary expressions, including mathematical, string, and boolean operations. As we have seen with the `stats` command, eval provides several functions. Among of the standard functions are case, if, several math functions such as abs, ceiling, floor, log, round, and so on, and string functions including rtrim, split, substr, tostring, and so on.

We will explore the eval command with an interesting real-world use case: identifying users who are window-shopping as opposed to those who have made purchases. This use case is applicable to both brick and mortar and online retail stores. Because MyGizmoStore.com is an online retail store, we will have customers who browse the web site and look at the products but who don't make any purchases. It would be interesting to find the ratio of users who are just viewing the pages to those who are actually making purchases.

You can use the eval command to find the basic distribution of HTTP status code that ended up in 200, which means that the requests have been successful, and see how that number compares with the total for the rest of HTTP status codes in our sample MyGizmoStore.com data. This doesn't get us the detail about the ratio of users purchased, but it does allow us to get started at a broader level of success and failure across all events.

Splunk also allows you to use the wild char * to signify all. Typically, you will have different types of access to log files and most likely you will be processing and analyzing them all. To do that, change the search so that sourcetype is equal to access_*; this would pick all sources that start with access_. Use an eval command to evaluate a condition that looks at whether an individual event has status code 200; if it does, it is captured into OK and all other events are captured into FAILED. The values are then stored into a field SuccessRatio, which is aggregated. The results can be seen in Figure 3-25. You can see that the search processed all of the 8,688 events of sample MyGizmoStore.com data that we started with and it shows that we got more events in the FAILED category than in OK.

```
sourcetype=access_*
| eval SuccessRatio = if (status==200, "OK", "FAILED")
| stats count by SuccessRatio
```

Figure 3-25. *Number of status code 200 versus rest of status codes*

Now that you have a feel for the eval command, let us figure out the ratio of page views to purchases. To do that, use the events that have method GET and pipe those events into the stats command to find out the actual page views count. You then use the eval command within another count where we are looking if the event has an action that is equal to purchase. You will also use the AS argument of the stats command to rename the resulting fields Page_Views and Actual_Purchases. Type the following into the search bar:

```
sourcetype=access_*    method=GET
| stats count AS Page_Views,
count(eval(action="purchase")) AS Actual_Purchases
```

Figure 3-26 shows the results that 6,176 events have been processed that had the GET method, resulting in 6,176 page views, and only 144 are actual purchases.

Figure 3-26. *Page views versus actual purchases*

Splunk's eval command can be used to calculate the percentage difference between the values as well. Extend the search that you have just done and pipe the results into another eval command that calculates the percentage of actual purchases made compared with page views. This percentage is stored in a field called PCTofPurchases.

```
sourcetype=access_* method=GET
| stats count AS Page_Views,
        count(eval(action="purchase")) AS Actual_Purchases
| eval PCTofPurchases =((Actual_Purchases*100)/Page_Views)
```

Additional commands in this category of filtering, modifying, and adding fields are lookup, rex, replace, and fields. We will make extensive use of the lookup command in Chapter 11 where you will work with the Airline On-Time Performance project.

Grouping

The last category of commands is grouping results. The idea of grouping is to identify patterns in data that is grouped together. Transaction is the SPL command that allows to group events into a transaction. Events are grouped together if the definition constraints described along with the command are met. The constraints specified in the transaction command identify the beginning and the end of the transaction. This is very similar to transaction demarcation in the traditional database world.

We can make use of the transaction command to further analyze the use case of window-shopping versus purchase. In many cases, there will be users who add items or products into their shopping cart and never make a purchase, or users who spend a fair amount of time browsing, adding items to their shopping cart but never closing the transaction.

For our example, we will look at grouping the events related to a particular session id (JSESSIONID) and coming from the same IP address (client), where the first event contains the string addtocart and the last event contains the string purchase. Before we start work on our search, let's see what this would look like if you took a small set of sample events as shown in Figure 3-27. The highlighted areas in the rectangle show that you got the same IP address and session id and you also have addtocart and purchase actions in the related events. Splunk would find these relationships across the events and group them as transactions. In the sample set, we got two transactions that involve IP addresses 237.15.107.90 and 60.66.3.96. You could also group the transactions using the cookie values if they are present in the requests coming from a client.

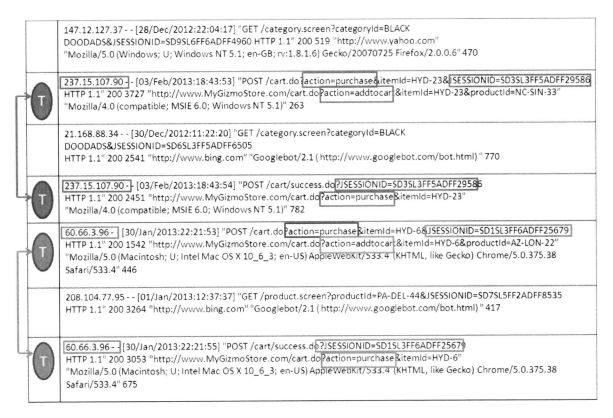

Figure 3-27. *Transaction grouping*

Now enter the following in the search bar and see how Splunk would group the events into transactions:

```
sourcetype=access_combined_wcookie
| transaction JSESSIONID clientip startswith="addtocart" endswith="purchase"
```

As seen in Figure 3-28, 609 events match the grouping model you have made and you can see that each grouped event has addtocart and purchase as the actions and the same IP address and session id.

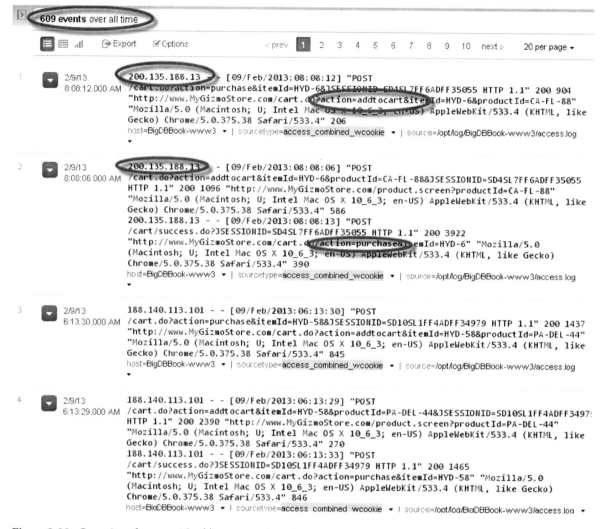

Figure 3-28. *Grouping of events with* addtocart *and* purchase

Another use case for the transaction command is grouping the events that have purchase action from the same clientip where each session is no longer than 10 minutes and includes no more than three events. These kinds of combinations and event groupings can be used to identify users who are spending either the shortest time to conclude their purchases or making long-running transactions. In the following search, we are using maxspan and maxevents attributes along with clientip field to group the events:

```
sourcetype=access_combined_wcookie    action=purchase
| transaction clientip maxspan=10m maxevents=3
```

Figure 3-29 shows that 519 events matched the grouping, and you can see that all of those events have purchase action.

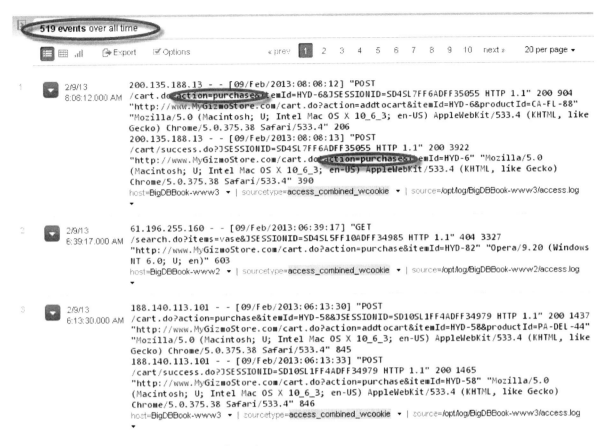

Figure 3-29. Transactions within a span of 10 minutes

We will wrap up this chapter by looking at one last use case that is very useful when analyzing combined access log data, and it also helps us to exercise some of the commands and concepts that we have looked at in this chapter. The use case is related to how much business we are losing because of failed transactions in our online applications. In the case of MyGizmoStore.com, this would mean that users have failed to complete the purchase process due to errors in the application or the server.

We can make use of HTTP status codes and analyze our sample data set. In Table 3-2, HTTP status codes for the classes of 4xx and 5xx were listed. HTTP status codes in the range 5xx are related to server errors. Any 5xx error means that something did not go right on the server end. A couple of common examples include status code 500, which indicates a generic internal server error, and status code 503, which means that the server is unavailable to process the request. From the operational standpoint, error codes 5xx mean that there are issues with the server that may need immediate attention. For an online retail store such as MyGizmoStore.com, status 5xx means that the site is unable to perform customer transactions, so the online retail store has probably lost business as customers failed to complete transactions. This could result in customers going to competitor stores or, at the very least, escalated customer support calls.

Look at the 5xx status codes in the sample data or the number of events where the status code is greater than or equal to 500. Use the status field and apply a condition where the status is greater than or equal to 500:

```
sourcetype=access_combined_wcookie    status >= 500
```

Figure 3-30 shows that there are a total number of 1557 events that have a status code greater than or equal to 500.

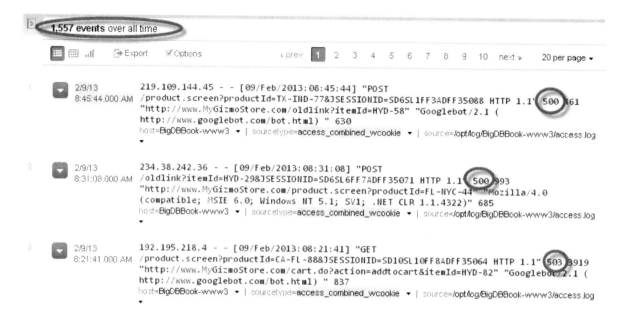

Figure 3-30. Events with status code greater than or equal to 500

You can use the Splunk statistical command `stats` to find exactly what the distribution of events is across the different 5xx status codes. Pipe the previously qualified events into the `stats` command and use the `count` function to count the different status codes. Figure 3-31 shows how the distribution of 1554 events looks between different 5xx status codes. In this case, we only have status codes 500 and 503.

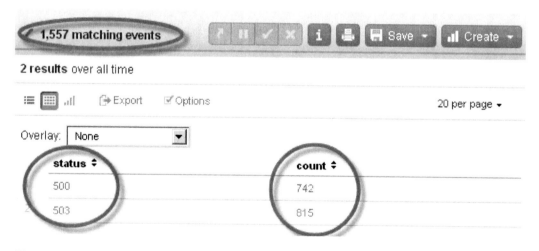

Figure 3-31. Events distribution for status code in 500 range

You can also make use of the `sort` command to sort the status count in ascending or descending order:

```
sourcetype=access_combined_wcookie    status >= 500
| stats count by status
```

Now that we have identified and determined that we have an issue with MyGizmoStore.com servers, we need to identify which of these events where the status code was greater than or equal to 500 were also involved in events that had a purchase action. If there are any qualifying events for that condition, that would translate into loss of business for MyGizmoStore.com. We will take the previous search and add an additional field condition, action=purchase, and find out the number of qualifying events:

```
sourcetype=access_combined_wcookie  status >= 500  action=purchase
```

Figure 3-32 shows that there are 175 events that have matched our condition of purchase action.

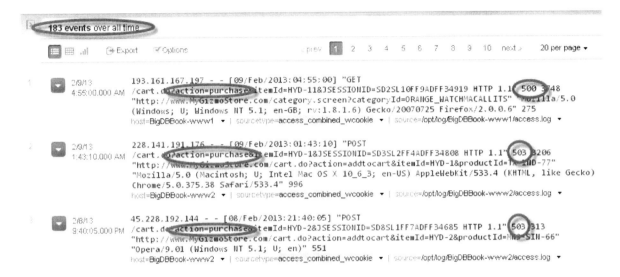

Figure 3-32. *Events having status in 500 and also purchase action*

We can further analyze these 175 events and see what the product distribution is across failed transactions. Splunk provides the *rename* command, which allows you to change or alias the column names to be more business-friendly, as logs may have cryptic field names.

Take the previous search and pipe that into the stats command to aggregate the results by product. The result set will be sorted in descending order. Finally, rename the count field "Product Units In Failed Transactions":

```
sourcetype=access_combined_wcookie   status >= 500  action=purchase
| stats count by productId
| sort -count
| rename count as "Product Units In Failed Transactions"
```

Figure 3-33 shows that the same 175 events are aggregated and the product count column is renamed accordingly.

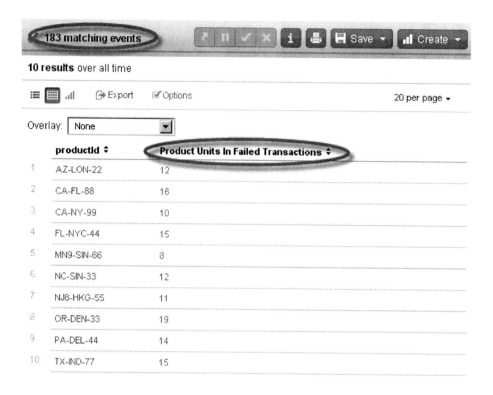

Figure 3-33. *Number of product units in failed transactions*

Summary

In this chapter, we examined the format of the combined access log to see what type of information is available in the log entries to process and analyze. You learned basic commands such as top, rare, stats, sort, addcoltotals, eval, rename, head, tail, and transaction, which helped you to report, sort, filter, and group events to find answers to the questions that enterprises typically need to ask. You also gained an understanding of using time picker and fields to validate data and process data in smaller subsets.

CHAPTER 4

■ ■ ■

Visualizing the Results

In this chapter we will focus on data visualization using reports and dashboards. You will learn about the default Splunk dashboards and new SPL commands that help to chart the data, and see how we can build reports using the Splunk report builder. We will also look at Splunk Apps to visualize data, get familiar with the default dashboards they provide, and work on creating custom dashboards that help to visualize multiple reports.

Data Visualization

Modern enterprises are looking at newer categories of data such as logs, network, clickstream, and social media as inputs to their mainstream data analysis to make better business decisions. What this means is that enterprises now have to deal with processing and analysis of big data. Apart from this, as enterprises expand and collect multiple sources of data, they are seeing huge data growth. Traditionally, this data was delivered in spreadsheets or tabular reports, making it very challenging to find the patterns, trends, and correlations necessary to take action. Not only that, the traditional spreadsheets or similar tools are no longer adequate to process and analyze multiple sources of data that enterprises want to look at holistically.

This is where data visualization comes into the picture. Visualization helps to communicate complex ideas or data patterns with clarity and precision. It helps the viewer focus on the substance of the data by presenting many numbers in a relatively small chart, enabling users to compare different pieces of data far more quickly and easily than by staring at page after page of raw numbers. One of the most important benefits of visualization is that it allows access to huge amounts of data in ways that would not be possible otherwise. Finally, and perhaps most importantly, visualizations give us access to actionable insight. Consider situations in which pricing changes over time for online stores such as MyGizmoStore.com. With visualization, we can quickly understand whether customers are buying more or if they have changed their buying patterns, and so on. Without that insight, we're effectively lost, and the odds of making a good decision are considerably reduced. Whether you are in IT or on the business side, visualization lets you quickly examine large amounts of data, expose trends and issues efficiently, exchange ideas with key players, and influence the decisions that will ultimately lead to success.

You saw in Chapters 2 and 3 how Splunk can work with multiple sources of data and process the data using SPL. The next question is: Does Splunk provide any tools to visualize data and make it easy to analyze big data to identify correlations, trends, outliers, patterns, and business conditions? Yes, of course it does! You will learn how to visualize data in Splunk in this chapter.

How Splunk Deals with Visualization

Splunk provides different ways to build reports and dashboards to visualize the search results coming from the data indexed into it. At the basic level, you can visualize all the data as a set of events in the timeline bar graph. In Chapter 3, you learned how to use the time picker to validate the data; it can also give us quick visual clues across events showing peaks and troughs in the activities. In the case of MyGizmoStore.com's sample data, the time picker lets us visualize any obvious patterns that can be drilled down into for more details. Let's do a quick search by entering `sourcetype=access_*` in the search bar. Figure 4-1 shows the activity across `MyGizmoStore.com` data; we can see that the activity is mostly constant except for some small spikes and drops in a few places. For real-world data, you will most likely see a very different graph with huge spikes at different places on the time line.

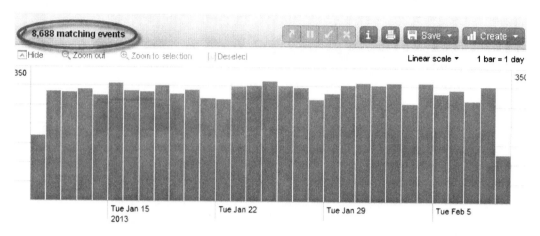

Figure 4-1. *Visualize data using time line*

Although the Splunk time picker is helpful, most of the interesting details come from statistical analysis done across different fields. The searches that we performed in Chapter 3 gave us the data in tabular format. For example, one of the searches we did helped us to find the top browsers for `MyGizmoStore.com`. If we take the same search and enter it into the search bar, it gives us the tabular report shown in Figure 4-2.

```
sourcetype=access_*
| top useragent
```

Figure 4-2. *Top browsers for* `MyGizmoStore.com` *in tabular format*

Splunk provides a very easy way to visualize the tabular data. All you have to do is click on the bar icon highlighted in Figure 4-2. Splunk will transform the tabular report that was showing the top browsers into a visual chart, as shown in Figure 4-3. You can now see the data visually instead of reading raw numbers. We can see that Mozilla 5.0 is the top browser, closely followed by Mozilla 4.0. The chart also shows the close proximity of the first two browsers and the next group of four browsers.

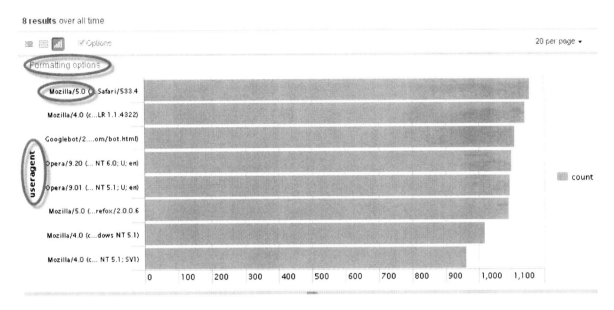

Figure 4-3. *Top browsers for* MyGizmoStore.com *in a column chart format*

Splunk provides different chart types to visualize the data; we will explore this later in this chapter. You can format the chart by clicking on the "Formatting options" link. Here you can select the type of chart that will best help to visualize the underlying data. Not all chart types are applicable for all data. For a Splunk chart to visualize data, it has to be in a structured format. Depending on the type of data structure that you have in the tabular format, Splunk will intelligently enable only those chart types and will gray out the rest of the options, as you can see in Figure 4-4. We will select a bar chart for this report, and we will go ahead and change the title for the chart to "Top Browsers," with X-axis title "Browsers," and Y-axis title "Browser Count."

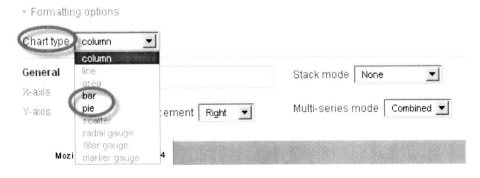

Figure 4-4. *Formatting charts*

The updated chart will look like Figure 4-5. Let's click on "Save" and save the search as "Top Browsers."

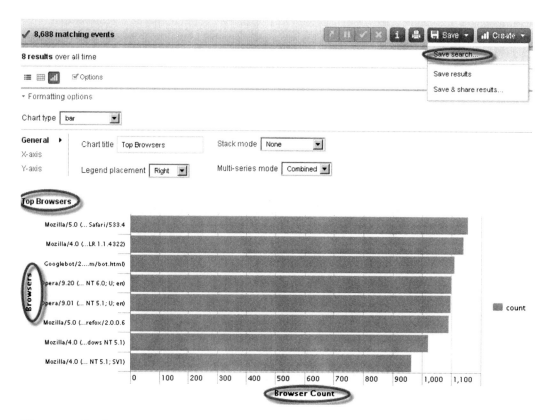

***Figure 4-5.** Top browsers*

The approach that we have taken so far is to build searches and view the data coming from search results. As you saw in Chapter 3, this approach requires a fair knowledge of SPL commands to build the searches. In the case of the top browsers search, we would have to be familiar with the top command. Splunk provides another way of building reports using reports builder. Let's explore report builder with the top five IP addresses search that we built in Chapter 3. With report builder, you don't need to have a thorough understanding of SPL reporting commands such as stats, top, rare, or the charting commands that we will explore in this chapter. You can build the reports using report builder in two steps.

- Define report
- Format report

You can access the report builder by selecting the "Create" drop down list and clicking on "Report . . ." (see Figure 4-6).

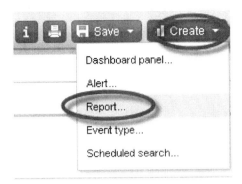

Figure 4-6. *Report builder*

In the Define report step, you can select the time range as you would in the time picker; in our case we will leave it as default, which is "All time." In the Report Data section, you will have to select the type of report that you would like to create. The three available options are:

- Values over time, used for visualizing the trends in a particular field over a selected period of time. Internally, the report builder will make use of the timechart command, which we will introduce to you shortly.

- Top values, used for reports that show the most common field values such as top browsers, top IP addresses, and so on. Report builder makes use of the top command, which you learned about in Chapter 3.

- Rare values, used for reports that want to display most uncommon or least common values. Report builder makes use of the rare command, which you learned about in Chapter 3.

For the top IP address report, we will select the "Top values" option in the report type. Based on the selection, report builder will automatically adjust the remaining choices for report definition. In our case, you will notice that the options for report display have now been hidden. The next option that we have to choose is the field in which we want to find the common or top values. Because we want to find the top IP addresses, we will select the field clientip from the "Fields" drop-down list. The Define report content page will look as shown in Figure 4-7. You will notice that the report builder also provides access to the search command that it is building. By default, the editing for the search is disabled, but for users who are familiar with SPL they can click on the "Define report using search language" link and edit the query. Report builder also defaults the values returned by top command to 1000 instead of 10, as you saw in Chapter 3. Because we only need the top five IP addresses, we will edit the search to return only five by changing the limit=5. Click on the Next Step: Format Report button.

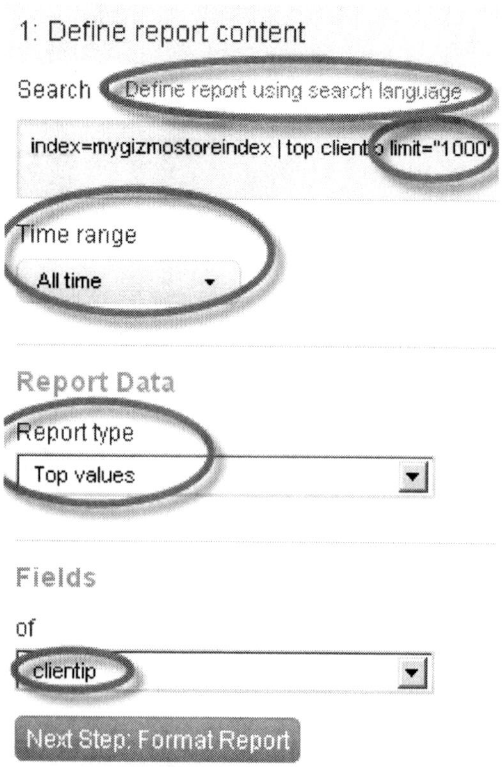

1: Define report content
Search
Define report using search language
index=mygizmostoreindex | top clientip limit="1000"
Time range
All time
Report Data
Report type
Top values
Fields
of
clientip
Next Step: Format Report

Figure 4-7. *Create report definition in report builder*

Report builder will show a column chart by default for the top five IP addresses. You will notice in the chart type options that Splunk has enabled all chart types; this does not mean that the data we got from the report can be visualized with all types of charts. For example, changing the chart type to a "radial gauge" and clicking on the "Apply" button will show the data as seen in Figure 4-8, and you can see that it doesn't make much sense.

Top 5 IP Addresses
0
10
20
30
40
50
60
70
80
90
100
243.189

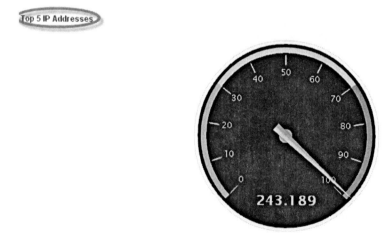

Figure 4-8. *Top IP addresses shown as radial gauge*

We strongly recommend that you experiment with different options and think about the audience to which you are showing these reports in order to choose the right chart type. Before we save this search, let's change the chart title to "Top 5 IP addresses," X-axis title "IP addresses" and Y-axis title "Count," change the chart type back to column and click the "Apply" button. The report should look as shown in Figure 4-9.

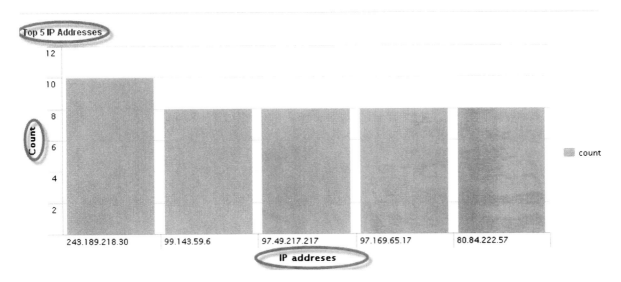

Figure 4-9. *Top IP addresses shown in column chart*

The key difference between using the report builder and working directly with SPL commands is that you would not need to know the SPL commands. You have also seen that report builder offers flexibility by allowing you to switch to search language mode if need be. You would probably still need to have some SPL knowledge to make changes to the default search generated by the report builder, as we did with the limit option for the top command. Save the report as "Top 5 IP addresses chart." You can always access the saved reports using the "Searches & Reports" menu.

You can use the stats, chart, and timechart commands to perform the same statistical calculations on your data. The stats command returns a table of results. The chart command returns the same table of results, but you can use the report builder to format this table as a chart. If you want to chart your results over a time range, you will be using the timechart command. Now that you have learned about the basic forms of visualization provided by Splunk, let's explore two powerful charting SPL commands:

- Chart
- Timechart

As in Chapter 3, we will take different use cases that are related to our online retail store MyGizmoStore.com to explore these commands.

Chart

The Chart command can be used to create a tabular data output that works very well for charting or visualization purposes. You have to specify the x-axis variable or field using the over or by clause. It is designed to work with all of the functions that are available with stats reporting command in Splunk. You can also use the eval command along with Chart, making it a powerful command to get the aggregated data that can be visualized. Different chart types can be created based on the structure of the data that is generated by the chart command. Chart types include:

- Column

- Bar

- Stacked column

- Stacked Bar

- Line

- Area

- Stacked line

- Stacked area

- Pie

- Scatter

- Gauges—Radial, Filler, and Marker

We will look at some of these chart types and how they work with the data that we are going to generate with the following use cases:

- Chart the number of GET and POST for each host

- Chart the purchases and views for each product category

- Which product categories are affected by 404 errors?

- Purchasing trend for MyGizmoStore.com

- Duration of transactions

Chart the Number of GET and POST for Each Host

All of the requests coming into MyGizmoStore.com have either GET or POST. We can start exploring the chart command to visualize the number of GET and POST requests for hosts BigDBBook-www1, BigDBook-www2, and BigDBook-www3. To do this, we will make use of the eval command to find out if the method is a GET or a POST. Once that is determined, the count function is used with the chart command to count the number of GET and POST. We are going to do it by host, which is going to be the field for the x-axis of the chart. Let's enter the following search into the search bar:

```
sourcetype=access_*
| chart count(eval(method="GET")) AS GET,
  count(eval(method="POST")) AS POST
  by host
```

Splunk shows the results in a tabular format that lists GET and POST for each of the hosts. One of the interesting things that you can do in the tabular format report is make use of the "Overlay" feature and have Splunk create a heat map across the values in the tabular data. Heat maps help us to visualize the high and low values easily by using different colors. We can see the tabular report with heat map on in Figure 4-10; as GET has high numbers, we can see the cells with high values are shown in red.

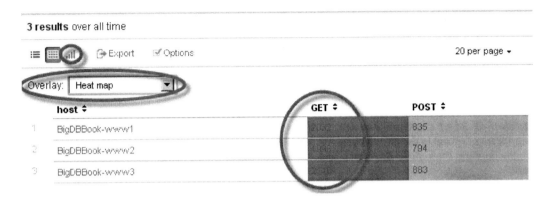

Figure 4-10. *GET and POST requests in tabular format with heat map*

To visualize the tabular data in the chart, you will click on the bar icon, highlighted in Figure 4-10. By default, Splunk picks bar chart; because we want the x-axis to be plotted by host, we will change the chart type to column instead of bar. You will see that the column chart works better when comparing the GET and POST with the amount of data that we have. As mentioned earlier, you will have to explore different charting options to see which one will be better for the generated data structure. Let's change the **chart title to "GET-POST-Chart"** and the y-axis title to "Number of Requests." The updated chart can be seen in Figure 4-11. Save the chart as "GETPOSTRequestsChart."

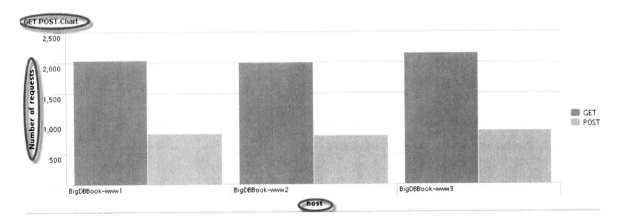

Figure 4-11. *Get and Post Requests in a column chart*

Chart the Purchases and Views for Each Product Category

We can reuse the work we did in Chapter 3, where we found the comparison between actual purchases and product views by the users of MyGizmoStore.com. Instead of using a stats command, we will use the chart command. We will give the resulting columns a more meaningful name. Let's enter the following into the search bar:

```
sourcetype=access_* method=GET
| chart count AS views,
  count(eval(action="purchase")) AS purchases
  by categoryId
| rename views AS "Views",
  purchases AS "Purchases",
  categoryId AS "Category"
```

Figure 4-12 shows the default chart for the search. As you can see in this chart, the number of purchases is very low or almost negligible compared to the number of views. What we are seeing here is a mix of very small and very large values. This is where you could make use of unit scale setting and making it log instead of linear to improve the clarity of the charts.

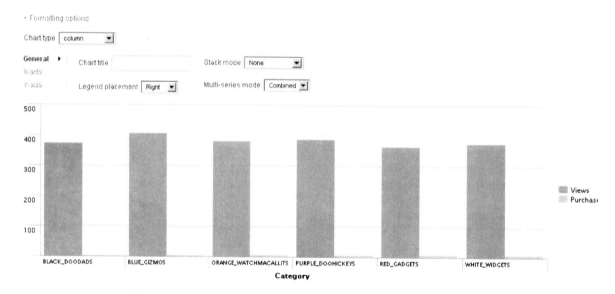

Figure 4-12. *Purchase versus views for product category*

Let us go ahead and click on the y-axis and change the value of the axis scale to log. We will rename the chart title to "Purchases to Views chart" and save the chart as "PurchasestoViewsChart." The updated chart can be seen in Figure 4-13. Again, depending on the audience for these reports, your choice of chart type and scale would vary. As we can see in this particular example, the charts in Figures 4-12 and 4-13 convey a different message, while the data set is exactly the same.

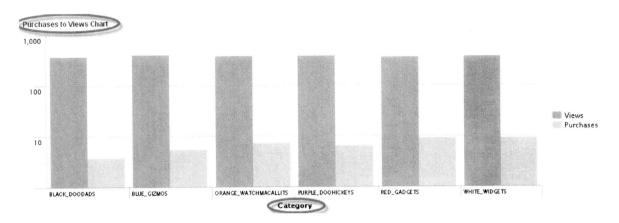

Figure 4-13. *Purchase versus views for product category using log scale*

Which Product Categories Are Affected by HTTP 404 Errors?

The HTTP 404 error message is shown on the client browser when a web server such as Apache or Microsoft IIS responds to a client request that the requested resource is not found or unavailable. This resource could be a HTML page, an image, or something similar. Any online retail store or web site would want to keep track of all 404 error messages so that they can fix the web site and make sure that users are finding what they want; otherwise, the business might lose money by letting users drop off or even go to a competitor. In Chapter 3, we used the stats command to find the number of events that had 404 status. In this use case, we would want to create a chart to find the number of 404s for each of the product categories. Let's enter the following search into the search bar:

```
sourcetype=access_* status=404
| chart count by categoryId
```

You can make use of a pie chart to show the relationship of parts of your data to the entire data set. Because we are getting a single value for each item in the product category, we can create a pie chart to see which categories have the highest number of 404 errors. In the chart type, select pie and name it "404 Errors Chart." Figure 4-14 shows the resulting pie chart. You will notice that the fields for the x-axis and y-axis formatting options are disabled, as this is obviously a pie chart.

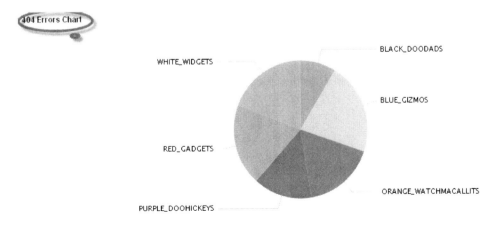

Figure 4-14. *404 Errors Pie chart*

One of the neat features that Splunk provides for these visual charts is drill-down values for each field, and this happens when you hover the mouse on a particular part of the pie. Figure 4-15 shows us the count and percentage for the WHITE_WIDGETS products category. We will save the report as "404 Errors Chart."

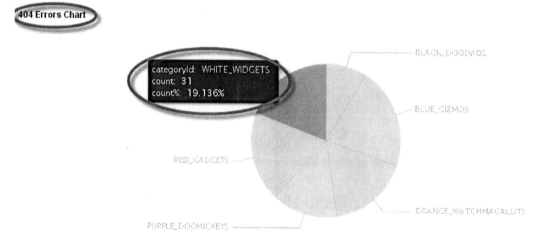

Figure 4-15. *Drill-down information for pie chart*

Purchasing Trend for MyGizmoStore.com

We want to visualize how the purchasing trend looks for each of the product categories. Trends are best visualized using sparklines. Sparklines show progression over a period of time very well, and this visualization is widely used in the medical industry and to show patterns such as earthquake activities. Sparklines show hidden patterns such as big spikes or big drops that IT or business users, doctors, or scientists might be looking for. Splunk provides a sparkline function that can be used with the stats or chart commands. In the search shown here, we are finding the requests that have purchase action using the eval command. We will count the events that match that condition and pass them to the sparkline function, which is used with the chart command. We also change the column name to "Purchases Trend" and computation is done for each product category. Let's enter the following search:

```
sourcetype=access_*
| chart sparkline(count(eval(action="purchase")))
  AS "Purchases Trend" by categoryId
```

On hitting enter, if you see a blank chart, that means that you are in the chart mode. You can click on the small table icon and that should bring you to the tabular report, as shown in Figure 4-16, with sparklines shown for each of the product categories.

6 results over all time

Figure 4-16. Purchasing trend for product categories

Depending on the audience for these reports, your choice of chart type would vary. Sometimes the sparklines may not be best suited for users in a Network Operations Center (NOC), but they could work very well for a management report. A line chart may serve the users of NOC better, as the scale for line chart is much bigger than that for sparklines. In this tabular report, we did not get the total number sold for each product category. We can do that by adding an additional count function into the search and renaming it "Total." We can also rename the categoryId field "Category Name," which is more business-friendly.

```
sourcetype=access_*
| chart sparkline(count(eval(action="purchase")))
  AS "Purchases Trend"
  count(eval(action="purchase"))
  AS Total
  by categoryId
| rename categoryId AS "Category Name"
```

Figure 4-17 shows the new tabular report with sparklines and "Total" column for each product category. We can see that RED_GADGETS was the best-selling product category. We will save the report as "Purchases Trend."

6 results over all time

Figure 4-17. *Purchasing trend with total for each category*

Duration of Transactions

In Chapter 3, you learned how to group a set of events as a transaction. We will now make use of the transaction command and create a chart to show the number of transactions based on their duration in minutes. In this search, we find all of the events that have purchase action and will group the events that have the same IP address. The transaction happens within 10 minutes. For the grouped transactions, we are using the chart command to count them based on the duration. Let's enter the following search:

```
sourcetype=access_* action=purchase
| transaction clientip maxspan=10m
| chart count by duration span=log2
```

We can use the default column chart and change the title to "Transaction Duration." Figure 4-18 shows the resulting chart. We can see that a large percentage of the transactions happened in less than a minute, followed by transactions in the span of two to four minutes, and then during the span of one to two minutes. Let's save the chart as "Transaction Duration Chart."

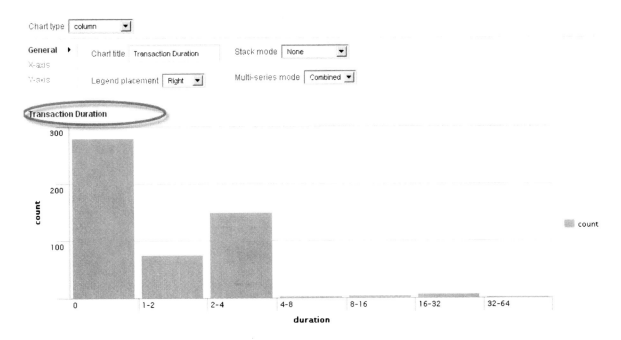

Figure 4-18. *Transaction duration*

Timechart

The Timechart command is similar to the chart command that we have used so far, except that the timechart creates a chart with aggregated values against time shown in the x-axis. The time on the x-axis can be controlled using the span option.

We will take a couple of use cases to explore the timechart command using MyGizmoStore.com sample data:

- Top purchases by product
- Page views and purchases

Top Purchases by Product

We would like to see what the chart looks like for purchases made for each of the products mapped against time on the x-axis. In the search, we make use of the count function to get the count of events with purchase action for each product and we eliminate any nulls using the usenull option. The time scales of the charts (and timeline) are calculated based on the context of the data. For example, if you are looking at data at the minute level, the scale will automatically set to minutes. If you are looking at hourly data but it's a lot, the scale can be daily, and so on, unless you force it with the span argument. Let's enter the following search:

```
sourcetype=access_*
| timechart count(eval(action="purchase"))
  by productId
  usenull=f
```

As you can see in Figure 4-19, the chart shows multiple column lines for each of the x-axis time variables, making it difficult to get the essence of the data. As we saw earlier, Splunk sometimes intelligently grays out the chart types that are applicable to a given data structure. In this case, the only other options available for the chart types are line and area.

Figure 4-19. Top purchases by product

Instead of changing the chart type, we will keep the column chart but adopt the stack mode option and select "stacked." This will update the chart as shown in Figure 4-20. Rename the chart "Top Product Purchase Chart," the x-axis title "Timeline," and the y-axis title "Count," and then save it all as "Top Product Purchase Chart." You can hover on each item in the stack and the count and percentage for the individual part of the stack will be displayed.

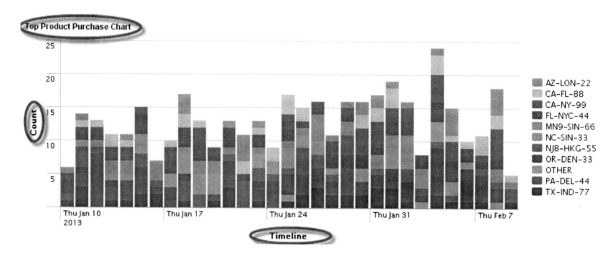

Figure 4-20. Product purchases in stacked mode

Page Views and Purchases

We can take the same use case and make use of the `timechart` command to show the data against a timeline on the x-axis. In the following search, we make use of the `eval` command to find if an event has a GET method or a purchase action. We use the `per_hour` function in `timechart` to lay out the data over a timeline. `Per_hour` is an aggregator function and works well to get a consistent scale for the data:

```
sourcetype=access_*
| timechart per_hour(eval(method="GET"))
   AS Views,
   per_hour(eval(action="purchase"))
   AS Purchases
```

We will change the chart type to "area" for this chart. When you compare the area chart to the default column chart (by switching the values in the chart type drop-down box), you will notice that the **area chart is much clearer for the data structure that we got for this use case. We will make the chart title "Purchases and views area chart," select stacked mode, and switch the legend placement to the top. The resulting visual area chart is shown in Figure 4-21. Let's save the chart as "Purchases and views area chart."

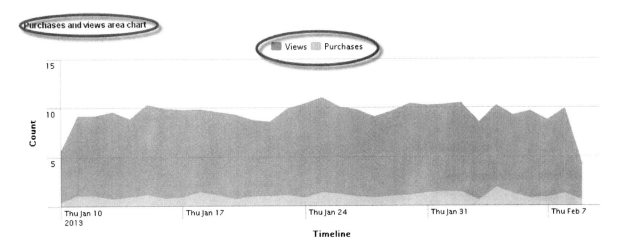

Figure 4-21. *Purchases and views area chart*

So far, you have learned how to use the `chart` and `timechart` commands and have become familiar with different chart types. In general, column or bar charts are better for comparing the frequency of values of fields. Most charts require search results structured as tables with at least two columns, where the first column provides the x-axis values and the rest of the columns provide the y-axis values. Apart from charts, Splunk also allows us to visualize the resulting data through gauges. Gauges work best when you have a single resulting value from the search that can be used to map against the range provided in gauge. Splunk provides radial, filler, and marker gauges. You will learn more about the `chart` and `timechart` commands in Chapter 10.

In Chapter 2, you learned about Splunk Technology Add-ons and Splunk Apps. One of the key differences is that Apps are used to visualize the data after it has been loaded into Splunk. Earlier in this chapter and also in Chapter 3 we looked at top IP addresses for `MyGizmoStore.com`, but what would be more useful and compelling from a business standpoint would be to find out the geolocation of these IP addresses. That way, we could spot patterns such as where most of the MyGizmoStore.com customers come from. There are two Apps available in Splunkbase to visualize the geolocation of IP addresses, and we will explore both of them:

- Google Maps
- Globe

Visualization Using Google Maps App

We will have to install the Google Maps App on our Splunk instance. You can do this using the same technique you learned in Chapter 2, making use of the "Find apps . . ." link under the "Apps" menu. Search for Google Maps on Splunkbase and install it. Once the App is installed, you will be able to see Google Maps in your "App" menu, as shown in Figure 4-22.

Figure 4-22. *Google Maps App*

Clicking on "Google Maps" will bring up the familiar Google maps. This App provides a geoip command, which takes IP addresses and plots those IP addresses on the map across the globe. In our sample MyGizmoStore.com data, we have the field clientip, which has the IP addresses. Let's enter the following search:

```
sourcetype=access*
| geoip clientip
```

Figure 4-23 shows the resulting Google map, which is zoomed to the U.S. level. We can see that MyGizmoStore.com has customers in 180 distinct locations; in the United States, Kansas, Kentucky, and New York are the top three areas.

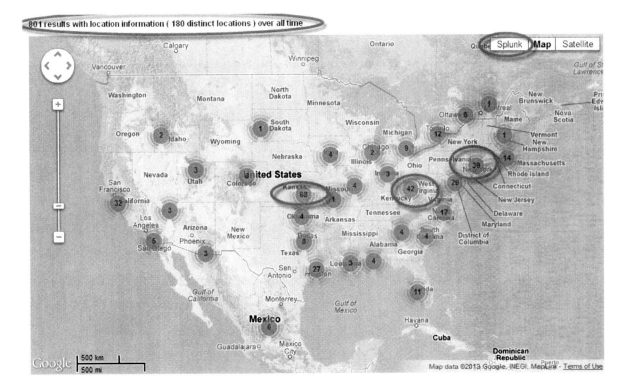

Figure 4-23. *Visualization of* MyGizmoStore.com *IP addresses in United States*

We can zoom out using the zoom control to see the distribution across the globe. If you select the Splunk overlay, you will get a darker background of the map with a brighter display of numbers on top, as shown in Figure 4-24. As you can see, North America is the most popular region for MyGizmoStore.com.

801 results with location information (180 distinct locations) over all time

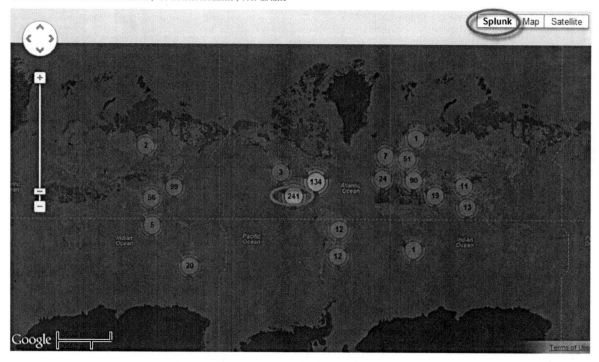

Figure 4-24. *Visualization of* MyGizmoStore.com *IP addresses across the globe*

We can save the search using Save search under the "Actions" menu, as seen in Figure 4-25. We will save the search as "Customer Location Map." You can click on the "Settings" menu to edit settings such as geolocation database as well as specify how the reverse lookup for IP addresses will work. Under the "Views" menus, you will find a couple of demos that you can watch to familiarize yourself with the App.

Figure 4-25. *Saving search with Google Maps App*

The second App that we are going to look at to visualize IP addresses is Globe.

Globe

Globe is very similar to the Google Maps App except that the visualization of geolocation is on a rotating globe and the locations are shown as shining beams coming out of the globe. The longer the beam, the more activity. You can install the Globe App from Splunkbase; once you have installed it successfully, you will be able to find it under the "App" menu, as shown in Figure 4-26.

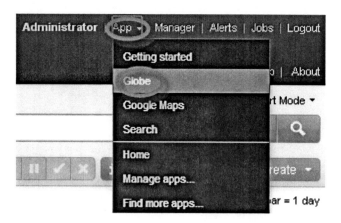

Figure 4-26. Globe App

After clicking on "Globe," you will select "ChromeGlobe," as shown in Figure 4-27.

Figure 4-27. Globe view

Once the view is selected, we can make use of the same search that we used with Google Maps:

```
sourcetype=access*
| geoip clientip
```

This will bring up a rotating globe with light beams, and you can tilt the globe to see it from different directions. Figure 4-28 shows the customer activity from the North Pole side. At the time of writing, the Globe App is experimental, and it is not publicly available for download. We expect it to be available on Splunkbase by the time that you are reading this chapter.

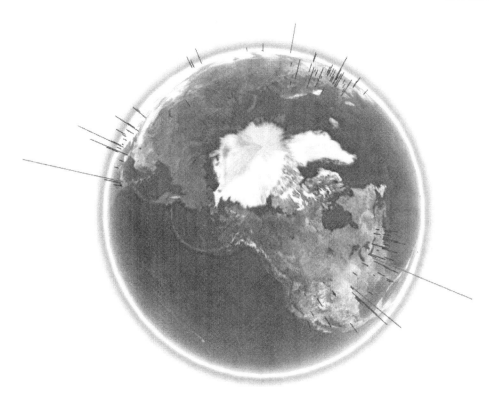

Figure 4-28. *MyGizmoStore.com customer activity in Globe*

As with Google Maps, you can save the search using the "Save Search..." option under the "Actions" menu as shown in Figure 4-29.

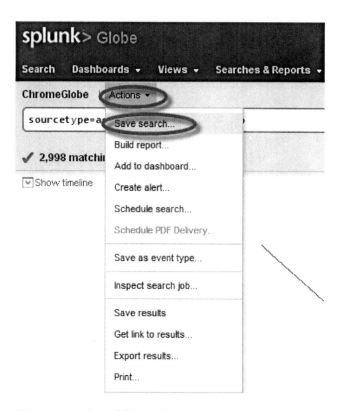

Figure 4-29. *Save Globe search*

So far, we have focused on building reports that can be used to visualize the data loaded into Splunk and on making use of Splunk Apps to help visualize the data. We will now explore how we can put these reports together as a dashboard.

Dashboards

Dashboards are popular for measuring and tracking company performance, and they represent a useful way to manage specific metrics that are the best success indicators. A dashboard allows sharing of key information between different departments in a company and can be considered a one-stop place to look at the current state of affairs. A dashboard can pinpoint strengths and challenges; users of dashboard do not have to weed through pages of unnecessary, unmeaningful data. Instead, a dashboard can identify and display data based on advanced analysis. Managers can make well-informed, evidence-based decisions based on the information in a dashboard.

Enterprises are looking to move beyond historical and descriptive data by using forecasting and predictive measures for their decision making. Only a few key metrics should be introduced in a dashboard. Your audience does not want tons of data; they want disciplined thinking and well researched information. It is important to remember that the main objective of a dashboard is communication; you should not distract viewers with elaborate graphics, gauges, and dials. The data in a dashboard should be clean, usable, and integrated in a way that is meaningful to your enterprise.

Splunk provides several ways of creating meaningful dashboards that are built from the searches and visual charts. The easiest way would be using the Splunk Web user interface. We have been using Splunk's search App to make searches; however, you may not have realized that the search App already has a basic dashboard (Figure 4-30). This provides information about the data loaded into Splunk, the number of events indexed, different sources of data, source types, and the hosts. It gives a holistic snapshot and lets the users take action either by doing drill-downs or by creating searches to find specific data.

Figure 4-30. *Splunk Search App Dashboard*

Splunk also provides a default collection of five status dashboards, which can be accessed by clicking on the "Status" menu, as shown in Figure 4-31.

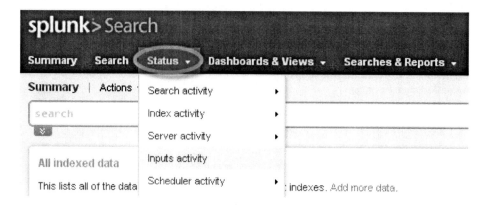

Figure 4-31. Splunk status dashboards

- The Search activity dashboard provides information about search activities for the Splunk instance. You can see the peak load times for searches, the most popular searches, and so on.

- The Index activity dashboard provides several useful statistics breakdowns for indexes and index size, utilization of resources such as CPU per index, top five sources that have been indexed in the last 24 hours, and so on.

- The Server activity dashboard provides information about recent browser usage, Splunk web errors, and information about how Splunk is performing. Figure 4-32 show a nice server activity dashboard and good usage of Gauge charts to show information about errors, access delays, and uptime. You can control the threshold and range values for the gauge charts to make the data more meaningful.

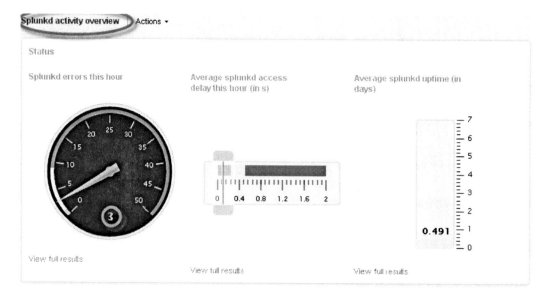

Figure 4-32. Splunk activity overview with gauge charts

- The Inputs activity dashboard provides information related to inputs or processed and ignored log files.

- The Scheduler activity dashboard provides information about the search scheduler, with charts showing the started and skipped searches, average execution times, average time taken to run scheduled searches, and so on.

Now that you have learned what Splunk provides as default dashboards, and the wealth of information that can be used, let's explore how to create a custom dashboard. Throughout this chapter we have created different searches and visual charts that are related to MyGizmoStore.com sample data. We will handpick some of them to create a dashboard that provides snapshot information about MyGizmoStore.com. We will take the following reports:

- Chart of purchases and views for each product

- 404 Errors

- Purchases trend

- Transaction duration

- Top purchases by product

To get started, click on the "Create dashboard . . . " link under the "Dashboards & Views" menu, shown in Figure 4-33.

Figure 4-33. *Create dashboard*

This will bring up the dialog box "Create new dashboard." We will enter MyGizmoStore as ID and MyGizmoStore.com for Name, as shown in Figure 4-34. Click on the Create button.

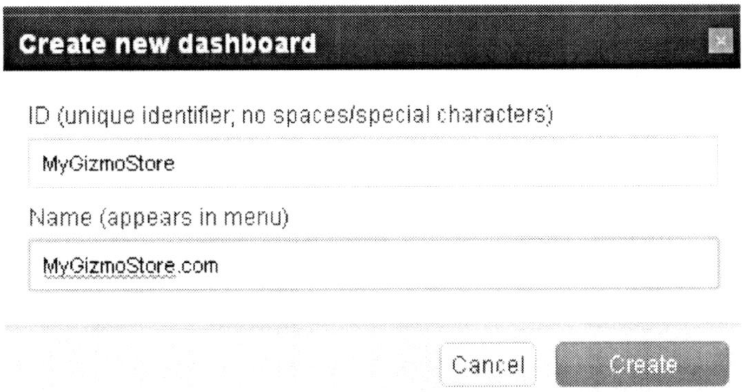

Create new dashboard

ID (unique identifier; no spaces/special characters)

MyGizmoStore

Name (appears in menu)

MyGizmoStore.com

Cancel   Create

Figure 4-34. *Create MyGizmoStore dashbaord*

An empty dashboard has now been created. To make it editable and add previously created reports, click "On" for Edit, as shown in Figure 4-35.

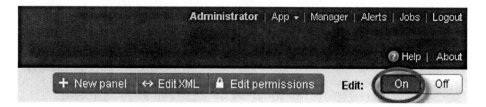

Figure 4-35. *Edit MyGizmoStore.com dashboard*

To start adding the reports into the dashboard, we will have to create a new panel that will hold the report. Click on "New panel;" that will bring up a dialog box, as shown in Figure 4-36. We will start creating a panel for the Purchases and Views chart. Enter "Purchases and Views" for the title and select the radio button for "Saved search." The drop-down box will show the list of saved reports. Select "Purchases and view area chart." Click on the Save button.

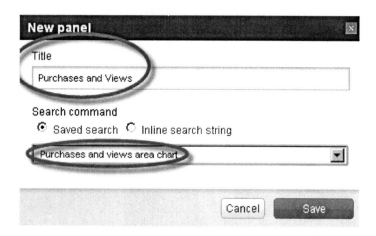

Figure 4-36. *Purchases and views panel*

This will bring up the report in tabular format. In order to visualize it as a chart, you have to click on "Edit" and select "Edit Visualization," as shown in Figure 4-37.

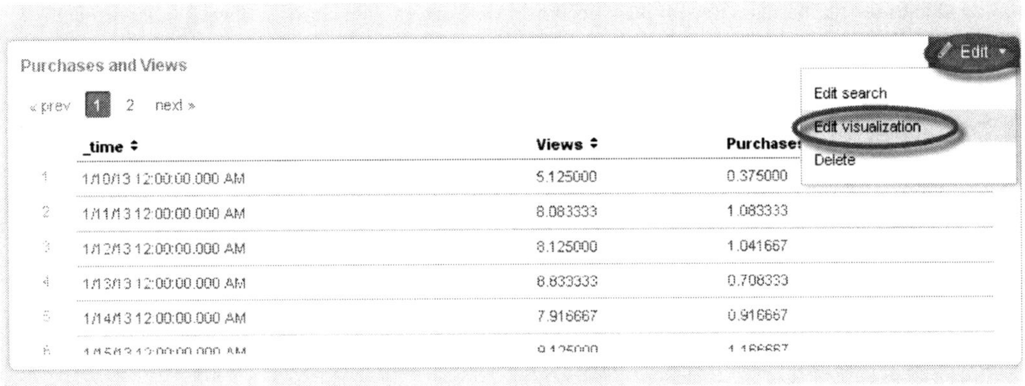

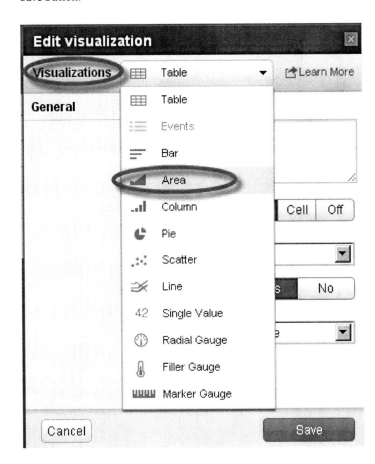

Figure 4-37. *Edit visualization for purchase and views*

This will bring up the dialog box, where you can make changes depending on how you want to visualize the tabular data. Let's select the "Area" chart as shown in Figure 4-38, which we saw worked well for this data. Click on the Save button.

Figure 4-38. *Edit visualization*

You will now be able to see the tabular data for purchases and views as an area chart, as shown in Figure 4-39.

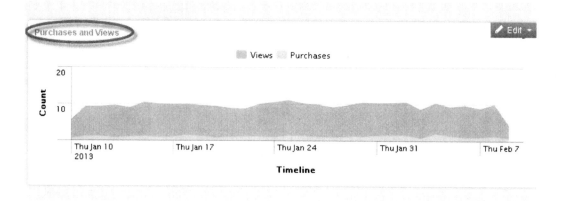

Figure 4-39. *Purchases and Views*

Now that you have learned in a step-by-step manner how to add a report to the dashboard, we will add the rest of the reports to the dashboard.

For 404 errors report:

- Click on New Panel; name as 404 Errors and select "404 Errors Chart" report from the drop-down box. Click Save.

- Edit the panel, and select "Pie as Visualization." Click Save.

For Purchases trend report:

- Click on New Panel; name as Purchases Trend and select "Purchases Trend" report from the drop-down box. Click Save.

For Transaction duration report:

- Click on New Panel; name as Transaction Duration and select "Transaction Duration Chart" report from the drop-down box. Click Save.

- Edit the panel, and select "Column as Visualization." Click Save.

For Top Product purchases report:

- Click on New Panel; name as Top Product Purchases and select "Top Product Purchases Chart" report from the drop-down box. Click Save.

- Edit the panel, select "Column as Visualization," and select stacked as stack mode option. Click Save.

Now that we have added all of the reports that we want to be part of the MyGizmoStore.com dashboard, we can arrange them properly to make the dashboard easy to understand. You can drag and drop each panel in the dashboard to rearrange them. Once the dashboard is rearranged, go to the "Dashboards & Views" menu and click on MyGizmoStore.com link to bring up the new dashboard, as shown in Figure 4-40.

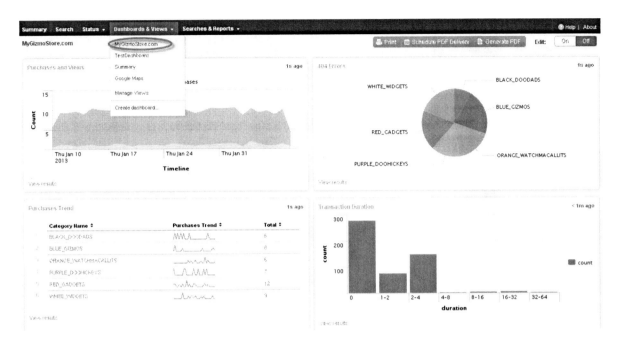

Figure 4-40. *MyGizmoStore.com dashboard*

You have already seen how Splunk Apps such as Google Maps or Globe help us to visualize the data. In Chapter 2, you learned about and installed Splunk Technology Add-ons for Windows and *Nix to collect the data in Windows or Unix environments. These Add-ons also have full-fledged Splunk Apps, which means that you can not only collect the data but also visualize it through prebuilt dashboards.

Let's see how we can make use of Splunk's *Nix App. You can download and install the *Nix App the same way as you have installed other Apps. If you have installed the *Nix Add-on at the same time, you will have to disable it before you can get the *Nix App to work. To disable a particular App, use Splunk Manager and click on "Apps." Once you have installed *Nix App, it will be listed under the "App" menu, as shown in Figure 4-41.

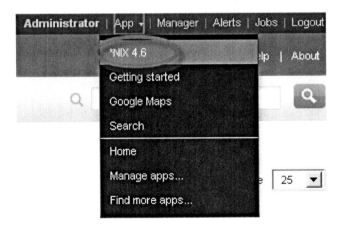

Figure 4-41. *Nix App*

On Unix, you can enable the sources of data that you want to load into Splunk. To show the *Nix dashboard working we have enabled /var/logs as file and directory inputs and cpu, memory, top, and who in scripted inputs. *Nix provides a comprehensive set of dashboards that let you visualize the information across CPU, memory, disk, network, users, and different log files. You can build custom dashboards on top of it or create reports in different charting formats using the data loaded into Splunk by *Nix.

We will wrap up this chapter by looking at sample dashboards in *Nix. Figure 4-42 shows the CPU overview dashboard that you can access from the CPU menu, which provides visual information about CPU consumption by user, process, and so on. You can see that the App makes use of the timechart command to create reports that are part of the dashboard.

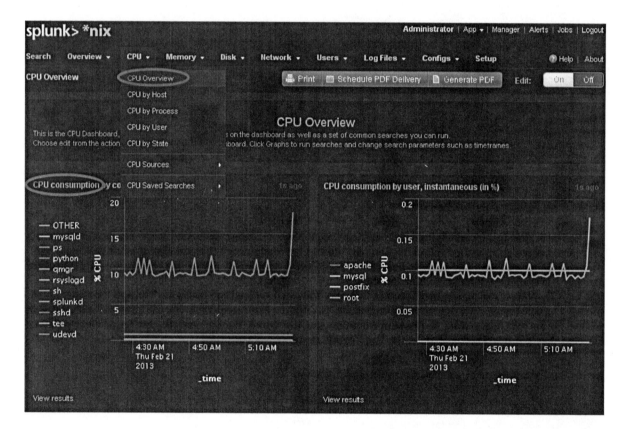

Figure 4-42. *Nix CPU overview dashboard*

Figure 4-43 shows the memory dashboard that visually shows the memory usage by process, usage by top 10 users, and so on.

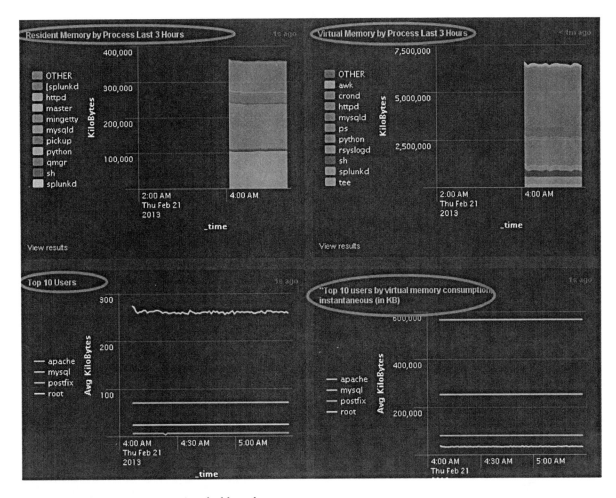

Figure 4-43. *Nix memory overview dashboard*

Figure 4-44 shows the logging dashboard, which visually shows throughput across different log files. Splunk App for Windows also provides a comprehensive set of dashboards similar to *Nix. You can install this in the same way as *Nix.

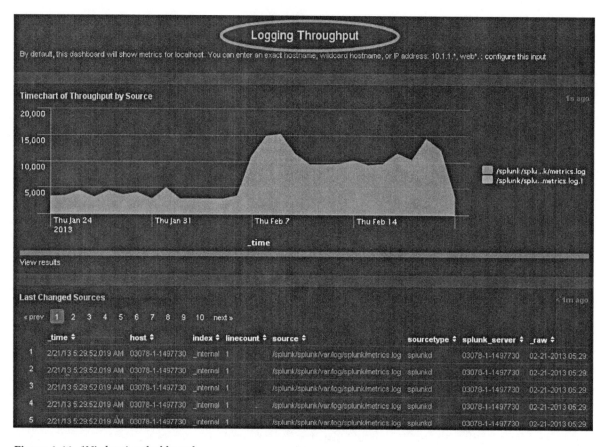

Figure 4-44. **Nix logging dashboard*

Summary

In this chapter, you have learned how to visualize data indexed into Splunk. You have seen the reporting capabilities of Splunk using report builder and SPL commands `chart` and `timechart`. You learned about different charting types and how to use them for different types of data structures, using the `MyGizmoStore.com` sample data. Finally, you learned how to build dashboards and explore Splunk Apps that help in visualization and provide prebuilt dashboards.

## CHAPTER 5

Defining Alerts

In this chapter, you will learn about different types of alerts that Splunk provides and how to create alerts and take action. You will make use of the searches and data from `MyGizmoStore.com` and *Nix and take different use cases to trigger different types of alerts.

What Are Alerts?

Alerts are just short messages or notifications that help individuals keep informed about certain things that have happened or potentially will happen. We are all used to alerts in our daily lives: for example, calendar notifications that alert us about meetings we should attend. Those of us who travel by air in the United States are familiar with Homeland Security's advisory system, which issues alerts in a color-coded threat level. As online customers, we are used to e-mail alerts that we get from online stores or online ticketing sites when the price of an item or a ticket or a hotel room rate drops. Enterprise IT teams are interested in security, network, CPU, memory, and types of alerts that have huge impact on the performance of applications, security breaches that may be happening, or if some Service Level Agreement (SLA) is not going to be met. Even without a computer, we all are used to weather alerts on the radio, TV, mobile phones, and tablets.

Alerts have become part and parcel of our lives now whether we explicitly recognize them as alerts or not. For enterprises, alerts about network, hardware, applications, web sites, and so on are key indicators to what is going on with the IT infrastructure, how they are doing compared to SLAs; and from the business point of view, how operations are performing whether from inventory, shipping, call center, customer support, or sales. So now the question is, what does Splunk provide for alerts? We are going to explore this in the next few sections of this chapter.

How Splunk Provides Alerts

It is extremely important to know what is happening with all the data that is getting collected and how you can take advantage of being informed about the positive and negative types of events. We saw Splunk processing, analyzing, and visualizing different sources of data; now we need to see how, from the IT and business standpoint, users can be alerted to abnormal events. You can use Splunk as a monitoring tool for the enterprise. Splunk provides three different types of alerts:

- Scheduled alerts—these types of alerts are used with searches that are scheduled at a specified interval. The searches work on a historical set of events, and an alert can be triggered if a certain alert condition is met. For example, an online store such as `MyGizmoStore.com` will have a search that counts everyday product sales; an alert can be triggered if the count falls below a certain number. These types of alerts are designed for situations where an immediate action is not required but users want to know if the condition for the alert is met so that they can be informed and possibly take some action or probe further into the issue

- Per results alerts—these types of alerts are used when you want to know something as soon as an alert condition is met in the search. A classic example is having a search that would check for any HTTP status greater than or equal to 500 (i.e. server errors) and send an alert once every hour the first time that the condition is satisfied

- Rolling-window alerts—these types of alerts are used to monitor events in real time within a rolling time window, such as the previous 10 minutes. The alert gets triggered when the alert conditions are met by events that qualify in the rolling window. A classic example is potential fraud detection, in which someone makes more than three purchases with the same credit card in the rolling window of 10 minutes

Now that you have learned about alerts and different types of alerts provided by Splunk, let us explore how we can create alerts. As in previous chapters, we will take different use cases that are related to MyGizmoStore.com and also make use of the data we have collected using *Nix. Three use cases we are going to look at are:

- Alert based on product sales

- Alert on failed logins

- Alert on critical errors in log files

Splunk Web is the easiest way to create alerts. We will start with the first use case.

Alert based on product sales

In this use case we want to find out how many products were sold yesterday, and raise an alert if that number drops below 30.

To start with we will build a quick search so that we can make use of it with an alert. In Chapter 2 we made use of time picker to zoom into a particular range of dates. To find the total number of products sold yesterday, we could make use of the "Yesterday" option in the time picker. Splunk provides time modifiers that help to customize the time range for the search. Splunk provides three time modifiers:

- earliest—specifies the earliest for the search time range

- latest—specifies the latest for the search time range

- now—is the current time

Time modifiers support specifying time units in seconds, minutes, hours, days, week, days of the week, month, quarter, and year. In the search below, we make use of earliest and going back one day by specifying -1d, which means minus one day from the current date. To compute the total number of purchases, we use an eval command to find the events that have a purchase action and make use of the count function. Figure 5-1 shows that the search has used the time modifier and shows the total number of products sold for yesterday.

```
sourcetype=access_*
  earliest=-1d
  action=purchase
| stats count(eval(action="purchase")) AS "Total Products Sold Yesterday"
```

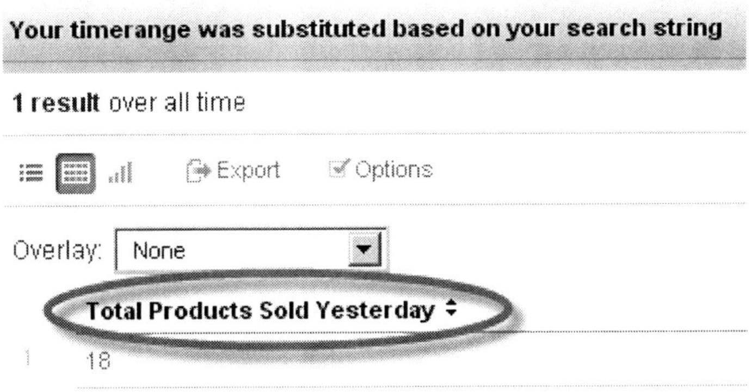

Your timerange was substituted based on your search string
1 result over all time
Export
Options
Overlay:
None
Total Products Sold Yesterday
1
18

Figure 5-1. *Total products sold yesterday*

Now that we have a search that can tell us the total number of products sold for yesterday, we can go ahead and schedule an alert using Splunk Web, which would also save the search for the alert. Select "Create" and click on the "Alert. . ." link, as shown in Figure 5-2. You can also add alerts to the saved searches using the Splunk Manager interface and select a particular search in the "Searches and reports" section. All of the alert options that we will explore in this chapter are available when you add an alert to the saved search.

All time
Save
Create
Dashboard panel...
Alert...
Report...
Event type...
Scheduled search...

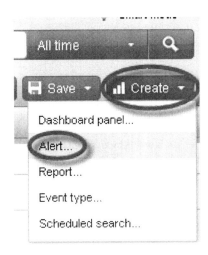

Figure 5-2. *Create alert*

In the Create Alert dialog box that comes up, enter "Yesterday Purchases Alert" as Name. In the Schedule drop-down, select "Run on a schedule once every". The dialog box will expand with additional options. Select "Day" for the schedule, meaning once every day. In the Trigger if drop-down box, select "Number of results" and "is less than", and add 30 as the value. This alert will get triggered if the number of products sold for yesterday is less than 30. Figure 5-3 shows the selected options. Click the Next button.

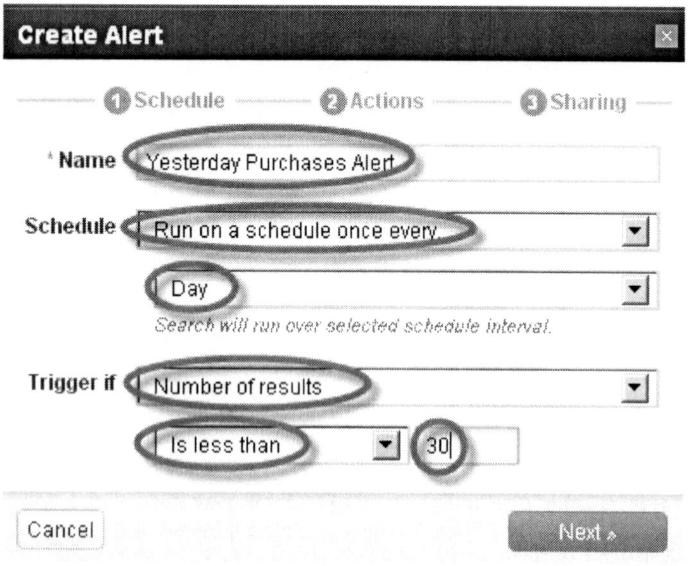

Figure 5-3. *Schedule an alert*

In the Actions page of the Create Alert dialog, you will see different options for the actions that need to happen. Three possible options for actions include sending an e-mail to a user or group of users, running a script that might take an immediate preventive action such as provisioning additional disk space, and making use of Splunk Alert manager to view the triggered alerts.

We will select the alerts manager option for this use case and explore the alerts manager once we have created the alert. Each alert can be labeled with a severity level that indicates the importance of the alert. For example, you would like to see the alerts that are related when CPUs are utilized above 90 percent or memory is being maxed out, and so on. Figure 5-4 shows a drop-down box for severity where the options include Low, Medium, High, and Critical. Because this is not a high priority alert, we will select "Medium". You can make use of the severity level as a way to filter the triggered alerts in alert manager. Alerts can be triggered on all results or for each result. Select the radio button for "All results". We will discuss throttling in our next two use cases. Click on the Next button. Figure 5-4 shows the enabled options.

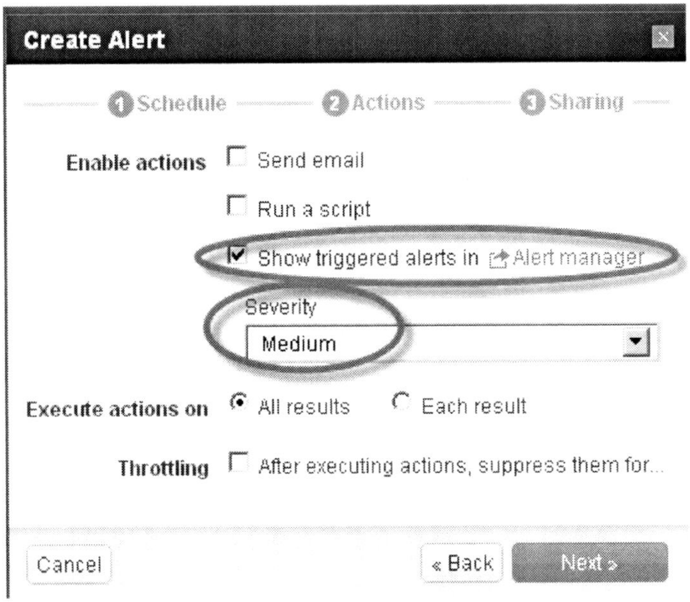

Figure 5-4. Alert actions

In the final step of the alert creation process, you can either share the alert with other users or keep it private. This is similar to sharing searches. Select the radio button for "Share as read-only to all users of current app". This will make multiple users view the alerts, and keep them informed about certain activity or performance indicators. Click on the Finish button. You will see the alert saved successfully popup box. Click OK.

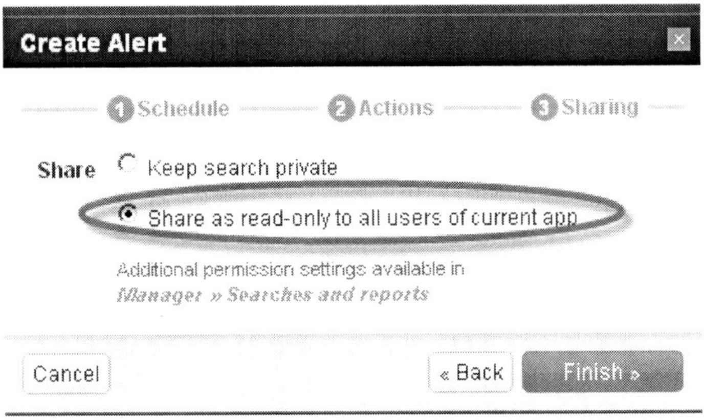

Figure 5-5. Share the alert

The Alert manager displays records of all alerts, and provides options to search, filter, and view alert based on app, owner, severity, and alert. Selected alerts can also be deleted as well. To access the Alert manager, click on the "Alerts" menu, as shown in Figure 5-6.

Figure 5-6. *Alerts manager*

In the Alert manager, select Yesterday Purchases Alert in the Alert drop-down box as shown in Figure 5-7. You will see the alerts getting filtered and records related to this alert are displayed. In this case, we only have one alert.

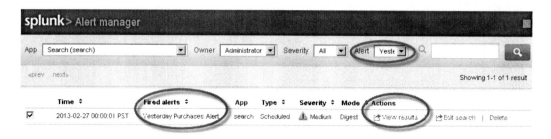

Figure 5-7. *Yesterday purchase alert in Alert manager*

You can click on the "View results" link to see the actual results for this search and find out why the alert has happened. As you saw in Figure 5-1, the number of products sold was only 18, and as our alert condition was set to trigger an alert when the sales were less than 30, this resulted in an alert getting triggered.

Alert on failed logins

We installed Splunk *Nix App in Chapter 4 and enabled data from different logs to be indexed into Splunk. Now we can make use of data loaded from security logs and find out how many login failures have happened in the last 24 hours to create a different type of alert that works in real time. Go the *Nix app using the "App" menu. *Nix creates its own index os, so we start the search specifying the index name. After the index name, search for any events that have failure in them where the sourcetype is linux_secure. Make use of the stats command to find the count of failures by remote host or rhost. Finally, sort the results.

```
index=os sourcetype=linux_secure failure
| stats count by rhost
| sort by count
```

Figure 5-8 shows the results from the above search that we have entered in the search bar of the *Nix app. The results would vary significantly, as the results are specific to the server that we are using at the time of writing this chapter. You can see the IP addresses of the remote hosts or rhosts where they are trying to login and the number of failures, sorted in descending order.

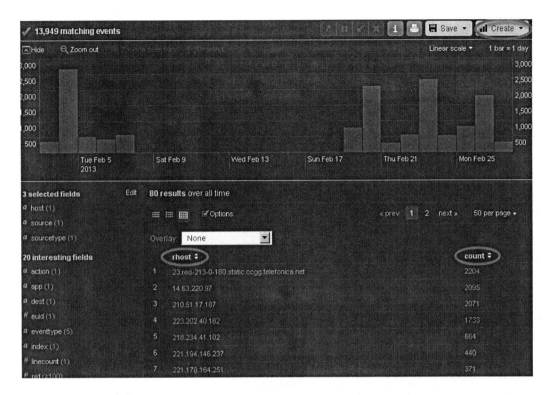

Figure 5-8. Login failures

We will create an alert for our search by selecting "Alert . . ." from the Create drop-down. In the Create Alert dialog box, we will name the alert as Login Failure Alert and select "Trigger in real-time whenever a result matches". This option is the second of the alert types that we discussed earlier, known as a per-result alert. This alert would get triggered whenever an event that comes in matches the search in real time. Figure 5-9 shows the selected options. Click on the Next button.

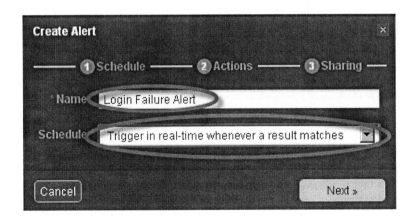

Figure 5-9. Create alert for login failures

In the actions page of Create Alert dialog box we will enable an action to make Splunk send an e-mail to a user at alert@mydomain.com. You can include the results either inline or as a CSV or PDF attachment. We will make it inline and also enable the alerts to be viewed via Alert manager as we have done before. Because this alert is related to security, we will make the severity High. Figure 5-10 shows the selected option. Click on the Next button.

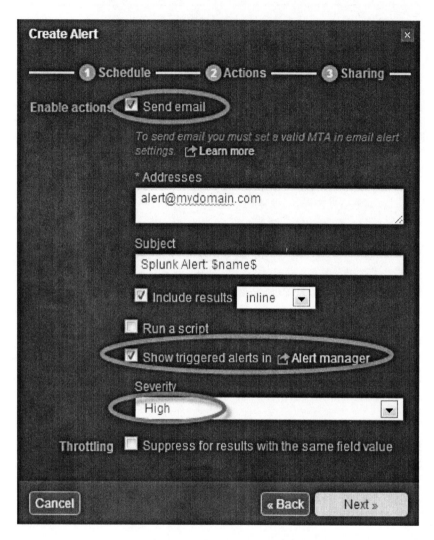

Figure 5-10. Alert actions for login failures

Once the alert has been successfully created, we can view the triggered alerts by launching the Alert manager. We can filter the alerts using the app name, which is *Nix in our case. As the *Nix app can have many alerts, we can further filter it down by selecting the particular alert, which in our case is a Login security alert from the alert drop-down box. Figure 5-11 shows the set of alerts that have been triggered. We can see the type for this alert is set to be Real-time and it is a Per Result type of alert. What you can also see is that users may be overwhelmed with the number of alerts that can potentially get triggered. This is where Splunk's throttling alert action capability comes in handy, and we will explore that in our next use case.

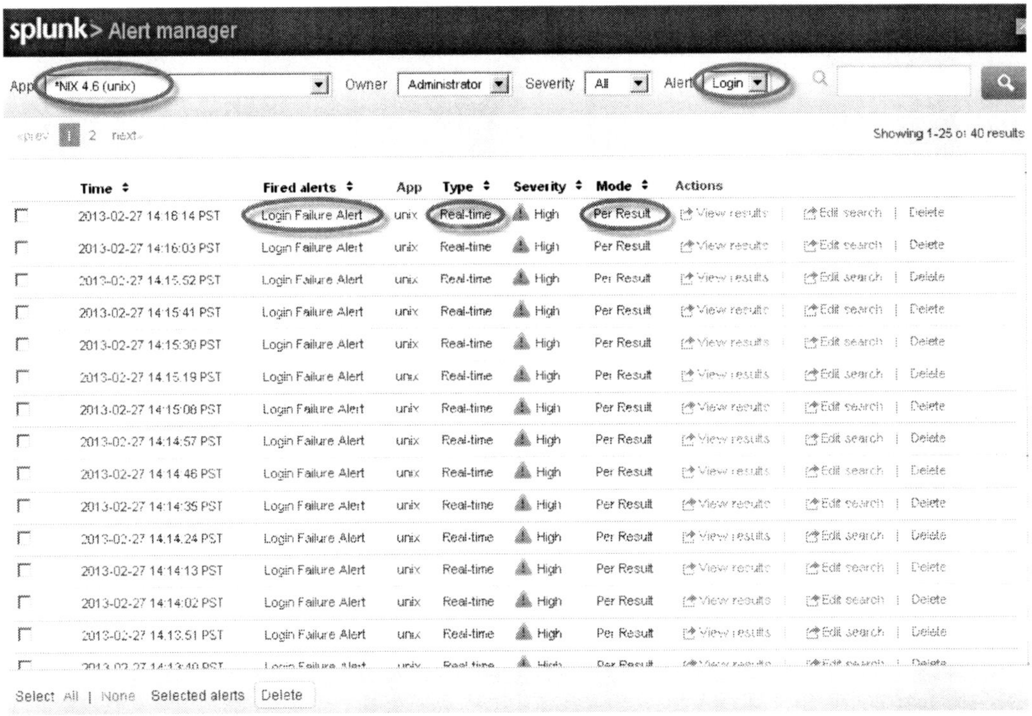

Figure 5-11. *Alert Manager*

Clicking on the "View results" link against the alert will show the particular set of events that have triggered the alert, as shown in Figure 5-12.

Figure 5-12. *Event that triggered login failure alert*

Alerts on critical errors in log files

In this use case, we will make use of the data from the log files in Unix, which are enabled in the *Nix app. We can set an alert that will look for any critical errors. We can do that with a simple search in which we indicate that the index is os and sourcetype is linux_secure and find any indexed events that have the word critical or error in them. In the *Nix app, enter the below search and create a new alert called Critical Errors Alert, as shown in Figure 5-13. We will select "Monitor in real-time over a rolling window of" option in the Schedule drop-down. In our case, we will specify a 30-minute window and trigger the alert if the number of results is greater than 15.

```
index=os sourcetype=linux_secure critical OR error
```

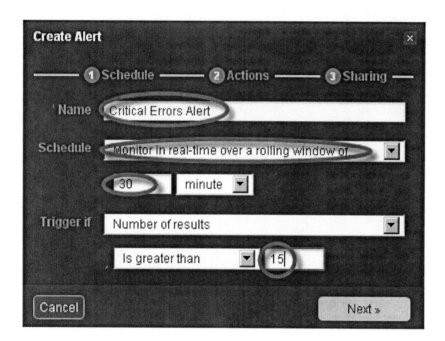

Figure 5-13. *Creating a critical errors alert*

In the actions page of the alert dialog box, we will enable sending e-mail to our `alert@mydomain.com` user and also enable the alerts manager option. One new option that we will select for this use case is throttling. The throttling option suppresses the alerts after the initial alert gets fired off for a specified period of time. As you saw in the previous use case for login failures, instead of getting repeated alerts, you can say that the alerts have to be suppressed for a specified time either in seconds, minutes, or hours. In this case, we will specify it to be 15 minutes. You will have to find the optimal time that would be best suited to the alert that you are creating so that the throttling does not suppress an important alert.

Splunk allows you to run scripts when alerts get triggered. This option can be enabled by selecting the checkbox for "Run a script" option in the Actions page of Create Alert. This feature can be used to take immediate action through an automated process that can done using scripting. Classic examples include: disabling a user login after three failures, blocking an IP address that is used in a DDOS attack, provisioning additional compute or storage on the cloud when the user activity touches a predefined threshold, or sending the details of the triggered alert to another application or system that would trigger the required action. The scripts that need to be executed have to be placed in the $SPLUNK_HOME/bin/scripts directory, where $SPLUNK_HOME is the directory where Splunk is installed.

Once created, the alert can always be edited using the "Manager" menu by clicking on the "Searches and reports" link in the Knowledge section. Figure 5-14 shows the specified options for the critical errors alert. Click the Next button.

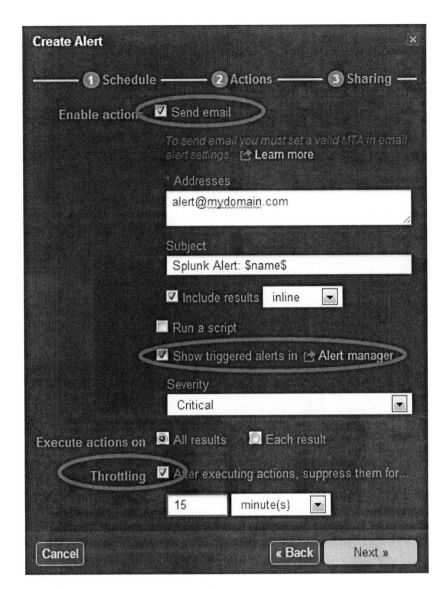

Figure 5-14. *Select throttling for the alert*

Once the alert is created, you can go to the Alert manager and filter based on the app and alert. We can select *Nix as app and Critical Errors Alert in the Alert and that would bring the triggered alerts so far. Figure 5-15 shows the alert records. You will notice that the Mode column is shown as Digest, meaning that the alert is related to a set of events.

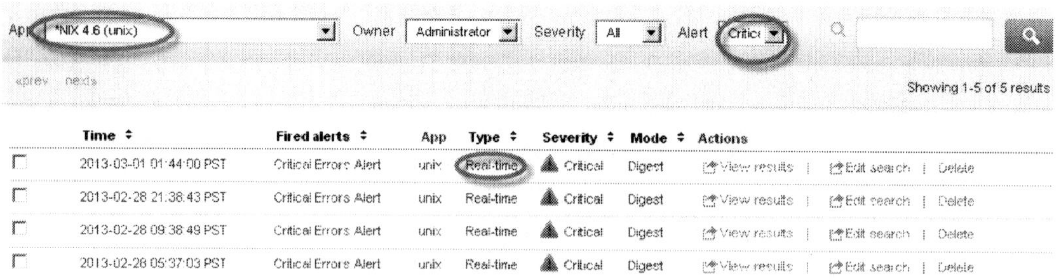

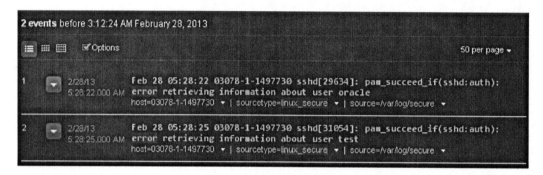

Figure 5-15. *Triggered alerts for critical errors*

Clicking on "View results" shows the list of events that pertain to the digest of the qualified alert. Figure 5-16 shows the sample set of events for this alert.

2 events before 3:12:24 AM February 28, 2013

☰ ⠿ ▦   ☑ Options   50 per page ▾

1   ▼   2/28/13   Feb 28 05:28:22 03078-1-1497730 sshd[29634]: pam_succeed_if(sshd:auth):
        5:28:22.000 AM error retrieving information about user oracle
                        host=03078-1-1497730 ▾ | sourcetype=linux_secure ▾ | source=/var/log/secure ▾

2   ▼   2/28/13   Feb 28 05:28:25 03078-1-1497730 sshd[31054]: pam_succeed_if(sshd:auth):
        5:28:25.000 AM error retrieving information about user test
                        host=03078-1-1497730 ▾ | sourcetype=linux_secure ▾ | source=/var/log/secure ▾

Figure 5-16. *Events related to critical errors*

Summary

In this chapter, you have learned about the three different types of alerts that Splunk provides. We explored how to create alerts using a common set of use cases, leveraging the searches from MyGizmoStore.com and *Nix along with different actions that are available with alert. You learned how to make use of the alerts manager to view the triggered alerts.

## CHAPTER 6

Web Site Monitoring

In this chapter, you will learn how to model and create a web site monitoring tool that contains reports and a dashboard to help IT and business in enterprises measure different aspects of online retail store web sites such as MyGizmoStore.com.

Monitoring web sites

Web site monitoring is often used by enterprises to ensure that their web sites are live and responding to user requests. Before updates are put into production for web sites or applications, IT typically simulate the actions of thousands of visitors to a web site and observe how the site or application responds. They also simulate visitors across multiple geographies and servers.

The sample data we have created for MyGizmoStore.com is quite similar to that approach with a well-defined set of patterns that are customizable. Monitoring tools send out alerts when pages or parts of a web site malfunction, allowing IT to correct issues faster. You saw how to use Splunk alerts to get notifications in Chapter 5.

Monitoring is essential to ensure that a web site is available to users and downtimes are minimized. Users who rely on a web site or an application for work will get frustrated or even stop using the application if it is not reliably available. This becomes even more critical if the web site is an online business like MyGizmoStore.com, where a nonfunctioning web site means loss of revenue as customers make their purchases from a competitor. Monitoring can cover many things that an application or web site needs to function: network connectivity, DNS records, database connectivity, bandwidth, and computer resources like RAM, CPU load, disk space, events, and so on. Commonly measured metrics on the IT side include response time, availability of servers, traffic patterns; and on the business side user demographics, time spent on web site by users, how many new users versus existing users, and so on.

In Chapters 2, 3, and 4, you learned how to process, analyze, and visualize data using different SPL commands. In this chapter, we will make use of different SPL commands and searches that you have learned previously and work toward creating a web site monitoring tool that would help us monitor the MyGizmoStore.com site. This chapter reviews some of the previous explanations and adds ideas about how to quickly put together some key reports from the IT and business perspectives that can be monitored through a dashboard. You will be able to expand on this monitoring tool as you work on more diverse data sets that come up in the later chapters of this book.

We will make use of the following use cases to create reports for the monitoring dashboard.

IT Operations

- Hits by host

- Hits by host without internal access

- Traffic with good HTTP status

- Traffic with bad HTTP status

- Top pages by bad HTTP status

Business

- User demographics by region

- Bounce rate

- Unique visitors

IT Operations
Hits by host

In this use case we would like to see the traffic coming into MyGizmoStore.com hosts for a specified period of time. Although you can make use of the timeline to adjust the period of time, you are familiar with the time modifiers that were used in Chapter 5; we will make use of them to specify the time range. In the search below, you are looking at the events over last seven days and counting the number of events per host and charting them using the timechart command. You will notice the time modifier earliest has an @ sign in the value assigned to it. Time modifiers allow the @ sign as part of the specified value to indicate how the time should be rounded down. In our search, we round it down to a day, but you can round the time modifier to any supported time unit such as second, minute, hour, day, week, particular day of the week, month, quarter, and year. Let's enter the following search into the search bar:

```
sourcetype=access_*
  earliest=-7d@d
  latest=now
| timechart count by host
```

Figure 6-1 shows the column type chart for this search, in which each column represents one of the MyGizmoStore.com hosts. On most dates, two of the three hosts are serving almost equal number of requests and the other host is less heavily loaded. IT can use this report and further analyze the data by increasing the number of days to see if they can find some patterns on which one of the three hosts is serving fewer requests. Then they can fine-tune the load balancer so that all user requests are serviced quickly and the hardware resources are properly optimized. We will make use of the "Formatting options" link to make changes to the chart and x- and y-axis titles, and save this report as "Hits by Host" (Figure 6-1).

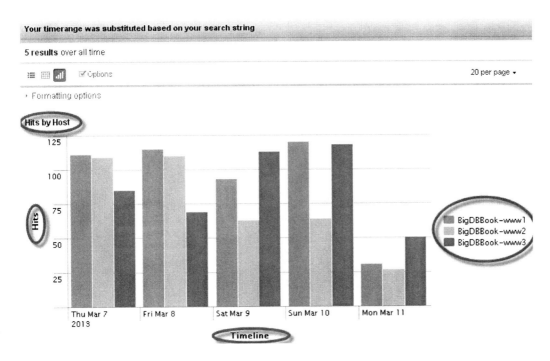

Figure 6-1. Hits by host

Hits by host without internal access

For online retail stores such as MyGizmoStore.com, there will be hits coming in from internal employees or users who are either browsing or making purchases. It would be interesting to see how different the hits by host chart looks if we remove the events that come from IP addresses that are internal to your network. To do that, we can take the search and update it, where we will only look at the events where the clientip field is not equal to 192*. You can add more subnets to the search if need be. Let's enter the updated search into the search bar:

```
sourcetype=access_*
  earliest=-7d@d
  latest=now
  clientip != 192*
| timechart count by host
```

Figure 6-2 shows the column type chart for the updated search. The sample data we have for MyGizmoStore.com doesn't have many internal access events, so we don't see any difference in charts, but in the real world it would be useful to compare these two charts to see if the internal requests are skewing the request numbers. IT could make use of this information to divert the traffic coming in from internal IP addresses to a different host so that it doesn't impact external customers. We will make changes to the chart and *x*- and *y*-axis titles, and save this report as "Hits by Host without Internal Access".

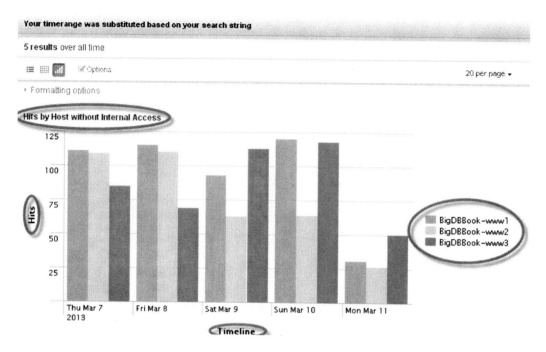

Figure 6-2. *Hits by host without internal access*

Traffic with good HTTP status

In our previous chapters we worked on searches that used HTTP status codes. From the operations standpoint, it is essential to know the success rate for HTTP traffic; this tells us how many requests are good and how many are bad. We will start by finding out the good traffic. We are using a 15-day window in the time modifier, but you can adjust it to the last 24 hours by making `earliest=-24h@h`. Let's enter the following search into the search bar:

```
sourcetype=access_*
  earliest=-15d@d
  latest=now
  status>100 status<300
| timechart count BY status
```

Although we specified the search to find any events that have the `status` field in the range of 100 to 300 exclusive, our sample data only has events with status 200 in that range, which is what the chart in Figure 6-3 shows. We will make appropriate changes to the chart and *x*- and *y*-axis titles, and save this report as "Traffic with good HTTP status".

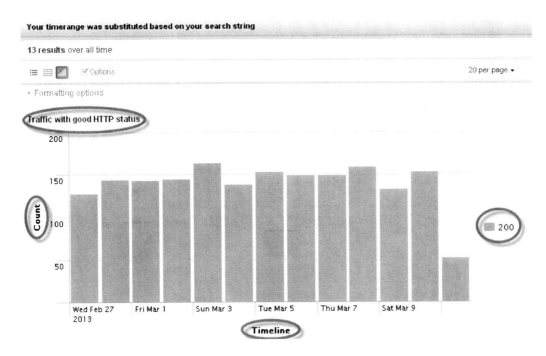

Your timerange was substituted based on your search string

13 results over all time

≡ ⊞ **iil** ☑ Options   20 per page ▾

▸ Formatting options

Figure 6-3. Traffic with good HTTP status

Traffic with bad HTTP status

Knowing about traffic with bad HTTP status is much more important than knowing about good traffic, as bad status could result in direct loss of revenue. We can update our previous search so that it will find all the events where the status field is greater than or equal to 300. HTTP status codes in the 3xx range can be used if the web site has moved or to provide temporary redirects (in which case, they don't necessarily mean bad HTTP traffic). In this particular example we have included status codes in the 3xx range, but the search can be refined to exclude them by changing status >=300 to status >= 400. This search could also be used as a report that shows non-200 HTTP status in the last 24 hours by changing the time modifier. Let's enter the following search into the search bar:

```
sourcetype=access_*
  earliest=-15d@d
  latest=now
  status>=300
| timechart count BY status
```

Figure 6-4 shows the chart with bad HTTP status; unlike our previous search, which only had HTTP status 200, we can see that the events are distributed between 400, 404, 406, 408, 500, and 503 HTTP status. Although we discussed what these codes mean in Chapter 3, it makes more sense to show the descriptions of these status codes than the numbers. We can achieve that using the lookup tables, which you will learn about in the next few chapters. As this chapter provides a basic framework to create a monitoring tool, by the end of this book you will be able to morph these reports into something that would serve enterprise requirements. We will make appropriate changes to the chart and x- and y-axis titles, and save this report as "Traffic with bad HTTP status" (Figure 6-4).

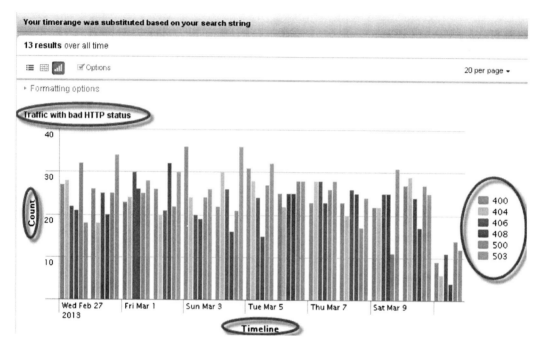

Figure 6-4. *Traffic with bad HTTP status*

You can always drill down into individual columns in the chart or click on different status legends on the right-hand side. For example, clicking on 404 status will show the drill-down events list as seen in Figure 6-5. You will notice that the search also has an additional field clause, which is status=404.

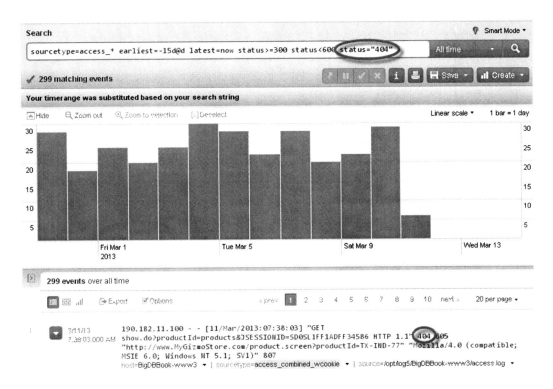

***Figure 6-5.** Drill-down chart for HTTP 404 status*

Top pages by bad HTTP status

One possible next step after finding the information about bad HTTP status traffic would be to find out the top pages that are part of the bad HTTP status. We can do that using the following search, in which we find the bad HTTP status events, distinct IP addresses, and get a count using the `uri` field, which is used as pages that the user is trying to access. Let's enter the following search into the search bar:

```
sourcetype=access_*
  earliest=-3d@d
  latest=now
  status>=300
| stats dc(clientip) as "unique ips"
  count as "total count" by uri, status
```

Figure 6-6 shows the resulting chart. This would look very different based on the data set you have and what values the time modifiers have. In Chapter 4, you saw that all chart types do not lend themselves very well to all types of data. In this particular use case, we got a chart in which the values of each fields are at different scales. What this means is that we have a mix of very small and very large values. To make the chart more meaningful, we will change the chart type to be "line" and choose a log instead of linear unit scale setting, to improve the clarity of the chart. To set the unit scale to log, click on "Formatting options", click on *Y*-axis, and change the value of axis scale to log. We will make appropriate changes to the chart and *x*- and *y*-axis titles, and save this report as "Top pages by bad HTTP status".

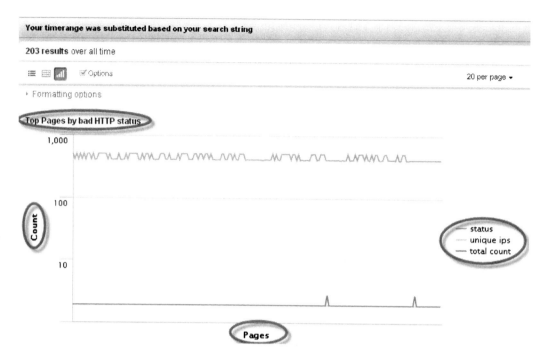

Figure 6-6. *Top pages with bad HTTP status*

Business
User demographics by region

From a business standpoint, it would be interesting to know what regions or places have most users/customers. In Chapter 4, we used the Google Maps app to visualize IP addresses. We can leverage the geoip command from the Google Maps app to find out the top five countries where users come from. The geoip command takes IP addresses as input and plots those IP addresses on the map across the globe. In our sample, MyGizmoStore.com data, we have the field clientip, which has the IP addresses that we can input. Other than plotting IP addresses in Google Maps, the geoip command also outputs the following fields:

- <field>_ country_name
- <field>_country_code
- <field>_region_name
- <field>_city
- <field>_latitude
- <field>_longitude
- _geo

The field names are prefixed with the name of the field that is given as input. Because we use the clientip field as input, geoip will produce new fields that have names clientip_country_name, clientip_country_code, clientip_region_name, clientip_city, clientip_latitude, and clientip_longitude. The _geo field has the combined latitude and longitude information. We can start with a simple search to see the new fields produced by geoip. Let's enter the following into the search bar:

```
sourcetype=access_*
| geoip clientip
```

Figure 6-7 shows the left bar of Splunk search, and we can see the fields produced by the geoip command.

Figure 6-7. *Fields produced by geoip command*

Now that you are familiar with the geoip command, you can extend this simple search by adding the time modifiers to find all of the events for the last 15 days, and piping the results from the geoip command to find the top five countries using clientip_country_name field. You can use the clientip_city field if you want to look at it by city instead of country. Let's enter the following search into the search bar:

```
sourcetype=access_*
  earliest=-15d@d
  latest=now
| geoip clientip
| top limit=5 clientip_country_name
```

We will make appropriate changes to the chart and *x*- and *y*-axis titles, and save this report as "User demographics by region". Figure 6-8 shows the column chart for the top five countries where users come from; in this case, we can see that the United States is at the top of the list.

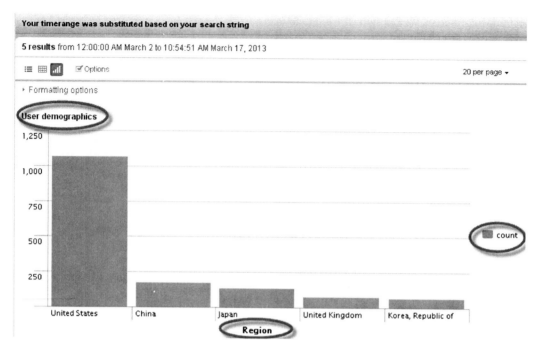

Figure 6-8. *User demographics by region*

Bounce rate

From a business standpoint, it would be interesting to know what the bounce rate for customer visits is on web sites such as MyGizmoStore.com. A visit is considered to be a bounce when the visitor enters and exits on the same page without visiting any other pages on the site. To do this, we make use of the transaction command to group the events based on how many pages have been visited for a particular IP address. In the following search, we make use of clientip as the common identifier for the group of events, and the maxpause option is used to specify that the pause between the events is not greater than one hour. The transaction command produces two fields:

- duration, which is the difference between the timestamps for the first and last events
- eventcount, which is the number of events in the transaction.

We make use of the eventcount field with the eval command to categorize the transactions into:

- Bounced
- 2-5 pages
- 6-10 pages

To categorize the transactions into these three buckets, we use a case function that compares the value of the event count per transaction and puts them into one of the three buckets. Let's enter the following search into the search bar:

```
sourcetype=access_*
| transaction clientip maxpause=1h keepevicted=t mvlist=t
| eval user_type=case(eventcount=1, "Bounced",
       eventcount<=5, "2-5 pages",
       eventcount<=10,"6-10 pages")
```

```
| top limit=5000
  user_type
```

Figure 6-9 shows the pie chart of the distribution of bounced rates and the percentage of visits that are bounces and visits of various depths, such as 2-5 pages and 6-10 pages. We will make appropriate changes to the chart and *x*- and *y*-axis titles, and save this report as "Bounce Rate".

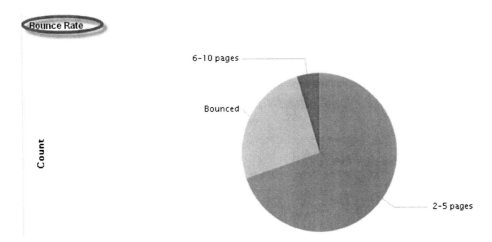

Figure 6-9. *Bounce rate*

You can hover the mouse over the pie to find the count of users and the percentage of the bounce rate, as shown in Figure 6-10.

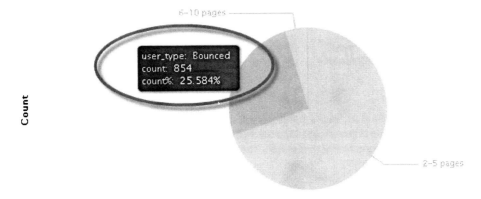

Figure 6-10. *Bounce rate with count and average*

Unique visitors

From a business standpoint, it would be interesting to know how many unique users are coming to the web site. We can use a distinct count function with the clientip field to see how many unique visitors there are for a given period of time. In the following search, we look for unique visitors over last 10 days, served by different hosts of MyGizmoStore.com. Let's enter the following search into the search bar:

```
sourcetype=access_*
  earliest=-10d@d
  latest=now
| timechart  dc(clientip) AS unique_visitors by host
```

Figure 6-11 shows the column chart for unique visitors across the hosts BigDBook-www1, BigDBook-www2, and BigDBook-www3.

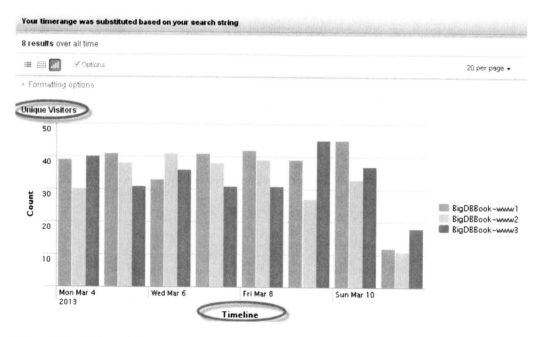

Figure 6-11. *Unique visitors*

We will make appropriate changes to the chart and *x*- and *y*-axis titles, and save this report as "Unique visitors".

In Chapter 4, we worked on a few reports that were added to the MyGizmoStore.com dashboard. Some of those reports are good candidates for the web site monitoring tool as well. The reports that we added in Chapter 4 that are worth going back and looking at include:

- Chart of purchases and views for each product

- Purchases trend

- Transaction duration

- Top purchases by product

- Get versus post

Now you can create a dashboard called "Web Site Monitoring". Although you learned how to create a new dashboard in Chapter 4, we will do a quick recap on how to add a report to the dashboard. Click on the "Create dashboard ..." link under the "Dashboards & Views menu", as shown in Figure 6-12.

Figure 6-12. *Create a dashboard*

This will bring up the dialog box "Create new dashboard". Enter WSM as ID and Web Site Monitoring for Name, as shown in Figure 6-13. Click on the Create button.

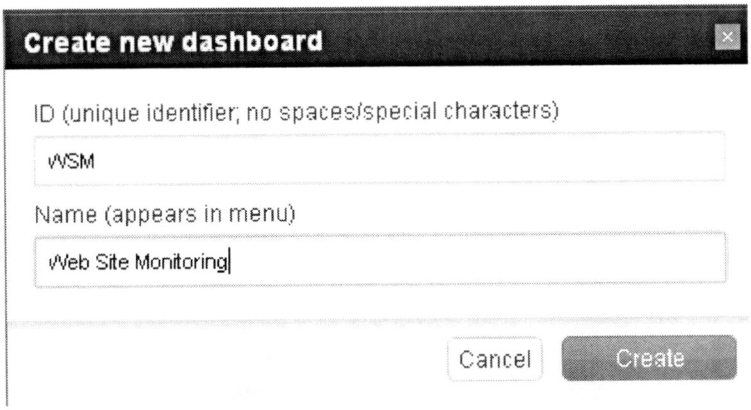

Figure 6-13. *Create web site monitoring dashboard*

You will see that an empty dashboard has been created. To make it editable and add previously created reports, click "On" for Edit, as shown in Figure 6-14.

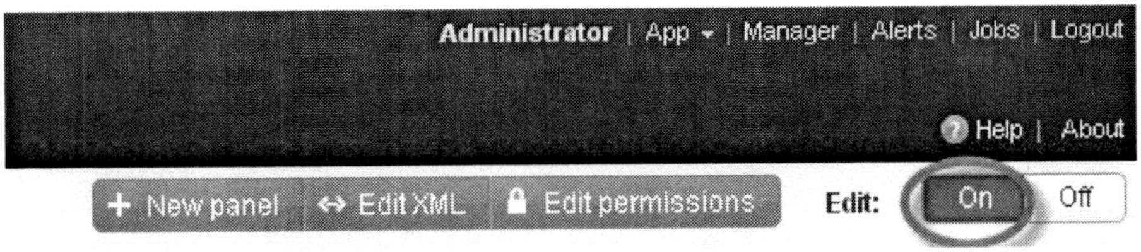

Figure 6-14. *Edit web site monitoring dashboard*

To start adding the reports into the dashboard, you have to create a new panel that will hold the report. Click on the "New panel" button; that will bring up a dialog box as shown in Figure 6-15. Start creating a panel for the Hits by Host report. Enter Hits by Host as Title and select the radio button for "Saved search"; the drop-down box will show the list of saved reports. Select Hits by Host report and click on the Save button.

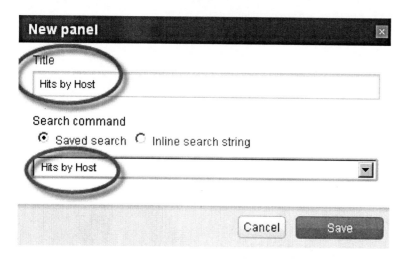

Figure 6-15. Hits by Host panel

This will bring up the report in a tabular format. In order to visualize it as a chart, you have to click on "Edit" and select "Edit Visualization", as shown in Figure 6-16.

Hits by Host

_time ‡	BigDBBook-www1 ‡	BigDBBook-www2 ‡	Bi
3/10/13 12:00:00.000 AM	120	64	118
3/11/13 12:00:00.000 AM	31	27	50

Edit search
Edit visualization
Delete

Figure 6-16. Edit visualization for Hits by Host

This will bring up the dialog box, where you can make changes depending on how you want to visualize the tabular data. Select the "Column" chart as shown in Figure 6-17, and click on the Save button.

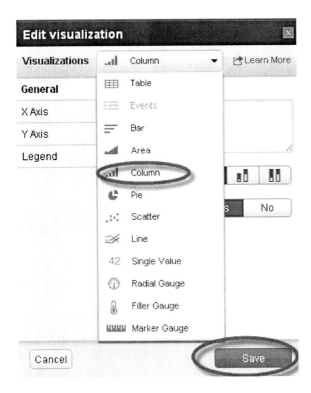

Figure 6-17. *Edit visualization*

You will now be able to see the tabular data for Hits by Host as a column chart, as shown in Figure 6-18.

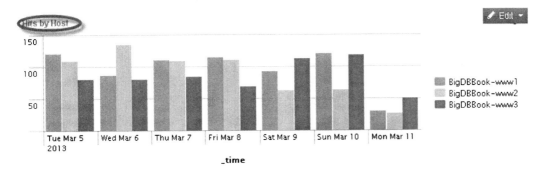

Figure 6-18. *Hits by Host*

For each of the remaining reports that we have created in this chapter, you can create a new panel in the "Web Site Monitoring" dashboard and edit the visualization for the panel to make it render the appropriate chart. Once all of the panels are added to the dashboard, you can move the panels around so that the monitoring information is grouped appropriately.

Once the "Web Site Monitoring" dashboard is created, it can be accessed from the "Dashboards & Views" menu, as seen in Figure 6-19.

Figure 6-19. *Launch web site monitoring dashboard*

Figures 6-20 and 6-21 show the dashboard panels for all of the reports that you have created in this chapter. The charts will look significantly different, as they will be based on the sample data that you have. In addition to the reports created in this chapter, you can add the reports from previous chapters to the Web Site Monitoring dashboard.

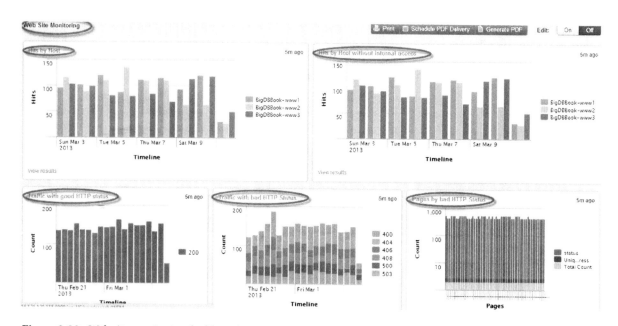

Figure 6-20. *Web site monitoring dashboard*

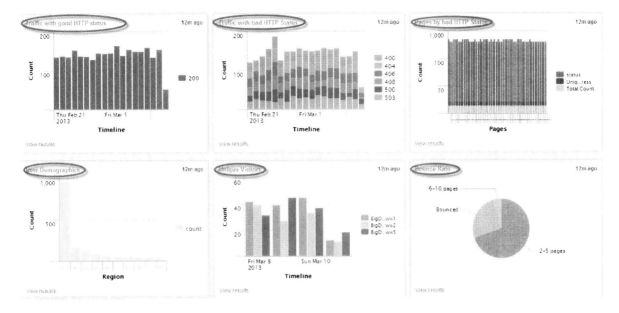

Figure 6-21. *Web site monitoring dashboard*

Summary

In this chapter, we created a web site monitoring tool that shows information about hits by host, good and bad HTTP traffic, user demographics, unique visitors, bounce rate, and other criteria.

CHAPTER 7

■ ■ ■

Using Log Files To Create Advanced Analytics

In this chapter we present a new paradigm, that of using log files to create advanced analytics, bypassing the route taken by traditional analytics. In doing this, we review the best way to create the analytics with a set of recommendations we call Semantic Logging. We'll go over an example that illustrates how easy it is.

Traditional Analytics

Traditional analytics cover a number of functions, which are typically known by various names such as business intelligence, data mining, online analytical processing (OLAP), or just plain analytics. In general, companies use analytics to get a better sense of their operations, cut costs, improve decision making, and identify inefficient processes, which can lead to identify new business opportunities and reengineering their processes.

The challenges with developing analytics are that most of the raw information lives in data stores that are usually decoupled or spread across distributed systems. This makes it very difficult to consolidate. Probably the most difficult challenge is that the information needed to perform analytics has to be made available for this purpose intentionally, and such an effort involves going through the typical software development life cycle, which takes time. It also becomes another task of the Information Technology (IT) department that will go in the queue of things to do, and usually it will have a lower priority.

At the heart of the issue is the fact that the development cycle of traditional analytics is based on what is called Early Structure Binding, where you need to know beforehand what questions are going to be asked of the data. The typical development steps can be summarized as follows:

- Decide what questions to ask

- Design the data schema

- Normalize the data

- Write database insertion code

- Create the queries

- Feed the results into an analytics tool

This is a process that can take days, weeks, or even months, depending on the complexity of the procedures to obtain the necessary data. More often than not, the data required for performing the analytics will be placed in a data warehouse or data mart, which will then be accessed by the various analytical tools. This architecture, along with the typical cases for data collection, can be seen in Figure 7-1.

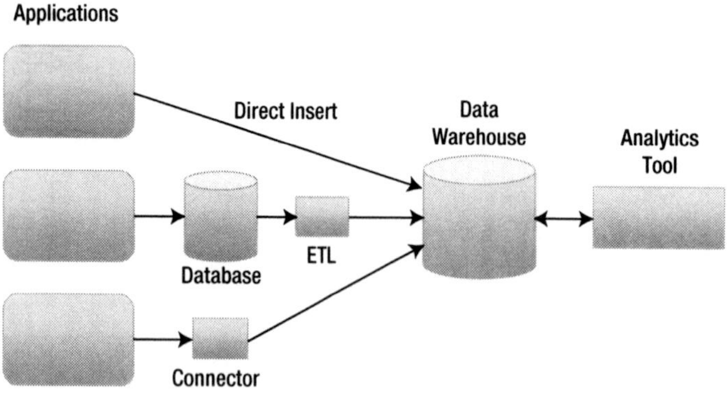

Applications

Direct Insert

Data
Warehouse

Analytics
Tool

Database

ETL

Connector

Figure 7-1. *Typical analytics tool architecture*

As can be seen in Figure 7-1, the data can be collected by including code in the systems to directly insert the desired data into the data warehouse, or can be extracted from an existing database, which in many cases requires some sort of transformation using an Extract-Transform-Load (ETL) program. Finally, many data warehouse systems include connectors that allow for an easier collection of data from applications. As mentioned earlier, no matter how the data is collected, this is an additional step that has to be done specifically for this purpose.

A Paradigm Change

We propose that by using log files in combination with Splunk you can obtain all of the information typically required for performing traditional analytics. Better still, the process is faster and simpler. First of all, pretty much any program or computer-based device produces at least one log file; granted, in some cases, for example, in Windows, the typical information associated with log files is stored in an event management system. The fact that the log files are generally ignored does not make them any less important.

As you have seen in the previous chapters, log files contain a gold mine of information, with a wide range of data that can be used to ensure the system security and to meet compliance mandates. They are a definitive record of activity and behavior in your systems. Because of this, they can provide insight for the IT department and the business in general, presenting customer behavior, product and service usage, and, ultimately, end-to-end transaction visibility.

As can be seen in Figure 7-2, when you collect machine data from your systems infrastructure, such Windows servers, Linux/Unix servers, virtualization and cloud servers, and networks along with the log files of your applications and databases, you can obtain a complete picture of all of your systems.

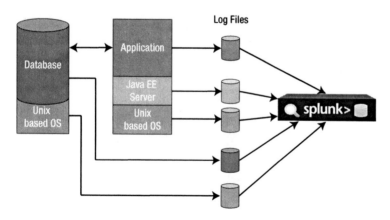

Figure 7-2. *Typical log files of a corporate system*

Using a tool like Splunk makes it easy to collect all of these log files, even though they are not considered structured, as there is no upfront schema for the data, there are no custom connectors, and in general you don't have to use a relational database. The marketing literature on Splunk refers to the information collected and processed as operational intelligence. The primary capabilities of Splunk to provide operational intelligence include:

- Searching, which allows drilling down into the data to troubleshoot issues and incidents that facilitate the root cause analysis

- Real-time visibility, by which you can monitor and alert your system as a whole, allowing tracking of Service Level Agreements (SLAs) and performance issues

- Historical analysis, by which you can find trends and historical patterns, behavior baselines and thresholds, and produce compliance reports

The key to why it is easier and faster to perform analytics using log files with Splunk is that it uses an alternate development cycle to Early Structure Binding. With Splunk you use what is called Late Structure Binding, which has these simple steps:

- Write data (or events) to log files

- Collect the log files

- Create searches, graphs, and reports using Splunk

This process takes minutes or hours instead of days, weeks, or even months, as compared to the traditional development life cycle. What makes this proposition more attractive is that you are not changing any behavior of the developers, as they already log key information. Although this is done mainly for debugging and auditing reasons, most of the code out there has a healthy amount of logging instructions scattered throughout.

In general, it can be said that from the perspective of the IT organization, collecting a bunch of log files is less complicated than having to deal with an ever-growing data warehouse, as most operating systems have decent log file management allowing for rotation based on size or time.

Semantic Logging

We define semantic logging as data or events that are written to log files explicitly for the purpose of gathering analytics. As mentioned earlier, most existing code already has logging statements. With semantic logging, we formalize the process by adding or modifying logging instructions throughout the code. In general, it is something very easy; for example, this high-level snippet of pseudocode that is called every time a commercial transaction is to be processed:

```
void submitPurchase(transactionID)
{
    log.info("action=submitPurchaseStart, transactionID=%d", transactionID,\
            " productId=%d", productId,\
            " listPrice=%d\n", listPrice);

    // Each of the following calls throw an exception on error, which is also logged
    submitCreditCard(...);
    generateInvoice(...);
    fulfillOrder(...);

    log.info("action=submitPurchaseComplete, transactionID=%d\n", transactionID);
}
```

The logging statements, emphasized in boldface in the previous pseudocode, call a method that logs the corresponding information and also includes a timestamp. By just adding those two logging statements, all at once we have enough information to answer the following questions:

- What is the hourly, daily or monthly purchase volume?

- How long are the purchases taking during different times of the day and different days of the week?

- Are the purchases taking any longer than they did last month?

- Are the underlying systems getting slower over time, or are they stable?

- How many purchases are failing? Graph these failures over time

- Which specific purchases are failing?

A couple of years ago, we generated some fake data that would have been created by the code sample presented earlier. The file contains 5 weeks' worth of transactions starting on Monday, August 29, 2011, until Sunday, October 2, 2011, with almost 4 million events. You can find this file in the download package of the book under the name c7sampledata.log.gz.

Before we load the data into Splunk, we create a new index called c7 following the steps we presented in Chapter 2. It is in this index that we will load the sample data following these steps:

- From any screen in the Splunk user interface, click on the Manager link, which is located on the top right corner

- Select the "Data inputs" option on the manager screen

- To the right of the "Files & directories" option, under the "Actions" column, click on "Add new"

- Select the c7sampledata.log.gz file by clicking on the "Browser server" box under the "Preview data before indexing" option and select the "Continue" button on the bottom right corner

- As Splunk is not familiar with this data source, it presents a pop-up window giving you the option to start a new source type or use an existing one. Chose to start a new source type

- At this point, Splunk presents a preview of the data, where it highlights the timestamp within the event and it shows the timestamp it interprets as can be seen in Figure 7-3

Data Preview   /root/c7sampledata.log.gz

If your data looks correct, continue
If it looks incorrect, adjust timestamp and event break settings.

	Timestamp	Event
1	8/28/11 9:00:02.355 PM	Mon Aug 29 2011 00:00:02.355 EDT action=submitPurchaseStart transactionID=1000000 productId="Blue Gizmo" listPrice=100
2	8/28/11 9:00:02.508 PM	Mon Aug 29 2011 00:00:02.508 EDT action=submitPurchaseStart transactionID=1000001 productId="Blue Gizmo" listPrice=100
3	8/28/11 9:00:02.764 PM	Mon Aug 29 2011 00:00:02.764 EDT action=submitPurchaseComplete transactionID=1000000
4	8/28/11 9:00:02.972 PM	Mon Aug 29 2011 00:00:02.972 EDT action=submitPurchaseComplete transactionID=1000001
5	8/28/11 9:00:05.939 PM	Mon Aug 29 2011 00:00:05.939 EDT action=submitPurchaseStart transactionID=1000002 productId="Black Doodad" listPrice=90

Figure 7-3. *Data preview*

In Figure 7-3, you can see that the timestamp assigned by Splunk is different than the timestamp that is highlighted in the actual event. The difference is because the event has the timestamp based on the Eastern Daylight Saving time zone, but the Splunk user has defined the Pacific time zone so it presents the timestamp accordingly.

■ **Caution**   Even though Splunk stores the correct timestamp, it displays it according to the time zone defined in the user profile. This also affects the time periods selected in searches.

■ **Note**   As an aside, to see or change the time zone of the user profile you are working with, you can go to the Manager and select the "Your account" option. You can see in Figure 7-4 that the screen presented allows managing a number of items in the user profile, including the time zone.

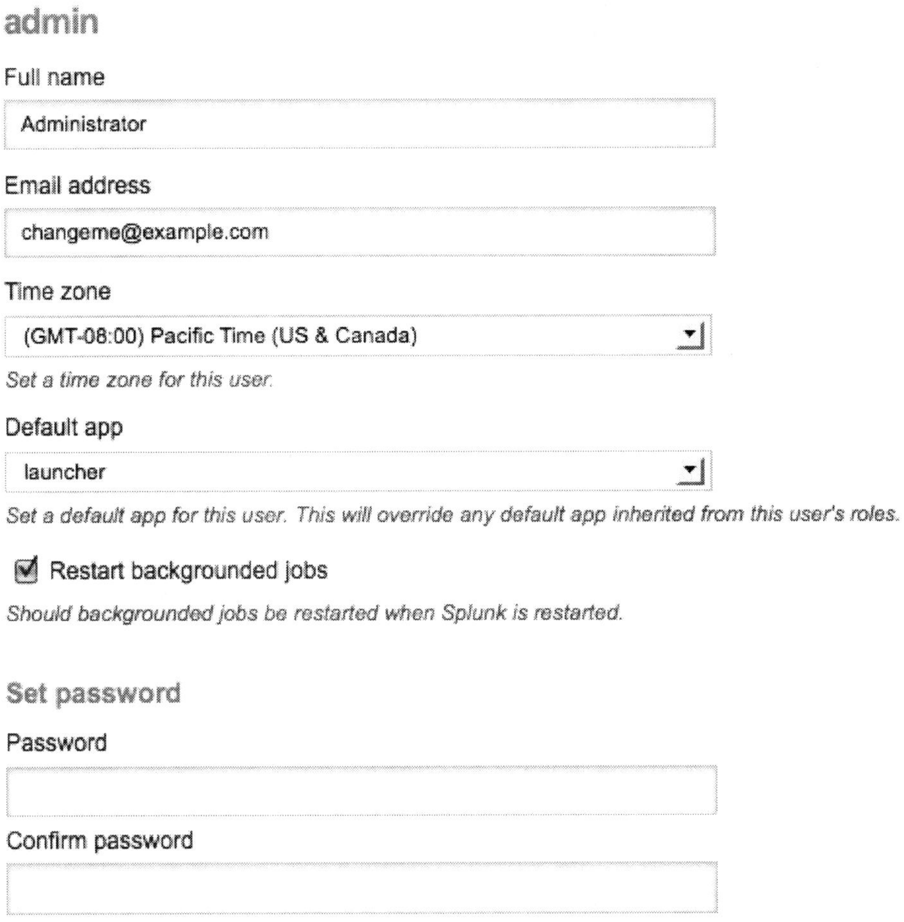

admin

Full name

Administrator

Email address

changeme@example.com

Time zone

(GMT-08:00) Pacific Time (US & Canada)

Set a time zone for this user.

Default app

launcher

Set a default app for this user. This will override any default app inherited from this user's roles.

☑ Restart backgrounded jobs

Should backgrounded jobs be restarted when Splunk is restarted.

Set password

Password

Confirm password

Figure 7-4. *User profile*

Back to loading the sample data. Now that you understand the reason for the difference in the timestamps, and all the rest of the data looks correct, we click on the "continue" link above the data preview. This action presents a popup window asking to name the new source type. We choose to call it c7example and click on the "Save source type" button on the bottom right corner. You will get another popup window stating that c7example was successfully created, and then you click on the "Create input" button.

You are now presented with the screen shown in Figure 7-5, where you select "Index a file once from this Splunk server". Not visible in Figure 7-5, but under the "More settings" section, select the index where Splunk will store this data from the corresponding pull-down menu, which for this example is c7.

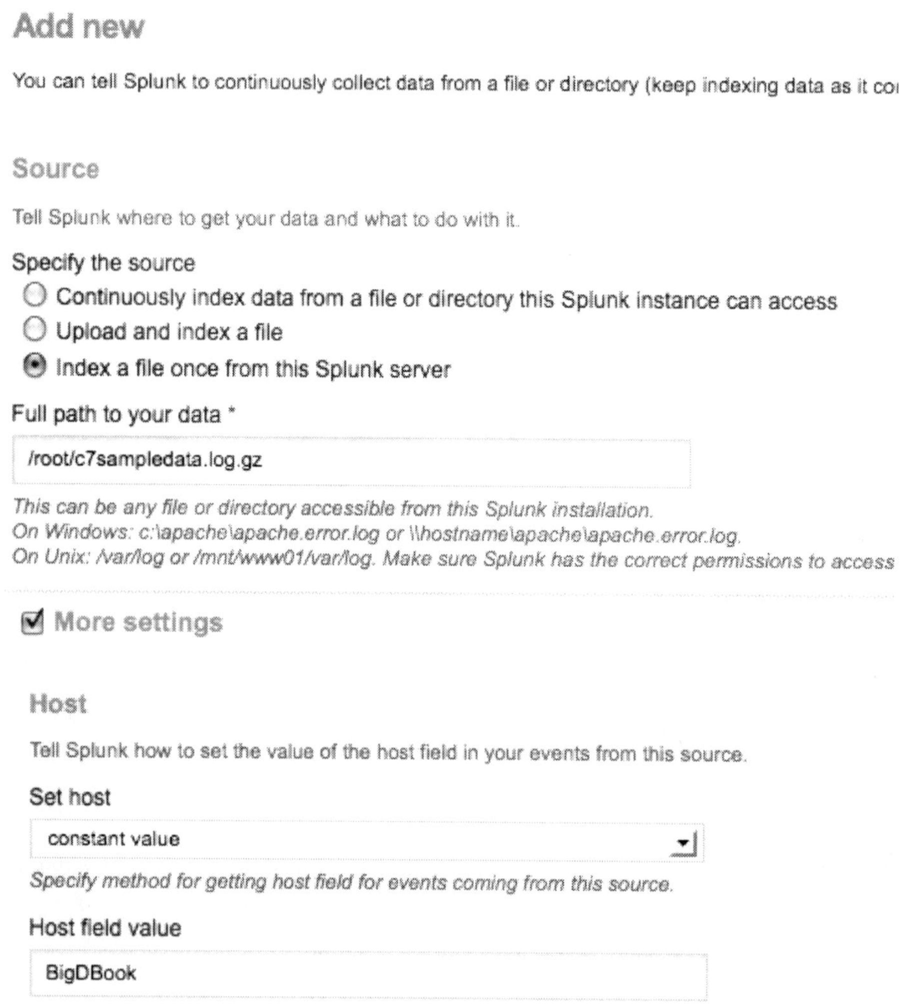

Add new

You can tell Splunk to continuously collect data from a file or directory (keep indexing data as it co

Source

Tell Splunk where to get your data and what to do with it.

Specify the source

○ Continuously index data from a file or directory this Splunk instance can access

○ Upload and index a file

◉ Index a file once from this Splunk server

Full path to your data *

```
/root/c7sampledata.log.gz
```

This can be any file or directory accessible from this Splunk installation.
On Windows: c:\apache\apache.error.log or \\hostname\apache\apache.error.log.
On Unix: /var/log or /mnt/www01/var/log. Make sure Splunk has the correct permissions to access

☑ **More settings**

Host

Tell Splunk how to set the value of the host field in your events from this source.

Set host

```
constant value                              ▼
```

Specify method for getting host field for events coming from this source.

Host field value

```
BigDBook
```

Figure 7-5. *Loading a new file*

Once you save the new file definition, Splunk goes ahead and starts loading it. For this example, it takes a couple of minutes to upload. To know when loading has completed, you can go to the Search window in the user interface and select "Index activity" from the Status pull-down menu, and then choose "Index activity overview". This will show the information presented in Figure 7-6, where you can see that the c7 index has 3,815,666 events, which coincides with the number of events the original file has.

Figure 7-6. *Index activity overview*

The next step is to verify that the data was loaded correctly. Before we look at the search, you must remember to select the correct time range for the data, as it is from the year 2011. To do this, select "Custom time" from the pull-down menu of the time picker. This presents a popup window that is shown in Figure 7-7. In it, you can see that we chose the earliest date to start and October 3 as the latest time. Even though the sample data finishes a few seconds before midnight of October 2, by choosing October 3 you don't have to type 23:59:59.999 in the latest time field. If you don't do this, the results of the search will be empty.

Custom Time Range

Range Type

◉ Date   ○ Relative   ○ Real-time   ○ Advanced search language

Earliest time

○ Specific Date   ◉ Earliest Date

| 03/14/2013 | 00:00:00.000 |

Latest time

◉ Specific Date   ○ Now

| 10/3/2011 | 00:00:00.000 |

Figure 7-7. *Setting a custom time range*

Now you can type the following search to verify the data: index=c7 | head. This displays the 10 most recent events, which you use to verify that the fields in the events are recognized and show up in the left side bar. You also look for the values of the default fields created by Splunk for each event: these are the host, the source type, and the source file. In Figure 7-8, you can see that the action, transactionID, productId and listPrice fields are present in the left side bar, and that the default fields of each event show the correct information.

Figure 7-8. *Verifying the loaded data*

As the verification is complete, you can write the searches that will answer the questions we posed earlier. However, simply calculating the metrics by using the submitPurchaseStart event is not correct, because if any of the functions before the submitPurchaseComplete event fails, the metrics will be flawed. Given that a complete transaction is made up by both events, you will have to use the Splunk transaction command, which we already reviewed in Chapter 3, to correctly calculate the desired metrics.

The transaction command is pretty expensive because it keeps track of all of the ends of a transaction until it can match them to a beginning, assuming that they are paired, which might not always happen. Depending on the amount of events and the time between the first and last, this can consume a lot of memory. The tracking is done in reverse order because Splunk sees events in reverse time order, going from the most recent to the oldest.

We start by finding out which is the maximum time between the beginning and end of the transactions in our sample data. This can be done by typing the following command:

```
index=c7
| transaction transactionID maxspan=1m
| stats max(duration)
```

134

In this search, you specify that the transaction is delimited by the field `transactionID`, that is, only events with the same value in that field will be grouped. You also specify that the maximum time between the first and last events in a transaction cannot be greater than one minute. We do this because otherwise some transactions will not be counted, as they are evicted to make space for others. By specifying a relatively big amount of time, we can be sure that all the transactions are included in the count.

As you saw in Chapter 6, the `transaction` command produces two fields, `duration` and `eventcount`. The result of this search is 0.6 seconds, so from now on you can specify that the maximum amount of time for a transaction is 1 second to make it more efficient. Next, you will find out which transactions are incomplete, that is, the purchase failed. One search to obtain this information is:

```
index=c7
| transaction transactionID maxspan=1s
| search NOT *submitPurchaseComplete
```

This search is rather interesting, because we use the `search` command twice. The first time it is used to bring all the events from the c7 index. Then we use it again to find the new events created by the transaction command that do not have a string with the regular expression `*submitPurchaseComplete`. Note that the first line does not include the word "search", as it is always implicit that the first command is a search. The second time the command has to be stated as search. It is worth indicating that the results of this search will include any transactions that take longer than 1 second, which we now is not the case. For the sample data, the result of this search produces four transactions, which are shown in Figure 7-9. This answers the question regarding failed purchases and specifically which of them are failing.

Figure 7-9. Incomplete transactions

You can now find out the average transaction times, which should help to answer a few of the questions in our list. We start with how long the purchases are taking. In order to get a good resolution on the column chart that we use to present the results, we chose to calculate the hourly average of the duration of the transaction, for a period of one week, any longer period of time will produce an illegible chart. Remember to change the time period for this search.

```
index=c7
| transaction transactionID maxspan=1s
| timechart avg(duration) span=1h
```

The results of this search can be found in Figure 7-10, which presents a column chart with a minimum value for the *y*-axis of 0.3. We did this to have a better resolution on the chart. As we hovered over one of the columns, Splunk presents detailed information of that particular column. From this, we can deduce that the longer times of the transactions are between 4:00pm and 7:00pm, pretty much every day.

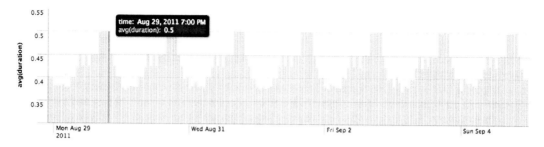

Figure 7-10. *Hourly average transaction times*

To find out if the underlying systems are getting slower, or if the purchases are taking any longer than they did the previous month, you can do a search similar to the previous one, but this time you calculate the average on a daily basis, thus we change the span argument of the timechart command to span=1d. This will reflect much better any issues over a longer time span, and we use a line chart, which will clearly show any trend there might be. After changing the time range to include all the 5 weeks' worth of data in the time picker, the results can be see in Figure 7-11.

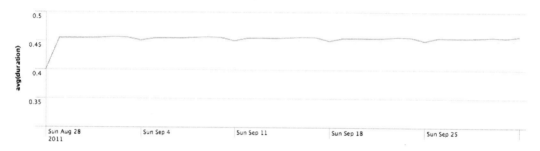

Figure 7-11. *Daily average transaction times*

As can be seen in Figure 7-11, except for the skew presented on the first day of the data, the transaction times seem pretty stable over the whole period covered by the sample data. If there had been any issues with the underlying systems, they would have shown up by presenting an increasing trend over time on this line. Very likely, the number of transactions is the same over the sample period. You can see this and partially answer the first question of the list by creating a graph that shows the daily purchases of all the products with the following search and the results, in the form of a stacked column chart, can be seen in Figure 7-12. In it you can see that the product sales seem to be very stable over time, and this is probably the reason why we don't see any degradation of the transaction time.

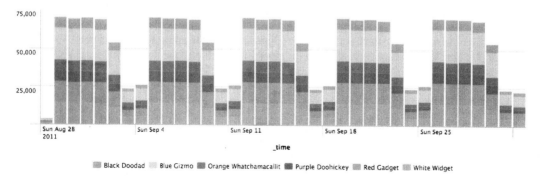

Figure 7-12. *Daily sales by product*

```
index=c7
| transaction transactionID maxspan=1s
| timechart count by productId span=1day
```

The skew on the daily average times observed in Figure 7-11 can be explained by reviewing the number of sales for Sunday August 28, 2011, the first day of the data in Figure 7-12. As there were so few sales that day, the transaction time was on average 0.1 seconds shorter than on the other days.

Admittedly, this is a simplistic example, but it clearly illustrates how easy it is to produce advanced analytics from log files, without having to mess up with data schemas and writing special code to insert data in a data warehouse or database.

Logging Best Practices

To take full advantage of semantic logging in Splunk, we have a few tips or best practices. The first one, which encompasses a few items, is to create human readable logs: specifically, to log in plain text. Although logging in binary might sound better because it is compressed, it requires decoding and does not lend itself to segmenting into partitions. You should also make it easy for humans, so do not use complex encoding for certain elements in the log entry that require lookup tables to decipher.

Another recommendation is to use categories for log entries. Most log systems, such as Log4J, offer a predefined set of categories that fit most needs; make use of them, as it will be a lot easier to control the amount of data that gets logged. Finally, as you will see in the social media part of this book, Splunk makes it very easy to use JSON, or XML for that matter, but do not use them unless you really require multidepth nesting.

The timestamp is another item to consider, and is probably the most important of all. Do not use time offsets, as that will unnecessarily lead you to all sorts of problems when doing comparisons and dealing with time zones. Use standard timestamps that are readable by humans and favor the beginning of the line. This is not only valid for log files, but as you will see in parts two and three of this book, for pretty much any data that makes use of a date and, or time. The bottom line: clearly timestamp every event.

Use clear key-value pairs. It might appear to take a lot more space by repeating so much information, but remember that Splunk stores the raw data in compressed format, and key-value pairs have a very high ratio of compression. The following example is very much the traditional statement most developers use to create a log entry:

```
log.debug("Error 1454 - %s %d\n", userID, transID)
```

The issue with this customary way of logging is that searching for the word "Error" is too vague. Although Splunk will find it very quickly, the operations required to make sense of the accompanying information afterward might be rather complicated. Usually, this kind of log entry will need regular expressions to be parsed, whereas a key-value pair can be processed using a single consistent rule. The following logging statement would be the recommended equivalent for the previous one:

```
log.debug("orderstatus=error, errorcode=1454, user=%s, transactionid=%d\n", userID, transID)
```

If you don't have the ability to create key-value pairs, but you do have a header line with the names of the columns or keys, that will also work well with Splunk. Unix commands such as ps and vmstat produce valuable information, but the column names are provided only as a header line. Splunk knows how to take these column names and use them as the key to the corresponding values. This will be reviewed in detail in the project in Part Two of this book.

It is somewhat popular to log fields with multiple values as a single line, as there is the perception that it is the easiest for the developer. We strongly suggest that multivalue information be broken down into separate entries in a log file. Have a look at this example:

```
Mar 10 2013 04:12:02 phonenumber=669-555-1212, app=angrybirds,facebook
```

It is more difficult to parse the app values in Splunk than using straight key-value pairs. Additionally, there are limitations regarding adding data to each multiple value. If you break each multivalue into separate lines in the log file it will be a lot easier to parse and handle additional data. From the developer's standpoint, this recommendation should not add complexity to the logging statement or set of statements. The previous example would be a lot better if logged as follows:

```
Mar 10 2013 04:12:02 phonenumber=669-555-1212, app=angrybirds, installdate=Jan-02-2012
Mar 10 2013 04:12:02 phonenumber=669-555-1212, app=facebook, installdate=Jun-29-2012
```

Another best practice is to log unique identifiers at every opportunity. With this you will be able to effectively span over multiple log files from different components in your system and it allows you to track transactions in detail. You can use the `transaction` command of Splunk to tie different log entries and convert them into a single transaction. This is also known as Transitive Closure, so that if you have some log entries as the ones in the following example, you will be able to link them all together:

```
transID=135889976
transID=135889976, otherUniqueID=qwas543
otherUniqueID=qwas543
```

The information that you are logging right now can be revealing and present more information than you expect. The effort of refactoring code to take advantage of this new paradigm can seem daunting at first, but start gradually and grow organically. In many cases, all it involves is modifying the existing logging statements so that they follow the best practices already described. Develop future applications with this new paradigm in mind and you will see that the return on investment will be priceless.

On the operations front, remember to use log file rotation policies, where you either destroy or backup the logs. You are probably better off destroying them, as all the information will now be indexed in Splunk. We also suggest that you log locally, in the server where the log files are being created. More details on this can be learned in Chapter 15, in which we review the use of Splunk forwarders.

To sum up, log what is evidently required, that is, log anything that can add value to the analytics, especially when aggregated and visualized, so log more than just debugging events.

Summary

In this chapter, you learned about a new paradigm to obtain advanced analytics from log files instead of the traditional method using a data warehouse. We went over an example that illustrated how simple this is and reviewed the best practices for semantic logging.

CHAPTER 8

■ ■ ■

The Airline On-Time Performance Project

This chapter presents a project to get you used to using Splunk outside the context of machine data, using CSV files, and interacting with relational databases. It discusses an embryonic methodology that we shall use for the project, and explains what will be done in the next three chapters.

The Airline On-Time Performance project is based on publicly available data, which keeps track of all the domestic flights within the Unites States of America, including Puerto Rico and the Pacific Territories and Possessions. This data set is particularly well suited for explaining how to use Splunk as an analytical tool for the following reasons:

- The amount of data is small enough that you do not need a large hardware configuration to work with it. Actually, it can be done on a typical laptop if you wanted to.

- Almost all of us have travelled on a commercial flight, so we have some experience with the domain from where the data is coming; that is, we do not need any special training to understand the majority of the data we are going to handle.

- It is based on real data, which makes the results of the searches we formulate all the more interesting. Whereas we detail various searches to explain multiple Splunk commands, there are still plenty of appealing questions left for your enjoyment.

- The source of the data is different than the traditional machine data we have been working so far in this book. We will be introducing how to import data using a CSV file and from relational databases.

The flight data, as we will refer to it, contains one record or event for every single scheduled flight leg. Sometimes the airlines assign the same number to a flight that stops in more than one airport. Each stop is considered a flight leg, so every event in the flight data only refers to a flight between two airports, sometimes referred as city pairs, which represent an origin and a destination. Note that the flight data is based on scheduled flights, not on the actual flights. If a flight was cancelled, diverted, or delayed, the event will reflect this in the corresponding fields. For example, if you count the number of flights for September 11, 2001, you will see the results show about 17,000 scheduled flights. As we know, most of those flights were cancelled, and there is a field that reflects this.

Each event has 109 fields. Not all of them are populated, either because they did not have the information available or the field was an addition sometime after the original release. An important part of any project that has to do with analyzing data sets is exploring the fields to gain a good understanding of the information they contain. The flight data is ideal for this, as there are some fields that contain similar information with subtle variations. Understanding the contents of a field and choosing the right one is quite important, as the quality or expectations of the results will change depending on the selected field.

With so many fields available containing such a variety of information for each single flight in the United States, there are lots of interesting and compelling questions that can be answered with Splunk searches that go from the very simple to the complex. You will often find that the greatest complexity lies not in the search itself, but in formulating the search. This is demonstrated in some of the searches we do.

The flight data we use has over 147 million events, which start on October 1987 and go up to September 2012. This abundance of data also provides a nice context to present various ways of optimizing searches and reporting. Even though there are tens of millions of events, the size is pretty manageable at approximately 65GB, and because the flight data is distributed by its provider as monthly files, you can download just the time periods you want.

This project also allows us to present a rather informal set of steps to follow when tackling a typical data analytics project like this one. More than anything else they are common sense; therefore, we do not dare to call it a methodology just yet. Every time we work on a new project we perfect it a little more and there is still plenty of room left for improvements. On a high level, these are the steps we follow:

- Obtain the data. This is rather obvious, but there are some issues that have caused us a lot of headaches. So we want to stress the importance of getting your data, and chiefly the reliability of the data source. This can be measured in three areas:

    - The accuracy of the data. Needless to say, the quality of data drives the quality of your search results. We have found that not all data sources are as accurate as we would like. Minor errors can be expected, but consistently unreliable information is not good. Even with the flight data used in this example, we found on a quick check that a huge amount of tail numbers or registrations of airplanes are invalid for American Airlines, and this goes back to April 1999. It does not affect us directly, so we decided to ignore it. But it is an example of why you need to carefully verify the accuracy of the data with which you work, especially the data that is critical to your business process.

    - The reliability of the servers. We have found that many providers of public data do not have the best infrastructure and downloading the data becomes a painful process fraught with multiple crashes, which require many efforts to download a single unit of data.

    - The regularity of the data. Not all providers of publicly available data make it available in a consistent fashion, so data that should be updated on a regular basis is spotty at best. Lags due to data collection and preparation are understandable, but beware of irregular updates as they will affect your business process.

- Load the data. Another obvious step, but it can be quite challenging the first time you do it, especially when you have to deal with data elements that define timestamps. Of course you can always choose to ignore timestamps, but that will limit the powerful features of Splunk.

- Verify the loaded data. Assuming that the data you loaded is correctly indexed in Splunk will lead you very quickly into trouble. Doing it only the first time and taking for granted that it will work fine after that is begging for problems.

- Build the searches step by step. The Splunk Search Language is unique. Whereas having SQL experience will help you a lot in formulating searches, we strongly suggest that you build your searches step by step and verify that the partial results you obtain match your expectations. Do this with small sets of data. Either restrict the time periods or use the head or tail commands to limit the number of events you use whilst building the search, so it is easy to verify the results.

- Verify the results. Use alternate search commands or external tools, such as databases or spreadsheets to verify the results of your searches. The more important the results are in your business process, the more critical it is that you feel comfortable and certify them.

Following this rather incipient methodology we have broken down the project into three chapters that will take you through the whole process of a typical analytics project, which also helps us to give you a tour of some of the most used Splunk features and commands, as well as some that are not as popular but very useful. This methodology is also used in the social media chapters of the book.

Chapter 9 is focused on obtaining the data. It explores the data and its fields at the necessary level to make sure that you have all what is necessary for successfully indexing the data into Splunk. In particular, it goes at length into the issues that relate to the timestamp. The flight data has the components of a useful timestamp spread over two different fields, which are not next to each other. Two different strategies on handling this issue are presented and used. This chapter also goes in detail on how to load data from CSV files and directly from a relational database. To summarize, this chapter focuses on the steps to successfully load the data and verify it is correctly loaded.

In addition to being the fun part of the project, Chapter 10 is the core of it. Here we go in detail into the actual analysis of the data set. It starts by exploring the fields of interest and understanding their contents. Using as an excuse the analysis of airport traffic, flights, delays, and reasons for the delays an interesting variety of searches are formulated, and a set of Splunk commands are introduced. These are:

- join, which allows you to do SQL style joins with some limitations.

- delete, as an option to get rid of events that have inaccurate information.

- Macros, which are parametrized chunks of searches that can be reused and can take arguments, if needed.

- Report acceleration. Most reports are summaries that handle all involved events and tend to be expensive. We explain how to accelerate these kinds of reports.

- Accelerating statistics. Using the TSIDX feature of Splunk and the associated tscollect and tstats commands to dramatically accelerate the searches. And by dramatic we mean from hours to seconds.

- Additionally, we delve into new attributes of some known commands, such as:

    - The limit attribute of the timechart command, and

    - The earliest and latest attributes of the search command

We also discuss how to visualize the results obtained from the various searches we create in this chapter. Although this is not a book dedicated to visualization, it is an important component of any analytics project, probably the most important one right after the accuracy of results. As such, we discuss visualization options available in Splunk throughout the searches we create and use.

In Chapter 11 we go in detail into lookup tables. Just as with relational databases, we have the ability in Splunk to use lookup tables, either based on CSV files or relational databases. We start by creating a lookup table based on a CSV file and then we automate it, to make the lookup transparent, a very handy feature of Splunk. We also create a new lookup table based on the results of a search, showing that not only you can use external lookup tables but also produce them. Asking for the airplane model used on a specific flight number over the years forces us to deal with another set of files provided by a different source, the Federal Aviation Administration (FAA), and create a search that does a double lookup. For this example, we use lookup tables that exist in a relational database. We finally venture outside the realm of Splunk to attempt to visualize the results of this search.

In Chapters 10 and 11 we extensively use commands that have been explained in previous chapters, such as stats, dedup, chart, timechart, top, eval, sort, fields, and head. The truth of the matter is that Splunk makes it so easy to handle analytics that you do not need much more than the basic search commands.

Summary

In this chapter we review the basics of a typical analytics project and propose a set of steps to follow when working on such a project: a methodology of sorts. We describe how the project is broken down and explain what will be done in each of the next three chapters.

CHAPTER 9

Getting the Flight Data into Splunk

This chapter discusses in detail the data used by the Airline On-Time Performance project. It introduces two ways to get structured data into Splunk: Using comma-separated value CSV files and directly from a relational database. This chapter also covers various ways to handle complex timestamps based on data spread over multiple columns.

Working with CSV Files

CSV files are probably the most popular and easiest way to import and export data to and from a relational database or other systems and applications. In this section we are going to use a public data set that contains information about all the scheduled commercial domestic flights in the United States. This data set will be the basis of our examples in Part II of this book.

The Flight Data

The Bureau of Transportation Statistics has a web site dedicated to TranStats, which is the Intermodal Transportation Database. This database contains information that is regularly updated on aviation, maritime, highway, rail, and other modes of transportation. As described earlier, we will focus only on the Airline On-Time Performance data, which is made available as a simple table that contains departure and arrival data for all the scheduled nonstop flights that occur exclusively within the United States of America. The data is reported on a monthly basis by U.S. certified carriers that account for at least one percent of the domestic scheduled passenger revenues. The flight data and a description are available at the following URL:

http://www.transtats.bts.gov/DL_SelectFields.asp?Table_ID=236&DB_Short_Name=On-Time

Alternatively, begin by going to www.transtats.bts.gov. Then click on the Aviation link, which can be found on the left side bar under Data Finder, By Mode. This selection is indicated in Figure 9-1.

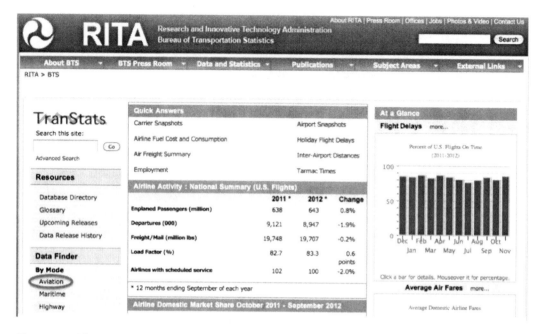

Figure 9-1. *The TranStats web site*

Selecting the Aviation link will take you to a page that presents a list of aviation related databases. On this list, select the Airline On-Time Performance Data by clicking on the name of the database. This will take you to the next page, which shows more detailed information about the database we are interested in. Here you will click on the download link as shown in Figure 9-2.

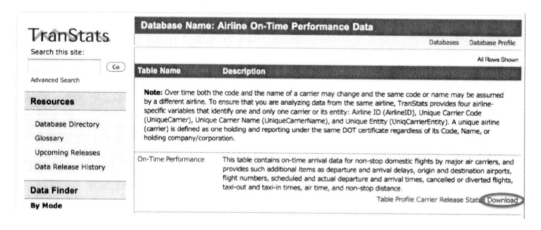

Figure 9-2. *The Airline On-Time Performance page*

Downloading the Data

The download page presented in Figure 9-1 offers a comprehensive set of options for download. You can filter the data by state, year, and month, making one month the standard unit of data for this example. The data is available starting with October 1987 and the web page clearly indicates which is the latest month available.

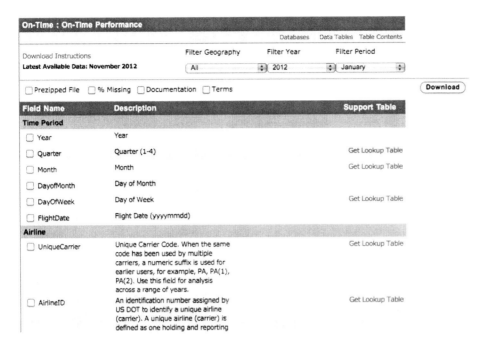

Figure 9-3. *On-Time Performance download web page*

Below the filter choices you will find four additional options related to the types of information. For our purposes, select the *Prezipped File* option. This option provides us with a zipped archive that contains a CSV file with all the available fields and an HTML file with a description of the fields. If you are interested in just selecting specific fields, you can choose them in this page as well. However, you should take the entire file, with all the fields, so that you can more easily follow along with our example in this book.

For starters, we will just download one month to get familiar with the data. However, in the next chapter we will be using all the data available. Downloading one or two month's worth of data is rather simple, but when you have to download all the data since 1987 month by month it becomes very cumbersome. You can automate the download by using either wget or curl. As of this writing the applicable URL to download the data with these programs is:

```
http://www.transtats.bts.gov/Download/On_Time_On_Time_Performance_YYYY_M.zip
```

In this URL format, YYYY is the year and M is the numeric month going from 1 (not 01) to 12. In the download package of this book, we include a script called *download_all_flight_data.sh* that automates the download of all the data. As an example, to download the month of October, 1987 using the wget command:

```
wget "http://www.transtats.bts.gov/Download/On_Time_Performance_1987_10.zip"
```

Getting to Know the Flight Data

One of the most important things to do when analyzing data is getting familiar with it. This statement applies for all types of data, be it big data, structured data, unstructured data, and so on. In general, data is a live entity that continually evolves. The contents are always changing as well as its format. This is the case for both real time and historic data. No matter the case, make sure you get really familiar with the data you are going to analyze. Particular care has to be placed on thoroughly understanding the meaning of the fields that are available. It has been our experience that many difficult problems arise by issues that can be traced directly to misunderstandings of the data.

Taking advantage of the fact that the download page presents all the fields available with a brief explanation, organized in groups, to gain familiarity with the data. The groups are as follows:

- **Time Period.** This group consists mostly of expected fields: year, quarter, month, day of the month, day of the week, and the actual flight date. This last field is one of the most important ones for defining the timestamp in Splunk.

- **Airline.** This is group that has to be studied carefully, as some issues can arise if the differences between these fields are not clearly understood. The differences between the *UniqueCarrier*, *AirlineID*, and *Carrier* fields are subtle, but they can have a big impact in the results of our analysis. We will discuss these fields in more detail in the next chapter. The other two fields in the airline group are the tail number of the airplane, which is the registration (the equivalent of the license plate for a car), and the flight number.

■ **Note**   Notice to the right of the *Carrier* field that there is a link to get a lookup table. Clicking on it downloads a table with the names of the airlines by unique code. You can see that many fields throughout the download page have such a link.

- **Origin Points.** Here we also find four fields that can be confusing: *OriginAirportID*, *OriginAirportSeqID*, *OriginCityMarketID*, and *Origin*. We will not go into the details of each field until the next chapter, but we want to reinforce the fact that you need to be acquainted with the data that you are analyzing.

- **Destination points.** These are similar to the origin point fields but indicate the corresponding destination.

- **Departure Performance.** These fields are mostly self-explanatory, except that we need to clarify the acronym CRS, which stands for Computer Reservation System. The field *CRSDepTime* is the official scheduled departure time of a flight. This field along with the flight date will be used to create the timestamp in Splunk, which will be explained later in this chapter.

- **Arrival Performance.** These fields are similar to those of departure performance but relate to the arrival performance.

- **Cancellations and Diversions.** Present if a flight was cancelled or diverted and gives the reason for cancellation.

- **Flight Summaries.** Information here includes things such as elapsed times and distances.

- **Causes for Delays.** Describes reasons for a flight's being delayed. This and the next two groups are interesting, because they illustrate how data can change in format over time. This group was added in June 2003.

- **Gate Return Information.** Contains information regarding the amount of time since the aircraft left the gate and returned. This is a newer group of fields added in October 2008.

- **Diverted Airport Information.** Describes detailed information for as many as five flight diversions. This group was also added in October 2008.

The fact that the number of fields increases over time is cause for concern. The issue at hand is that we potentially have to deal with three different types of CSV files: one with 56 fields, another one with 61, and a final one with 109 fields. We must examine carefully the CSV files we download.

We start by reviewing the first month available, October 1987. The easiest way to do this is to use a spreadsheet. We first look at the title line and notice that it has all the field names for the 109 fields. This is good because it means that we might not have to worry about dealing with three different CSV files. Now we look at the first line where we can see that there is no data starting in column 57 (CarrierDelay) to column 109 (Div5TailNum). A quick glance at all the data lines shows that pattern.

Although using a spreadsheet is the easiest way of examining a CSV file, it is not the most thorough. By definition, a CSV file contains Comma Separated Values. Even though the spreadsheet shows that there is no content in fields 57 through 109 in the data lines, we need to make sure that those fields are in the file and contain nothing. Each line should have a total of 108 commas, as the last field does not need one. Starting with field 57 there should only be commas with nothing between them or a pair of double quotes, one after another (""), which represents an empty string.

Although this is not exactly a fun endeavor, we counted the commas and found that the couple of random lines we chose have all the commas necessary to delimit 109 fields. This is really good news, because now we know that the October 1987 file has all 109 fields, even thought many of them are not populated. This means that there is only one type of CSV file to deal with, instead of three.

The next step is to make sure that the format is consistent throughout all the flight data. For this we choose to examine the months of June 2003 and October 2008, which are the ones where new fields are added. We reviewed these files and found the layout to be consistent. Now we are sure that all the files that contain the flight data have the same layout or format containing 109 fields, and we will not have to do any special processing for this reason.

When using a spreadsheet to examine the data, you also have to be aware that the spreadsheet will use default representations for data fields. For example, the FlightDate field is formatted as yyyy-mm-dd in the CSV file; however, the spreadsheet presents it as mm/dd/yyyy. Another example is the CRSDepTime field, which in the CSV file is a string with the following format: "hhmm". The string "0900" is presented in the spreadsheet as 900. You have to be aware of how data is presented in the spreadsheet as in many cases it will vary from the actual data.

Be thorough when you are getting familiar with the data you are going to analyze. Using a spreadsheet is a quick way to verify certain things, but it is not the best solution. You will always have to examine the raw data using a text editor.

Timestamp Considerations

Splunk is a time series indexer. This fact makes the timestamp of paramount importance in Splunk. All events are automatically timestamped. The search app user interface has a time range picker based on the timestamp that impacts the searches. The timeline column chart is based on the timestamp. By default, search results are sorted by the timestamp. There are also some commands like *timechart* that are based on the timestamp. In Splunk, data must have a timestamp, which is assigned based on the following precedence rules:

1. Splunk looks for a time or date in the event itself using an explicit TIME_FORMAT that you configure in the props.conf file.

2. If TIME_FORMAT was not defined, Splunk attempts to identify a time or date based on the event source type.

3. If it cannot identify a time or date, Splunk will use the timestamp from the most recent previous event of the same source.

4. If no events in a source have a date, Splunk will try to find one in the file name, which requires the events have a time.

5. For file sources, if no date or time can be identified in the file name, Splunk uses the file modification time.

6. If none of the above works, Splunk sets the timestamp to the current system time when indexing each event.

When we index the flight data into Splunk, it will automatically try to recognize a timestamp within all the fields of the CSV files. As a timestamp is not really obvious in the flight data, Splunk will follow the precedence rules and default to use the last modified date and time of the CSV file that contains the events.

This is not the ideal situation, as we will not be taking advantage of all the timestamp based features offered by Splunk and it could render incorrect search results. Admittedly we can still analyze the flight data using dates and times, as there are clearly defined fields that contain that kind of information, such as *FlightDate*, *CRSDepTIme*, and even *CRSArrTime*, if we want to analyze scheduled arrival times. But it will benefit us immensely to be able to combine some fields to create a Splunk timestamp.

Timestamps are usually represented in two ways: as an integer that contains the number of seconds since January 1, 1970, known as Unix time; or a string that contains a date and a time, which can be in various different formats. For example, December 21, 2012 9:15 AM is represented in Unix time as 1356081300.

As mentioned earlier, there are two fields in the flight data that, when combined, can provide us with a string that contains date and time. We will use the flight date and the CRS departure time, as almost everybody handles flights by the departure time.

Now we have a timestamp for our data, but we have an additional issue. There is no explicit definition of a time zone. All the times in the data are local. The time zone is implicitly defined because we know the city of origin of the flight. In this case, we have a couple of options. We can engineer a way to create a new field that contains the time zone or we can discuss the need of a time zone field. When you think about it, departure times of flights are rarely given associated with a time zone as everybody assumes the time is local to the departure city. Based on this we chose not to create a time zone for this data.

Mapping Fields to a Timestamp

We want to map columns 6 (*FlightDate*) and column 30 (*CRSDepTime*) to become the timestamp for the flight data. There are a couple of ways of handling this:

- Preprocessing the CSV files by moving those columns so that they become the new columns 1 and 2. Once this is done, we can specify in Splunk that these columns provide the timestamp.

- Modifying Splunk's timestamp processor to assemble the timestamp directly from the columns in their current positions.

As with everything, both options have their pros and cons. Telling Splunk to build a timestamp from contiguous columns is very easy, but it implies the extra step of preprocessing the CSV files to move the columns around. Modifying Splunk's timestamp processor requires defining a regular expression that handles the mapping. This regular expression can be expensive for the computer to process while Splunk indexes the data and can be a scalability concern, but there are no extra steps to be done. Because these options allow us to illustrate different features of Splunk, we will explain both.

Preprocessing the Flight Data

One of the advantages of working with structured data is that you always know its format; thus, moving the columns around is relatively easy to do. We have chosen to write an AWK program to do this. The program is based on counting commas, which is the field separator. All we have to do is read a line and write it with the columns rearranged. In AWK, the essence of the program looks like this:

```
BEGIN {
    FS = ",";
    OFS = ","
}
print $6, $30, $1, $2, $3, ..., $109
```

The field separator is defined as a comma with the FS variable for the input and OFS for the output. AWK will take each incoming line, separate fields and make them available as variables called $N, where N is the number of the field or column. Note that there is no need for a read instruction in AWK. By writing column 6 and 30 first, then the rest of the columns, we achieve our objective. The program, which we called *column_mover.awk,* is executed in Unix as follows:

```
column_mover.awk < original_data_file > modified_data_file
```

One of the issues that we bumped into using AWK is that it counts all commas, even those within a string, that is, within a pair of double quotes. In higher level languages, commas within a string are considered part of the string. This is relevant because two fields before column 30 are strings that always contain a comma. Column 16—*OriginCityName* and column 25—*DestCityName* always contain the city name and the state abbreviation separated by a comma, for example, "New York, NY." We solve this problem by accounting for those two extra commas. Now *OriginCityName* is made up of columns 16 and 17, shifting all the columns by one, and *DestCityName now starts at column 26 and is composed of columns 26 and 27, further shifting all the following columns by one for a total of two columns from this point on. The new AWK command looks like this:*

```
print $6, $32, $1, $2, $3, ..., $109, $110, $111
```

For this script to work correctly we need to verify that there are no other fields after column 30 that are strings that include a comma. After reviewing the data description, this is confirmed. This exercise of reviewing the data structure and its contents is a reminder of the fact that we need to be intimately familiar with the data we are going to analyze. After executing this program over the selected flight data, we are ready to load it into Splunk. We have included the AWK program in the download package of the book under the name *column_mover.awk.*

To define the desired timestamp combination, we will need to create a new source type. For this example we will be interacting with the user interface, but this can also be done by directly modifying the *props.conf* configuration file. Following the dialog for defining a new data input file, we select to preview the data before indexing and choose the file we want to load, as can be seen in Figure 9-4. As we are testing something new we will just use one month's worth of data. Once we are sure that all we do works properly, we can use all the data available. In this case, we will use the most recent data, which at the time of this writing is September 2012.

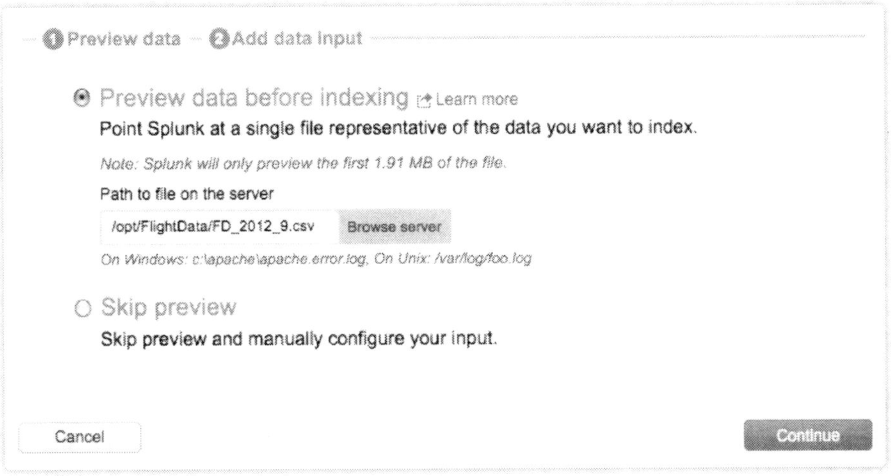

Figure 9-4. *Data preview dialog*

Once we click on Continue, the next dialog box gives us three options related to setting the source type. We choose to define a new source type, because we are going to have a special treatment of the fields. The next screen, presented in Figure 9-5, shows the first interpretation that Splunk has of the data. As you can see, it shows all the flight records as a single event. The default is to break the lines on the timestamp, but because Splunk cannot find it, it presents one single event that contains all the lines of the CSV file.

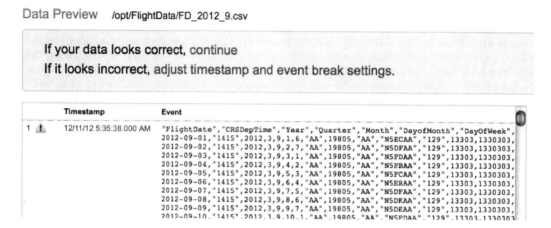

Figure 9-5. *Initial flight data view*

As the data looks incorrect, we click on "adjust timestamp and event break settings." This takes us to the next screen, where we have various options to define the event breaks. We select "Every line is one event" and click on the "Apply" button. After a little while processing the events, it displays them with the correct event breaks, as can be seen in Figure 9-6.

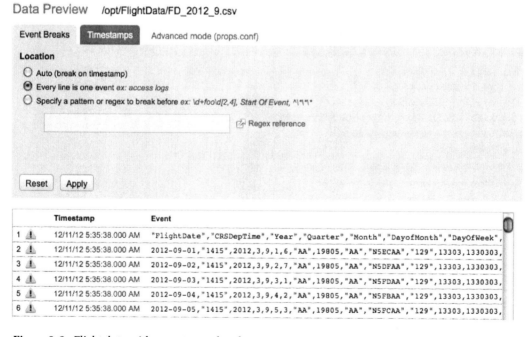

Figure 9-6. *Flight data with correct event breaks*

As can be seen in Figure 9-6, there is a warning triangle at the beginning of the line. When you hover the mouse over it, it states that it failed to parse a timestamp. In this case, following the rules of precedence, the timestamp it presents is the last modification timestamp of the CSV file.

The next step is to work with the timestamps. For this we select the corresponding tab. As you can see in Figure 9-7, there are three options under the Location title. Because the timestamp is located at the beginning of the event we do not have to specify a pattern that precedes it. This option is very useful if your timestamp is further into the event and has something like a string before it, for example:

```
Printed on 04/17/2012 Page 01
```

You would define the pattern as "Printed on". You can also use regular expressions to define the pattern. Because this is not our case, we select the last option where we define the maximum amount of characters into the event where the timestamp processor can find the timestamp. This number is 0 as the fields that are used to define the timestamp are always at the beginning of the event. The closer the timestamp fields are to the beginning of the event, the better for Splunk and the faster it will process the timestamp when it indexes the data.

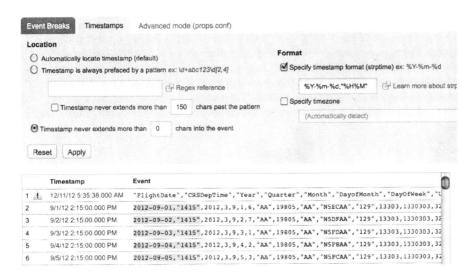

Figure 9-7. *Timestamp definition*

Under the Format title we can specify the timestamp format and the time zone. We already decided we will not be using the latter, so we can move on to define the former. This is done using *strptime()* expressions, which are really simple and well documented for multiple programming languages. In our case the expression is **%Y-%m-%d,"%H%M,"** which means the year using four digits, a dash sign, the month as a two-digit number, a dash sign, the day of the month, a comma, a double quote, the hour using a 24-hour clock (00 to 23), and the minutes as a two-digit number.

■ **Note**   Splunk does not currently recognize non-English month names (%B, %b) or weekday names (%A, %a). An alternative is to use numeric months (%m) and weekdays (%u).

Once we apply these definitions the data preview screen in Figure 9-7 presents us with a list of the events with the desired timestamp and it also highlights the section that matches the expression used for the timestamp. The only other thing to notice is that the header line still presents a warning related to the fact that it was not able to parse a timestamp, thus it is using the last modification time of the CSV file. In our case, this is not an issue as the title line of the CSV file does not have any flight information.

The next dialog box asks for the name of the new source type. We called it *FD_Source1* and saved it. From there it takes you to index data, where it automatically starts indexing once you have done your final selections on that screen. At this point you can move to the summary page of the search app and see how the count of indexed events increases until it stops. In our case September 2012 contains 490,200 events. Note that this count includes the title line.

Now we can verify if the new source type works correctly. We start this process by reviewing a few lines by typing the following search command in the search screen of the user interface * | head. As we have not specified any sorting, by default Splunk will sort using the timestamp in decreasing order, that is, from most recent to oldest. As expected, the header line is the first event to show up, followed by events with a timestamp of September 30, 2012 11:59:00.000 PM and continues with decreasing timestamps. The events contain the information we were expecting. We can also see that each of them correctly displays the name of the host, the source type and the source file.

Reviewing the left bar we notice that under Interesting Fields none of the field names of the header line are showing up. This is not good, as we will not be able to use the field names on the searches when we analyze the data. The problem is that we had incorrectly assumed that Splunk would automatically associate the field names in the header line with the data fields. This would normally happen if we use CSV as the source type, but in our case we defined a new source type. The solution is to set the CHECK_FOR_HEADER attribute to true in the *FD_Source1* stanza of the props.conf file.

Unfortunately, the CHECK_FOR_HEADER attribute works only at index time, so we will have to remove the data we have indexed and reindex it. After adding the CHECK_FOR_HEADER attribute in the *props.conf* file we stop Splunk. All changes to configuration files require a restart to be activated. The reason we stopped Splunk is that the command we will use to remove the events we already indexed only works when Splunk is not running.

The clean command, which is issued from the Command Line Interface (CLI) at the Unix prompt, deletes the data in one or all the indexes depending on whether you provide an index name as an argument. Because our instance of Splunk only contains the flight data on the main index, we will remove all the data by issuing the following command:

```
splunk clean eventdata
```

After we have deleted the data in the indexes we start Splunk again and we can proceed to reindex the data. We do it pretty much the same way we did before; we preview the data, but this time we select to use an existing source type and choose *FD_Source1* from the pull down menu. Splunk presents us with the data, which looks correct, and we proceed to index it.

Once the indexing is ready, we verify once again if the source type is working correctly. As we search for the first 10 events, we can see that the field names are on the left bar. Things are looking good. However, we also notice that the source type on each event is now *FD_Source1-2*. As suspicious as it looks, this turns out to be a side effect when CHECK_FOR_HEADER is true, as it causes the indexed source type to have an appended numeral.

■ **Caution** When CHECK_FOR_HEADER is set to true, the field names are stored in the server where the source type was defined. Because of this, this feature will not work in most environments where the data is forwarded.

Now we can try a few search commands to see if the field names are properly lined up with the contents. The first quick check is searching for the top 10 origin airports by number of scheduled flights by typing the following search command in the search screen of the user interface:

```
* | top Origin
```

This search goes over all the indexed events, which is one of the things we want as part of these quick verifications. The *Origin* field contains the code of the airport. The result shows the top 10 airports and contains the usual suspects: Atlanta, Chicago O'Hare, Dallas/Ft. Worth, Denver, and so on.

In the next check we will try to verify that if an event has the last field *Div5TailNum,* then it also contains an actual tail number. The search we use is:

```
* | where isnotnull(Div5TailNum)
```

The result of this search only includes the header line, so there are no flights with five diversions in this month. We try the search with *Div4TailNum* and *Div3TailNum* unsuccessfully, until we get a result with *Div2TailNum.* This shows only one event for Alaska Airlines on September 1 at 7:40 AM. We go to the left bar and click over *Div2TailNum.* Figure 9-8 presents us with a popup window that contains statistical information and a breakdown of the values of the field. There are two values here, Div2TailNum, which comes from the header line, and N768AS, which is a proper tail number and happens to be the same as the one contained in the *TailNum* field.

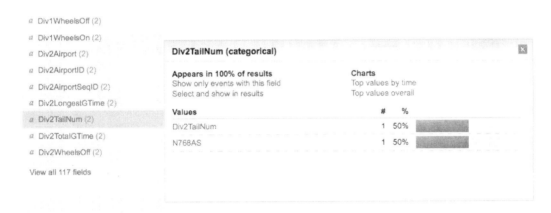

Figure 9-8. *Field pop-up with statistical information*

We do one last check with a field in the middle of the event. The chosen field is *Cancelled.* A value of 1 means the flight was cancelled, whereas a value of 0 means it was not. Execute the following search command:

```
* | chart count by Cancelled
```

This command goes over all the events and the field popup presents three values, 0, 1 and Cancelled. The last value comes from the header line as expected. Because there are no other values and the percentages of noncancelled flights (99.193%) and cancelled flights (0.807%) seem reasonable, this result and the ones of the previous searches make us feel comfortable that the data was correctly indexed into Splunk.

Modifying the Timestamp Processor

The second option for mapping the desired fields of the flight data into a timestamp is modifying Splunk's timestamp processor. Splunk automatically recognizes and extracts most of the obvious timestamps. It does this with a set of predefined regular expressions that can be found in a file called *datetime.xml* in Splunk's *etc* directory.

Using this option we will have to write a new regular expression that allows us to map columns 6 and 30 as the new timestamp. There are many different regular expressions that can be used to do this mapping. In this case we will do it in a similar fashion as we did with the AWK script that moved the columns around: we will count commas. As before we will have to consider that column 6 (*OriginCityName*) and column 25 (*DestCityName*) always contain the city name and the state abbreviation separated by a comma.

The idea is to skip the first five columns, capture the date, then skip the next 26 columns and capture the time. The regular expression we used is:

```
(?:[^,]*,){5}(\d+)-(\d+)-(\d+)(?:[^,]*,){26}"(\d\d)(\d\d)"
```

Without making this a regular expression tutorial we will explain what this strange combination of characters means. You can find various regular expression tutorials in the Internet by doing a search with your favorite search engine. In the first part, (?:[^,]*,), the enclosing parentheses mean that we are grouping this part of the regular expression. Groups are remembered for future reference, but this is quite expensive in processing costs. Because we are only using groups in this part of the regexp to skip fields, we use the ?: characters, which are a special directive telling the regular expression processor not to remember this group, which will speed up the processing.

Square brackets ([]) are used to define character sets, which tell the regular expression engine to match only one out of all the characters within that set. When a caret symbol (^) is used as the first character within a set, it means that the set is negated. In this case [^,] matches any character except for a comma. Please note that in all other contexts the caret symbol means the beginning of the line. The star (*) that follows it is a repetition directive that means zero or more times. Finally, the comma is a literal comma, which has to be matched. Thus (?:[^,]*,) means that this group matches any characters except for a comma, zero or more times, followed by a comma and do not remember this group for future reference. Groups can be repeated by enclosing the number of times with curly brackets ({}). In this case we repeat this first group five times.

In the next part, (\d+)-(\d+)-(\d+) we capture the date by defining these three groups. \d is a shorthand for the [0-9] character set, which describes all the digits. The plus sign (+) is a repetition directive that means one or more times. The dashes between groups have to be matched. Notice that these groups do not include the ?: directive; thus, they will be remembered so we can reference them later to extract the date.

The fourth part of the regexp skips the next 26 fields and then we capture the date as two distinct groups, hour and minutes, enclosed by double quotes (").

Using this regular expression, we create a new timestamp processor:

```
<datetime>
  <define name="flightdata_csv_timestamp" extract="year, month, day, hour, minute">
    <text><![CDATA[(?:[^,]*,){5}(\d+)-(\d+)(\d+)(?:[^,]*,){26}"(\d\d)(\d\d)"]]></text>
  </define>
  <timePatterns>
    <use name="flightdata_csv_timestamp"/>
  </timePatterns>
  <datePatterns>
    <use name="flightdata_csv_timestamp"/>
  </datePatterns>
</datetime>
```

Here we define a timestamp processor called *flightdata_csv_timestamp*, which extracts the year, month, day, hour, and minutes from those regexp groups we specified it should remember, in that specific order. The next statements tell that this processor will be used to process time and date patterns. Following best practices, instead of adding this XML code to the *datetime.xml* file, we create a separate file we call *datetime_flightdata.xml*, which can be found in the download package of the book.

Now that we have defined a new timestamp processor the next step is to associate it with a source type. This time, instead of using the user interface, we will directly work with the configuration file. In Splunk's *etc/system/local* directory we modify the existing *props.conf* file, and add the following stanza for a source type called *FD_Source2*:

```
[FD_Source2]
DATETIME_CONFIG = /etc/datetime_flightdata.xml
MAX_TIMESTAMP_LOOKAHEAD = 220
SHOULD_LINEMERGE = false
CHECK_FOR_HEADER = true
```

The first attribute specifies the file that contains the timestamp processor. Note that the filename is relative to the directory where Splunk is installed. The second attribute specifies how many characters into an event Splunk should look for the timestamp: in our case, how far away column 30 is going to be into the event. After reviewing the flight data we estimated that 150 characters would cover all the cases, but we decided to increase the limit to 220 just to be sure. Please note that if you do not get this number right, you can miss the timestamp altogether.

The line merge attribute is related to the way Splunk breaks the lines. As mentioned earlier, the default is to break at the timestamp. If there is no timestamp, it will create a single event that contains all the events, or flight records in our case. By setting this attribute to false, the behavior is that Splunk will create one event for every single line, that is, it will break an event where there is a line break. The final attribute specifies that Splunk should get the field names from the header line of the CSV files.

Now that we have defined the source type that uses the new timestamp processor, we can test it out. First, we delete all the events we indexed before using the CLI *clean* command. Then, with the user interface we add a new file using the preview option. Here we specify to use our new source type *FD_Source2*, which presents a preview of the events with the correct timestamp except for the header line.

After indexing the data we run the same set of quick tests we did for the preprocessing option and verified that this method works correctly.

Indexing All the Flight Data

Using the *download_all_flight_data.sh* script included in the download package of this book, we downloaded all the flight data into a directory called */mnt/flight_data*. Now we have to choose which option we want to use to index the data into Splunk. Because we are kind of lazy, we use the modified timestamp processor to avoid the extra step of preprocessing the data by moving the timestamp related columns around.

Before we go ahead and start indexing all the flight data, we have to add one attribute into the *props.conf* file where we defined the *FD_Source2* source type. As it turns out, by default Splunk only handles timestamps 2,000 days into the past from the current date. That is about five and a half years. Our data goes all the way back to October 1987. By setting the MAX_DAYS_AGO attribute, we define how many days into the past the timestamp is valid. The maximum value is 10,951 days, which is about 30 years, and a good number to cover our data.

For those events with a timestamp that goes beyond the limit defined by the MAX_DAY_AGO attribute the Splunk timestamp will show as the oldest valid timestamp. Thus, you will not be able to effectively use Splunk's timestamp, because it will show the wrong value. The alternative is to use a specific field or fields in your data that contain the date and time.

▓ **Note**　If the dates of your data are older than January 1, 1970 (the Unix Epoch), you will not be able to use Unix Time as an alternative timestamp.

As mentioned in previous chapters, one way to index a file one time into Splunk is using the CLI command *add* with the oneshot option. Given the large number of individual CSV files we need to upload into Splunk, the best way to index these files is using the following command:

```
splunk add oneshot /mnt/flight_data/On_Time_On_Time_Performance_YYYY_M.csv \
-sourcetype FD_Source2
```

However, as the *oneshot* option does not take more than one file, we created a quick script that specifies all 300 CSV files in the */mnt/flight_data* directory. You can find this script in the download package of this book under the name *fd_index.sh*.

As usual, we want to verify that all our data has been indexed correctly, so we run the following search command for September 11, 2001, a date for which we expect a high rate of cancellations, by first setting the date in the time picker of the search page in the user interface and then typing:

```
* | stats count by Cancelled
```

The results present 14,900 cancelled flights out of 17,430 scheduled flights. That is, only 14.5% of the scheduled flights were not cancelled. The next quick search is to verify that the data was indexed all the way back to October 1987, so we set the time appropriate time range in the time picker and execute the following search:

```
* | top Dest
```

As this search command will go through all the events for that month, we first verify that the number of events is the same as the number of flights. The numbers are off by one, which is correct. The header line in Splunk has a timestamp that is not in the month of October 1987. The actual results of the top destination airports are again the usual suspects: Chicago O'Hare, Atlanta, Dallas/Ft. Worth, Los Angeles, and Denver. Although not an exhaustive verification process, we feel comfortable that all the flight data is correctly indexed into Splunk.

Indexing Data from a Relational Database

One of the main reasons for using CSV files is that you do not have direct access to the database. But if you happen to have direct access to the relational database that contains the data you want to analyze, you can avoid that extra step of having to export that data in CSV format, even if the access is read-only. Using the DB Connect app from Splunk, you can create and manage connections to external relational databases, which can then be defined as data inputs in the same way you would any other file, raw TCP/UDP, or scripted input.

The main component of the DB Connect app is the Java Server Bridge (JSB). This bridge receives instructions from Splunk, which are then converted into SQL requests that are sent to one or more relational databases via a JDBC driver. When you install this app on Splunk you gain the ability to create and manage connections to external SQL based relational databases.

In this section, we will only explain the use of the DB Connect app to extract data from a relational database so that it can be indexed by Splunk. Using DB Connect to access lookup tables is explained in Chapter 11. To illustrate the process of indexing data directly from a database, we have created an instance of MySQL, which has a table that contains the flight data with exactly the same fields as the CSV file. There are two steps that have to be done to accomplish this objective:

1.   Define a database connection

2.   Fetch the data from the database

Defining a New Database Connection

Once you have installed the DB Connect app, you navigate to the Manager screen of Splunk. There, under the Data group, you will see a new choice called External Databases. When you select this option it takes you to a new screen, where you define the necessary information to connect with your database, as can be seen in Figure 9-9.

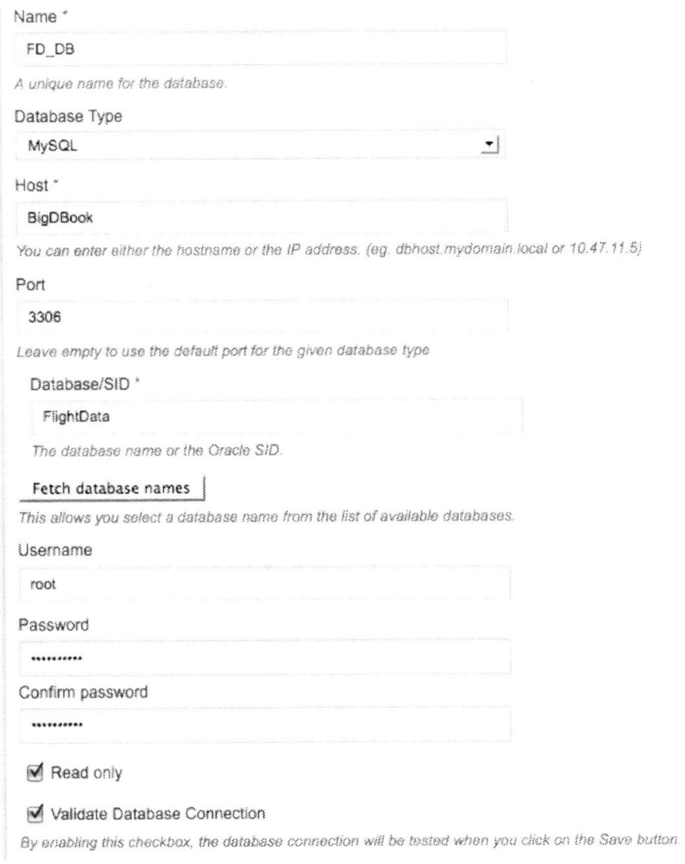

Name *

FD_DB

A unique name for the database.

Database Type

MySQL ▾

Host *

BigDBook

You can enter either the hostname or the IP address. (eg. dbhost.mydomain.local or 10.47.11.5)

Port

3306

Leave empty to use the default port for the given database type

Database/SID *

FlightData

The database name or the Oracle SID.

Fetch database names

This allows you select a database name from the list of available databases.

Username

root

Password

•••••••••••

Confirm password

•••••••••••

☑ Read only

☑ Validate Database Connection

By enabling this checkbox, the database connection will be tested when you click on the Save button.

Figure 9-9. *Defining a database connection*

As with any resource in Splunk, you have to give a name to this database connection. We called it FD_DB. From this point on, the information required is pretty much the same that is needed for any application to connect to a relational database. You must provide the hostname, port number, database name (or Oracle SID), username, and password, and (optionally) if you want this connection to be read-only. You can also validate the connection before saving it, which is quite useful to make sure that the connection is working correctly.

We did stumble on one problem when we tested the connection. It failed because the JDBC driver for MySQL was not present. As it turns out, DB Connect includes various JDBC drivers, but because of licensing issues it does not include one for MySQL. All you have to do is download the JDBC driver from the provider and place the jar file in $SPLUNK_HOME/etc/apps/dbx/bin/lib and then restart Splunk. The Java classpath is built dynamically when Splunk starts, and includes all the jar files in the lib directory just mentioned. That way the Java Bridge of DB Connect was able to use our new JDBC driver.

Database Monitoring

The DB Connect app mechanism for fetching data from a relational database is called database monitoring. There are two types of monitoring:

> **Dump**, *which executes the same SQL query every time it runs. If you do not specify an interval in the schedule, Dump will run only once. This is the most basic mechanism, and is equivalent to a oneshot load using the CLI for normal files.*

> **Tail**, *which works in a similar fashion to the Tail monitor that Splunk provides for file monitoring. This monitor type will determine which are the new records in the specified table and only output those. The query result must include a column that has a value that will always be greater than the value of any older record. You can use auto-incrementing values or timestamps for this purpose.*

As you are importing data from a relational database, you also get to choose the format in which it will be fetched. You have the standard options of CSV, CSV with header, and you can also define a template. Additionally, you can also define a key-value based format, or a multiline key-value format.

Independently of the output format, you can chose to have a timestamp value included, in which case the event will be prefixed with the timestamp value. You can also specify a column that contains the timestamp. If no timestamp is defined, the current time is used.

Another item that can be defined is the execution interval. You have the choice of defining a static value, which will be the amount of time that the monitor should wait in between executions, or a cron expression. For a fixed delay just specify a fixed number like 1h, for 1 hour, or 3000 for 3000 milliseconds (3 seconds). The cron expression uses the same syntax as the Unix crontab file.

Back to our example, we will use the dump option, as all we want to do is bring all the data from the database one time. (Unfortunately the flight data does not lend itself as a good example for using the tail monitoring mechanism, but this option is reviewed in Chapter 14.) Before we do that, we have to make a couple of decisions related to the format in which we bring the data and how to handle the timestamp.

We are really attracted to use the key-value format for a number of reasons. First of all, Splunk likes key-value pairs very much. Key-value is the default format and also the highest performing way to index data. Second, the key-value format maintains the richness and expressiveness of the data. As compared with the CSV format, with all its inconveniences such as those we have had to work with like counting commas, and having the header lines indexed as an event, the key-value format seems to be a lot easier and more powerful. We still have the issue of having the timestamp components in two different fields, but that can easily be solved by creating a view in MySQL that defines a new field with the timestamp, with all the other fields following:

```
CREATE VIEW ontime_ts AS
SELECT STR_TO_DATE(CONCAT(FlightDate, ' ', DepTime), '%Y-%m-%d %H%i')
AS FlightDateTime, ontime.* FROM ontime;
```

This approach of putting the timestamp first is similar to our earlier approach when working with CSV files. But because the new field has a timestamp that is easily recognizable by Splunk, we don't have to worry about defining anything related to the timestamp. Again, having the timestamp as one of the first fields in the event makes the indexing even faster.

The next step is to define the data input. We do that from the screen shown in Figure 9-10, which can be found under Database monitors, of the Data inputs screen of the Manager of the user interface.

A Database monitor will fetch data from a SQL database.

Name *

> SQL_FlightData

Monitor Type

> Dump (Always dump the full table/query)

Database

> FD_DB

☑ Specify SQL query

SQL Query *

```
SELECT * ontime_ts;
```

You can specify the SQL query that is executed against the database yourself. For
SELECT * FROM my_table {{WHERE $rising_column$ > ?}}

Sourcetype

Index

> fd_kv

Host Field value

> BigDBook

Output

Output Format

> Key-Value format

Specify how the event text content is generated.

Figure 9-10. *Defining a database monitor*

Defining the input is extremely simple. We name this database monitor SQL_FlightData, define it as a type Dump
from the pull down menu, and specify that we use using the FD-DB database connection. If you do not specify a query
to fetch the data, the input will default to selecting all the records of a table you define. In our case we want to use
the view we just defined, so our query is as simple as selecting all the records, but using that view, as can be seen in
Figure 9-10. Alternatively, we could have dispensed with creating a view in the database and written a SQL query in
Splunk that is essentially the same as the definition of the view:

```
SELECT STR_TO_DATE(CONCAT(FlightDate, ' ', DepTime), '%Y-%m-%d %H%i'
AS FlightDateTime, ontime.* from ontime;
```

Either way works and will achieve the same results, and we have no preference. As we are testing to make
sure our input works correctly, we specify fd_kv as the index into which the data will be loaded. We click on the
Save button, and the query starts executing, and Splunk starts indexing the data as it comes through the database
connection. Once it is done, we proceed to verify that data has been indexed correctly. We do this by checking the
output of the following search that can be found in Figure 9-11.

```
index=fd_kv | head
```

Figure 9-11. *Imported events from MySQL table*

In Figure 9-11 you can see that the events are now in key-value format, the timestamp is correct, and the fields on the left sidebar are populated correctly. On closer examination of the first event we notice that the FlightDateTime, Year, and FlightDate fields are in Unix time. This is because the latter two fields were defined as type year and date in the ontime table, and those types of fields are stored internally by MySQL as Unix time. The first field that we created with the view is automatically typed as date; thus, it also shows up as Unix time. As you can see, this is not a problem, as the Splunk timestamp was created correctly. However, having the other two fields as Unix time might be an issue, as any searches that use these fields would require us to convert them to something that is more human-friendly. Once again we have a couple of ways to handle this. We could convert the types at the import query, or we could use the Splunk convert command with the ctime function that converts from Unix time to a human-readable format:

```
... convert ctime(Year) ctime(FlightDate) ...
```

Before we decide on which approach we should use, we take a step back and think about how often we will use these fields. As it turns out, the fields in question are only needed to create the Splunk timestamp; therefore, it is a non-issue and we don't have to worry about it.

One more thing to note on the events is the assigned source type and source. As we did not fill these fields when defining the database monitor, it took default values that are quite illustrative. The source type is dbmon:kv, from which we can quickly infer a key-value format from a database monitor. The source itself is dbmon-dump://FD_DB/SQL_FlightData, from which we can also quickly deduce that it comes from a database monitor, specifically from a database called FD_DB and the monitor is called SQL_FlightData.

As you can see, connecting to a relational database using the DB Connect app is extremely simple. Even though we did not go through the other options such as tail and scheduling, using them remains as simple as what we have done here.

Summary

In this chapter we went through the process of importing data into Splunk using CSV files and direct connections to relational databases. We also dealt with timestamps and the various ways that they can be handled in order to create an effective Splunk timestamp and take advantage of it. We did all this using the airline on-time performance data, which is now fully indexed in our Splunk instance. In the next chapter, we will be analyzing this data and learning how to accelerate reports and searches.

CHAPTER 10

■ ■ ■

Analyzing Airlines, Airports, Flights, and Delays

In this chapter we analyze the flight data as related to the actual flights, delays, their reasons, and the performance of the airports. The analysis is done using Splunk search commands that were introduced in previous chapters as well as with new commands and features, such as analysis and reporting acceleration. We also discuss ways to visualize results from the searches we build, which otherwise produce results that can be very busy and difficult to understand.

Analyzing Airlines

One of the points we made in the previous chapter was that you should be intimately acquainted with the data that you are going to analyze. In this chapter we are going to take that knowledge of the flight data one step further. We start with some very basic searches to get an idea of the aggregated data. We ask simple questions, querying for such things as the number of airlines in the flight data. The airline business is very dynamic with airlines merging, going bankrupt, and every now and then a new one coming to market. Because of this we cannot just count the number of airlines in the most recent months, as we know that the number will vary over the 26 years of flight data.

The first thing we need to figure out is which field we should look at when counting the number of airlines. As you might remember, there are three fields that provide that information, each one in a different way. *UniqueCarrier* is a letter based code that can usually be understood by frequent travellers. When the same code has been used by various carriers over the years a numeric suffix is added to earlier carriers. For example, VX is today assigned to Virgin America, but before that it was used by Aces Airlines in Colombia, which ceased operations in 2003; thus, the code for Aces is VX(1).

The second field is *AirlineID*, which is a five-digit code based on the U.S. Department of Transportation certificate. The code is unique to each airline over time. If we would be doing a detailed analysis of airlines over time this would be the field to use, but it is not very useful at this stage of the analysis, as we would have to memorize a set of five-digit numbers until we implement a lookup for the airline name in the next chapter. A lookup table is used to display information from one table based on the value of another one. In this case we will display the actual name of the airline based on the code contained in one of these fields. The last field is *Carrier*, which almost always contains the same mnemonic as *UniqueCarrier*, but it does not add a numeric suffix for repeat uses; thus, the code is not always unique. Another issue appears when examining the corresponding lookup table, which can be downloaded from the TranStats web site; the field that contains the name of the airline includes the years in which the airline used that code. Additionally, the order of repeated codes in the lookup table is not consistent, as can be seen in a sample in Table 10-1.

Table 10-1. *Selected sample of the carrier field lookup table*

Code	Description
FL	Frontier Airlines Inc. (1960–1986)
FL	AirTran Airways Corporation (1994–)
PA	Pan American World Airways (1998–1999)
PA	Pan American World Airways (1997–1998)
PA	Pan American World Airways (1960–1991)
PA	Florida Coastal Airlines (2003–2005)
VX	Virgin America (2007–)
VX	Aces Airlines (1992–2003)

Because we will be using a lookup table to obtain the names of the airlines in the next chapter, the *Carrier* field has too many issues to be considered. The fact that the repeated codes are not always in order would force us to process the dates to find out the most recent one. Later we would need to strip those dates from the airline name field. It is just too much unnecessary work. We are better off using the *UniqueCarrier* field because its lookup table lends itself much better for our purposes. There is only one code for the airline and the description only contains the name of the airline.

▨ **Note**   For the kind of casual analysis we are doing in this book, we will not need a strict historic account of the names of the airlines over the years, so we feel comfortable using the *UniqueCarrier* field to represent the airlines with which we will be working.

The strategy we use for elaborating the searches is that we start by running them on a small sample of data, say the most recent month, just to make sure they work correctly. If the search uses arguments that span over various months, we test with just a couple of the most recent ones to verify that it handles the month boundaries correctly. The same applies to days, weeks, and years. The general idea is that you create the searches step by step and on a small sample of data so that you can quickly verify that they are producing the correct results.

Counting Airlines

We start by counting the number of airlines. As can be seen in Figure 10-1, we do this with the following command applied to just the month of September 2012, which we define using the time range picker in the user interface:

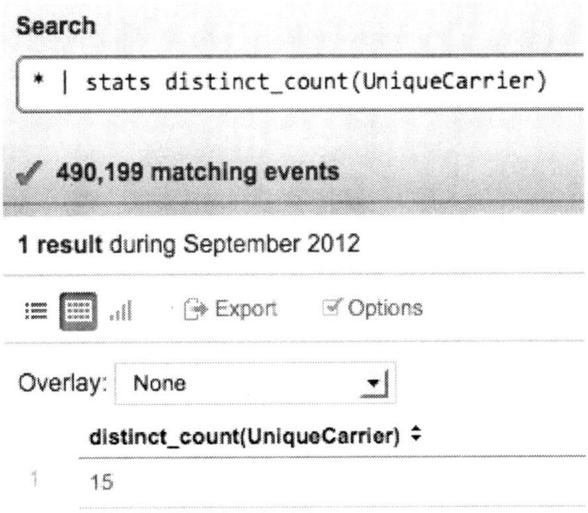

Search

```
* | stats distinct_count(UniqueCarrier)
```

✓ **490,199 matching events**

1 result during September 2012

:≡ ▦ .ıl   � Export   ☑ Options

Overlay: None ▾

distinct_count(UniqueCarrier) ↕

1   15

Figure 10-1. Counting airlines

The result is 15 airlines. There are various ways of verifying if the results are correct. We could load the particular month of flight data into a spreadsheet and perform various calculations over the 490,199 flight records, or we could take the easier road by formulating an alternate Splunk search command that should produce the same result. Because all we want to know is the number of carriers with scheduled flights in the month, we can use brute force and deduplicate all the events based on the *UniqueCarrier* field. That should produce only one event per airline. The way that Splunk works, going backward from the most recent date to the oldest, it should present the most recent event or flight record for each airline and then ignore the rest because they are duplicates of the first one. As we are not interested in the actual events but a count, we can use the following command, which also produces a count of 15:

```
* | dedup UniqueCarrier
| stats count
```

Now that we have verified that the search works correctly we can extend it. Because we have about 26 years worth of flight data the easiest way to review these results is on a yearly basis. If for any reason we want to drill down into a monthly level we can do that for the period of time of interest. Once again, keeping with our strategy of building the search step by step and with a small sample of data, we will modify the search so that it summarizes per year. Because of this we modify the time range to start at January 1, 2011, up to September 30, 2012. Note that *dc* is an abbreviation of *distinct_count*. The search and the results can be seen in Figure 10-2.

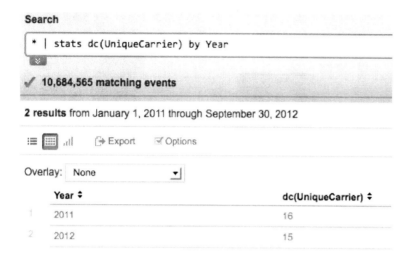

Figure 10-2. *Airline count for 2011 and 2012*

After scanning through 10,684,656 events the results show 16 airlines for 2011 and the already known count of 15 for 2012. This sounds about right as United Airlines merged with Continental Airlines in 2011; thus, the carrier count decreased by one. The next step is to run this search for all the flight data. The search took almost three hours on our cloud server. Later in this chapter we will discuss how to accelerate searches. In the meantime, we can review the results in the form of a line chart in Figure 10-3, which can convey the information much faster than a table of year and count.

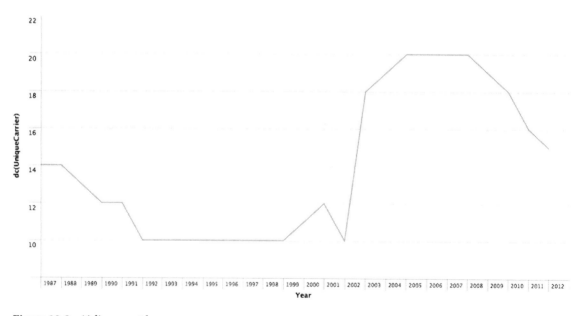

Figure 10-3. *Airline count by year*

Visualizing Results

As you can see in the chart in Figure 10-3, there is a drop from 15 airlines to 10 between 1988 and 1992 and then the number remains stable for almost a decade until there is a slight increase to 12. The effects of the attacks of September 11, 2001, can be seen as two carriers drop out of the picture. The industry rapidly grows again starting on 2003 to a peak of 21 airlines on 2006 from which it slowly drops to 15. We are interested in understanding what happened right after September 11, so we modify the search to count by month instead of year and we change the time period in the time picker to go from September 2001 to June 2002. The results show that one airline ceases to operate in October 2001 and another one in December 2001. It would be interesting to know which airlines ceased to operate. The first search that comes to mind is:

```
* | stats values(UniqueCarrier) by Month
```

The *stats* function *values* returns a list of all the distinct values of the *UniqueCarrier* field as a multivalue entry sorted in lexicographical order. The problem with this search is that even though the output contains the results we want, it is difficult to process. You can see a part of the total output in Figure 10-4.

Figure 10-4. *List of airlines by month*

The better way is using a graphic chart that will be easier to interpret. By clicking on the chart button, as indicated in Figure 10-4, we see that the results of the search produce only empty charts for any of the available chart types. Let us try with another search that will produce something that is more prone to be charted. Also, as we have pinpointed the demise of the two airlines before January 2002, we reduce the time period to five months, selecting from September 2001 to January 2002 in the time picker. We will use the *chart* command combined with sparklines:

```
* | chart sparkline(count,1w) by UniqueCarrier
```

By default, sparklines will be presented by the closest time unit of the data being processed. Because we are handling five months' worth of data in the search, sparklines defaults to monthly, but that is too small to be a useful visualization. Using *1w* as an argument expands the resolution from monthly to weekly. The output of this search can be seen in Figure 10-5.

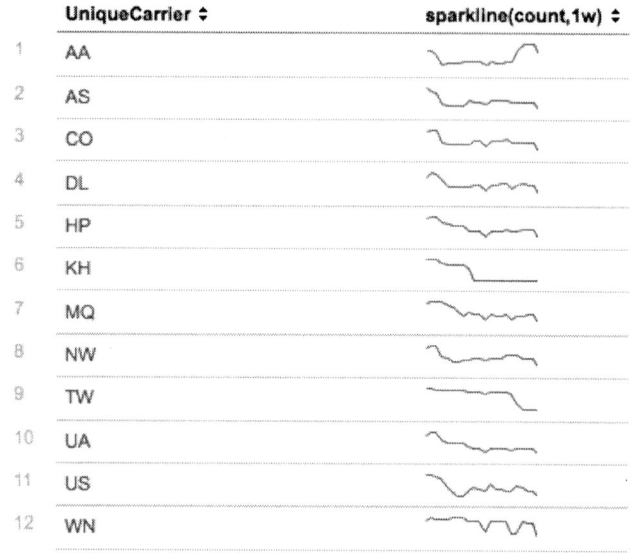

UniqueCarrier ⇕	sparkline(count,1w) ⇕
1 AA	
2 AS	
3 CO	
4 DL	
5 HP	
6 KH	
7 MQ	
8 NW	
9 TW	
10 UA	
11 US	
12 WN	

Figure 10-5. *Airlines out of business after September 11 using sparklines*

The output of the *chart* command is rather spartan. We can see that KH (Aloha) and TW (TWA) drop to zero at different points in time, so we have answers to our question. However, even though the sparklines provide a basic idea of what is happening it does not feel quite complete. Surely there has to be way to provide a more comprehensive and compelling visualization. To explore a better visualization we try using the *timechart* command:

```
* | timechart count by UniqueCarrier limit=0
```

In this search we count all the events (flight records) and group them by airline. It is a simple way of finding out which are all the airlines and also get a count of scheduled flights for each one of them. Because the span is over five months, the *timechart* command will present the results broken down by months, which is what we want.

The reason we use the *limit* argument is that by default the maximum number of items or data series to display is 10. There are 12 airlines in the time period we are analyzing, so we can set the limit to 12, or we can be lazy and set it to zero. By doing the latter we state that *timechart* should accommodate all the distinct items of the *UniqueCarrier* field, no matter how many there are. You have to be careful using an argument of zero as a large number of items will probably produce a rather chaotic and illegible chart. Had we not specified the limit argument, Splunk would have presented only nine items and grouped the additional ones under a category called *Other*. Which items fall in the *Other* category depends on the function used with the *timechart* command. In this example, it would have been those with the lower counts. The *useother* argument, which controls if the *Other* grouping exists, is turned on by default, but only has effect if *limit* has a value different than zero.

Communicating results in such a way that they are easily consumable by the intended audience is challenging. Charts with colors tend to be the favored method for this. The issue is finding the appropriate chart type. In this particular example, we will see that two different chart types clearly communicate two different pieces of information. The chart type options that Splunk offers for the results of this search are column, line and area. The column chart shown in Figure 10-6 clearly presents the airlines that went out of business during this period by the simple fact that the corresponding airline column is not there anymore. In this case the column for Aloha is no longer present starting the month of November and TWA is not there in January.

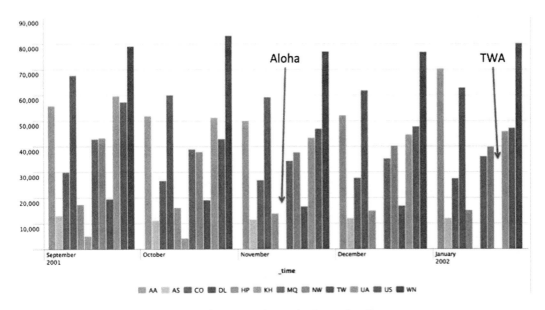

Figure 10-6. *Airlines out of business after September 11 (Column chart)*

Figure 10-7 presents the same results but using a line chart. As you can see the chart is pretty busy and does not convey the demise of the airlines as clearly as the column chart. However, in between December and January we can see a piece of information that was not obvious in the column chart. Whereas the traffic of TWA goes down to zero, the number of flights for American increases by about the same amount. What really happened was that American acquired TWA, thus a corresponding increase in flights. A completely different piece of information emerges by just changing the chart type.

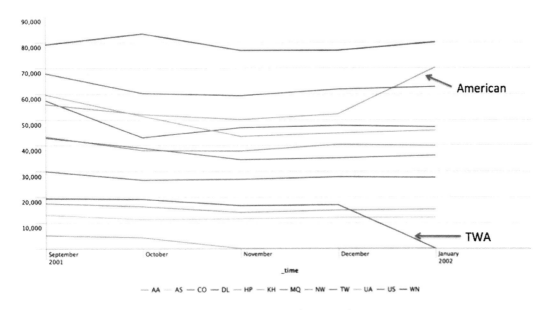

Figure 10-7. *Airlines out of business after September 11 (line chart)*

For completeness, we present in Figure 10-8 the area chart generated with the same results as the previous two charts. You can barely distinguish five items and nothing can really be deduced from reviewing it. Although this chart type is useless for the current type of results, it can be useful for results generated by other searches.

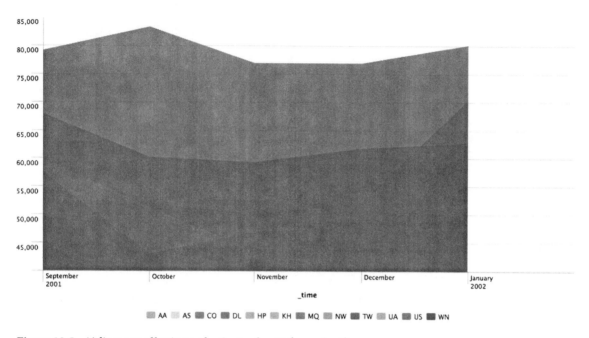

Figure 10-8. *Airlines out of business after September 11 (area chart)*

Now we have a pretty good idea about the airlines over the period of available flight data. The next step is to gain an understanding of the airports.

Analyzing Airports

The first question we have is similar to that when we analyzed airlines, how many airports are there? For this there are multiple fields that can be used, but before we get into that we have to note that the airports in the flight data are always paired by origin of a flight and the destination. The normal expectations are that there should be as many origin airports as destination airports. Let's start by checking this out for the month of September 2012. As can be seen in Figure 10-9, we use a similar search as that to count the airlines on the most recent month:

Figure 10-9. *Counting airports for September 2012*

The results show that there are exactly the same number of origin and destination airports, that is, 290 airports. We chose to use the *Origin* and *Dest* fields, as they contain a mnemonic of the airport name, which is quite familiar to frequent travelers. The other airport related fields contain codes that only make sense to specialists in the area. We should not use the *OriginCityName* field, as there are some cities that have more than one commercial airport. We can quickly find which these cities are with the following search:

Figure 10-10. *Cities with more than one airport*

Let us continue with the research on airports. We must check the number of airports over the 26 years of flight data. As airlines expand and contract, given the economic conditions, they will add or cancel service to airports throughout the United States. Just as we did with the airlines, let us try a search over the most recent couple of years:

```
* | stats dc(Origin), dc(Dest) by Year
```

Interestingly enough, for the year 2012 there are 301 origin airports and 302 destination airports, whereas for 2011 there are 299 origin and 301 destination airports. This is unexpected. We can only think this is caused by inconsistency in the data or an anomaly when crossing over years. Just to make sure the search is producing correct results we quickly run another search based on the *dedup* command as we did on the airline analysis, which generates the same results. In Figure 10-11 you will find the line chart that presents the number of airports, both for origin and destination. Once again, we reviewed the results using different chart types and decided to use the line chart as it was the one to present more clearly the results. However, to make the slight numeric differences between origin and destination airports more noticeable, we present only from 2000 to 2009 and change the minimum value of the y-axis of the line.

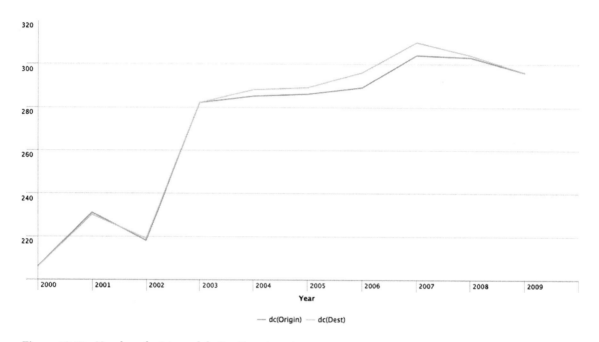

Figure 10-11. *Number of origin and destination airports*

We will bypass finding the most and least used airports as those searches are extremely simple. Instead, we will have a look at another visualization challenge. The search is rather simple; find the top five airports by airline. To make it easy we will just be using the origin airports:

```
* | top limit=5 Origin by UniqueCarrier
```

This search produces the output shown in Figure 10-12. This is a simple table ordered by airline code, but it spreads over eight pages; not exactly the best way to get a quick idea of the top five airports by airline.

	UniqueCarrier ↕	Origin ↕	count ↕	percent ↕
1	AA	DFW	12358	29.546922
2	AA	ORD	4045	9.671249
3	AA	MIA	3464	8.282128
4	AA	LAX	2450	5.857741
5	AA	LGA	1287	3.077107
6	AS	SEA	3704	30.902720
7	AS	ANC	1162	9.694644
8	AS	PDX	800	6.674454
9	AS	LAX	506	4.221592
10	AS	LAS	373	3.111964

Figure 10-12. *Top 5 airports by airline (table)*

Clicking on the chart button only offers column, bar and pie as options. The first two only display information for the first four airlines, but do not state the airports, showing only the individual count for the top five of each airline. The pie chart option does not display anything. As is the case when trying to visualize multiseries (or multiple items as we have been calling them in this chapter), there is no easy answer. A valid option is to break down the results into manageable units to make the results easier to digest. Using this principle, we consider breaking down the results by airline, ideally in a pie chart, as they are very easy to understand. The issue is that we cannot really generate as many pie charts as we want from a single Splunk command. The solution we come up with is to create a dashboard with each panel dedicated to display the top five airports for a specific airline. Needless to say, the searches for each panel are extremely simple:

```
UniqueCarrier=UA
| top limit=5 Origin
```

The easiest way is to save the searches to be used on each panel and create the dashboard. We have done this for the most recent month of flight data and with three randomly selected airlines, Southwest, Delta, and United. The resulting dashboard can be seen in Figure 10-13. This is an elegant and logical way of handling a set of results that is cumbersome to read when presented as a table and almost impossible to display as a single chart.

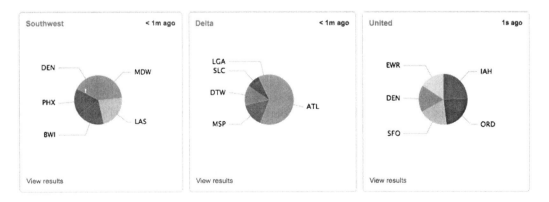

Figure 10-13. *Dashboard sample of top five airports by airlines*

Before we move on to analyze the flights we have to bring to your attention something that can create confusion. If you look at Figure 10-12, you will notice that the percentages do not quite add up. When you compare line 1, American Airlines at the Dallas/Fort Worth airport with 29.54% and a count of 12,358 and Alaska Airlines at the Seattle airport with 3,704 flights and 30.90%, things don't really make sense. How is it possible that Alaska with about a third fewer flights than American has a slightly higher percentage for their top airports? The answer is that those percentages are per airline for all the airports, as the search was done by *UniqueCarrier*. Thus, Dallas/Fort Worth represents 29.54% of the origin of flights for American, and Seattle represents 30.90% of the origin flights for Alaska.

Analyzing Flights

As all the searches we have done so far are simple, in this section we will start increasing the level of complexity. We start with a simple search, average flight time by airline and as usual so far we test the searches over the most recent month. The search, shown in Figure 10-14, is based on the *CRSElapsedTime* field, which contains the scheduled flight time in minutes, not the actual flight time:

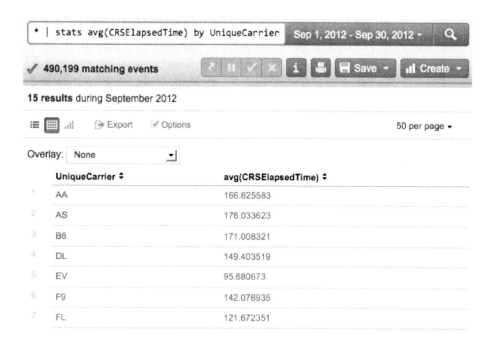

Figure 10-14. *Average flight time by airline*

The output of this search is a table that is ordered by airline and the average flight times, calculated as the arithmetic mean, have six decimal places. Before we determine the best way to present these results we want to polish the output a bit by rounding the averages and sorting them. We do this as follows:

```
* | stats avg(CRSElapsedTime) as AverageFlightTime
        by UniqueCarrier
| eval AverageFlightTime=round(AverageFlightTime)
| sort -AverageFlightTime
```

Every time we manipulate data to obtain partial results and pass them on to the next section of a search using pipes we like to define new fields, even if new fields are not explicitly necessary. In this case we create a new field that contains the averages in the second clause of the search. The output is rounded in the third clause of the search, which makes it more readable, and finally it is sorted in decreasing order. The round function defaults to an integer. If you want decimals, just specify the desired precision after the field separated by a comma.

Determining the best representation for a given data set can be difficult. The tabular form presented in Figure 10-15 provides a quick way to process the information if we are interested in the actual number of minutes, whereas a bar chart as seen in Figure 10-16 provides it without the exact numbers.

	UniqueCarrier ⇕	AverageFlightTime ⇕
1	VX	232
2	UA	195
3	AS	176
4	B6	171
5	AA	167
6	DL	149
7	US	146
8	F9	142
9	FL	122
10	WN	117
11	MQ	97
12	EV	96
13	OO	94
14	HA	93
15	YV	90

Figure 10-15. Average flight time in tabular form

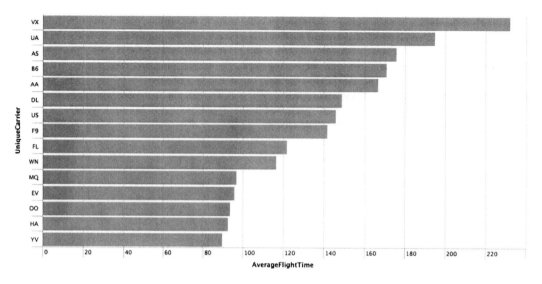

Figure 10-16. *Average flight (time—bar chart)*

It is interesting to note that all the regional airlines average around one hour and a half, confirming their regional status. Oddly, Hawaiian Airlines shows up in this group, but this can be attributed to the fact that the majority of their flights are short hops within the Hawaiian Islands, which skew the averages to the lower end as their flights to the continent are comparatively not as many.

The next search is looking for the longest flights, but that in itself would be very easy. We want to know which are the longest flights by airline and also from where to where. Presenting the origin and destination increases the complexity of formulating the search, which does not necessarily mean that the resulting search is more complex.

We cannot think of a way to do this search with only one command, so we will have to break it down in clauses using pipes. The first thought is to use the stats command to find the maximum scheduled flight time. As expected this will produce the longest flight by time for each airline. The next step is to understand what happens if we add the origin and destination to that particular search, as seen in Figure 10-17.

```
* | stats max(CRSElapsedTime) by UniqueCarrier, Origin, Dest        Sep 1, 2012 - Sep 30, 2012 ▾    🔍
```

✓ **490,199 matching events**   ↻ ❚❚ ✓ ✗ │ i │ 🖨 │ 💾 Save ▾ │ 📊 Create ▾

6,096 results during September 2012

☰ ▦ ⊿   ⟶ Export   « prev **1** 2 3 4 5 6 7 8 9 10 next »   50 per page ▾

☑ Options

Overlay: None ▾

	UniqueCarrier ⇕	Origin ⇕	Dest ⇕	max(CRSElapsedTime) ⇕
1	AA	ABQ	DFW	105.00
2	AA	ANC	DFW	375.00
3	AA	ATL	DFW	155.00
4	AA	ATL	MIA	125.00
5	AA	AUS	DFW	70.00

Figure 10-17. *Longest flight time by city pairs by airline*

The output of this search is a table with the maximum flight time for every origin-destination pair for each airline. This table contains 6,096 entries for the month of September 2012, which implies that there are as many unique city pairs for all the airlines. Just out of curiosity, we review the first table entry, American Airlines from Albuquerque to Dallas/Fort Worth, by doing a search to count all those entries, which totals 214 matching events.

```
UniqueCarrier=AA Origin=ABQ Dest=DFW
```

Clicking on the *CRSElapsedTime* field on the left side bar shows that there are two values, 100 and 105 minutes, as can be seen in Figure 10-18. This means that the airline itself has two different scheduled flight times. When clicking on the departure time field there are eight values and the scheduled arrival time has nine. Obviously, there is a wide assortment of choices for this city pair.

# CRSArrTime (9)	Min: 100 Max: 105 Mean: 101.425 Stdev: 2.262		
# CRSDepTime (8)			
# CRSElapsedTime (2)	Values	#	%
# DayofMonth (30)	100.00	153	71.495%
# DayOfWeek (7)	105.00	61	28.505%

Figure 10-18. *Scheduled elapsed times for American flights between Albuquerque and Dallas/Ft. Worth*

Now that we have a table with all the city pairs, we can try to sort them by the maximum flight time:

```
* | stats max(CRSElapsedTime) as MaxFlightTime
        by UniqueCarrier, Origin, Dest
| sort -MaxFlightTime
```

Because we are passing partial results to the next clause using a pipe, we created a new field, *MaxFlightTime*. The output of this search gives us a list of the longest flights independently of the airline, which is not exactly what we want. We want the longest flight for each individual airline. So let us try putting *UniqueCarrier* as an additional field of the sort command. This does not change the results because the only effect it has is to sort by airline for those entries that have the same flight time. The next logical step is to deduplicate the entries by airline. As the table has all the city pairs ordered from longest to shortest by airline the result should be what we want:

```
* | stats max(CRSElapsedTime) as MaxFlightTime
        by UniqueCarrier, Origin, Dest
| sort -MaxFlightTime
| dedup UniqueCarrier
```

As we verify that this search produces the results we want, which can be seen in Figure 10-19, we notice that it can be optimized from four to three clauses, as the *dedup* command has a sort option:

```
* | stats max(CRSElapsedTime) as MaxFlightTime
        by UniqueCarrier, Origin, Dest
| dedup UniqueCarrier
        sortby –MaxFlightTime
```

UniqueCarrier ⬍	Origin ⬍	Dest ⬍	MaxFlightTime ⬍
1 HA	JFK	HNL	660.00
2 UA	EWR	HNL	640.00
3 DL	ATL	HNL	580.00
4 AA	DFW	HNL	495.00
5 US	PHX	LIH	404.00

Figure 10-19. Longest flight by airline by city pair

So far we have worked with the maximum scheduled flight time, but we all know the reality is quite different, so let us throw in the actual flight time. We can do this by just adding it as part of the *stats* command and it should carry on throughout the rest of the sections:

```
* | stats max(CRSElapsedTime) as MaxSchedTime
         max(ActualElapsedTime) as MaxActualTime
         by UniqueCarrier, Origin, Dest
| dedup UniqueCarrier
       sortby -MaxActualTime
```

Note that we changed the names of the new fields to be more representative of their contents and that we sorted based on the maximum actual flight time. One of the issues of working with results that contain multiple fields is that they do not lend themselves to being presented as colorful charts. We tried the various chart types offered by Splunk, but whereas both flight times showed very nicely the names of the city pairs were not to be found. Figure 10-20 contains the results of this search in tabular form, which we feel is the best choice to present these results.

UniqueCarrier ⬍	Origin ⬍	Dest ⬍	MaxSchedTime ⬍	MaxActualTime ⬍
1 HA	JFK	HNL	660.00	670.00
2 UA	EWR	HNL	640.00	658.00
3 DL	ATL	HNL	580.00	597.00
4 AA	DFW	OGG	485.00	537.00
5 VX	JFK	SFO	390.00	519.00

Figure 10-20. Actual longest flights by airline

Interestingly enough, the longest flight by both scheduled and actual time for September 2012 is from New York to Honolulu with Hawaiian Airlines. This is the flight at the top of the output. It confirms what we saw in the previous search, that even though Hawaiian Airlines has the longest flight, the majority of the flights are short, thus skewing the average flight time. One thing we need to bring to your attention with this search is that if any of the actual flight times would have been the same or less than the scheduled time the order of the results would have changed and this could affect the way the audience perceives the results.

So that we cover all the cases, we will also look at the shortest flights, both by scheduled and actual flight time. Some readers might think that the easiest and fastest way of doing this is by reversing the sort of the longest flight search we just did, but this will not work as you will obtain the shortest of the longest flights because you are sorting by the maximum actual time, not the maximum scheduled time. If you reverse the sort by eliminating the dash ('-') and changing the field name to *MaxSchedTime* it will work correctly; however, it will have the wrong column titles on the table, as they should be Minimum Scheduled Time and Minimum Actual Time. The modified search follows along with a partial output in Figure 10-21.

```
* | stats max(CRSElapsedTime) as MaxSchedTime
          max(ActualElapsedTime) as MaxActualTime
          by UniqueCarrier, Origin, Dest
| dedup UniqueCarrier
        sortby MaxSchedTime
```

UniqueCarrier ⬍	Origin ⬍	Dest ⬍	MaxSchedTime ⬍	MaxActualTime ⬍
1	OO	SJC	MRY	18.00	
2	AS	WRG	PSG	27.00	47.00
3	YV	LIH	HNL	29.00	47.00
4	HA	KOA	OGG	31.00	38.00
5	EV	DEN	COS	39.00	58.00

Figure 10-21. Shortest flights by airline

The shortest scheduled flight is from San Jose to Monterey in California with 18 minutes, but it was scheduled only once during September 2012 and it was cancelled, so the actual time is null. As all the other flights do have actual flight times in the results we wonder about the effect of cancelled flights on the results. Reviewing other of the flights in the shortest group we find out that both minimum and maximum of either scheduled or actual flight times always result in a value. The San Jose to Monterey flight is a very special case, as there is no actual flight time and because there is only one flight the result of minimum or maximum actual flight time is null.

Analyzing Delays

There are plenty of fields in the flight data related to delays and we must state that we do not have much of an understanding of them. For details related to how delays and cancellations are reported you can visit http://www.bts.gov/help/aviation/html/understanding.html. Having said that, delays offer an opportunity to explore more complex searches and gain a better understanding of some of the search commands and their arguments.

Delay information is categorized by departure and arrival, and the specific causes are available since 2003. The flight data also includes reason for cancellation and diversion. Additionally, there is detailed information related to flights that return to the gate, which is collected since 2008.

Delays by Airline

We will start by finding out the delays by airline. Because delays are broken down by departure and arrival, we will focus on arrival delays as they are the ones that most affect the passengers. One of the fields in the flight data indicates if a flight has been delayed by 15 minutes or more; thus, we use *ArrDel15* as the base for our search. When there is a delay the value of this field is set to 1. The output of this search can be seen in Figure 10-22.

```
* | stats count(ArrDel15) as Total,
         count(eval(ArrDel15=0)) as OnTime,
         count(eval(ArrDel15=1)) as Delayed
         by UniqueCarrier
```

	UniqueCarrier ⬍	Total ⬍	OnTime ⬍	Delayed ⬍
1	AA	40409	24252	16157
2	AS	11869	10745	1124
3	B6	18093	15025	3068
4	DL	60843	54754	6089
5	EV	59543	48966	10577

Figure 10-22. *Delays of 15 minutes or more by airline*

The results of this search present the total number of flights by airline, as well as the number of flights that are on time and those that are delayed, all this sorted by the unique carrier code. Although in principle this produces what we wanted, it is kind of unfair. Comparing Southwest with almost 90,000 flights in September 2012 against Virgin America with around 4,600 flights, based on the resulting table is not a level comparison. The column chart of this search is a tad clearer as the proportions convey a better picture, but it is still not enough. The fairer way is to present the delays as a percentage of the total flights, which can be done by adding the percentage calculations to the previous search:

```
* | stats count(ArrDel15) as Total,
         count(eval(ArrDel15=0)) as OnTime,
         count(eval(ArrDel15=1)) as Delayed
         by UniqueCarrier
| eval PCTOnTime=round(OnTime/Total*100,2)
| eval PCTDelayed=round(Delayed/Total*100,2)
| sort - PCTDelayed
```

In addition to calculating the percentage we are rounding the result to two decimal points and by sorting the results on the delay percentage we have a better idea how the airlines did during September 2012. As can be seen in Figure 10-23, American Airlines had a bad month, with almost 40% of their flights delayed by 15 minutes or more on arrival. The next one with the highest number of delayed flights is Southwest with over 11,600, but given their high number of flights that only represents 13% of their total flights. We considered using a column chart for these results, but it's the same as the one produced with the previous search. The problem is that the scales of the values to be charted are very different, the totals in the thousands and the percentages in the tens. The chart still provides a good idea of proportion but the percentages are absent.

	UniqueCarrier ⬍	Total ⬍	OnTime ⬍	Delayed ⬍	PCTDelayed ⬍	PCTOnTime ⬍
1	AA	40409	24252	16157	39.98	60.02
2	EV	59543	48966	10577	17.76	82.24
3	UA	42456	35100	7356	17.33	82.67
4	B6	18093	15025	3068	16.96	83.04
5	OO	49518	41894	7624	15.40	84.60

Figure 10-23. *Arrival delays by airline*

Causes of Delays by Airport

The next search is the cause of delays by airport. There are five causes for delays, each with its own field, which contains the number of minutes of the delay:

- Carrier—*CarrierDelay*

- Weather—*WeatherDelay*

- National air system delay—*NASDelay*

- Security—*SecurityDelay*

- Late aircraft—*LateAircraftDelay*

Details on how these fields are reported can be found in the URL presented at the beginning of this section. As with the previous search, the results only make sense if they are presented as percentages. This means that we will have to calculate percentages for each of the causes of a delay:

```
* | stats count(ArrDel15) as Total,
        count(eval(ArrDel15=1)) as Delayed,
        count(eval(CarrierDelay>=15)) as CarrierDel,
        count(eval(WeatherDelay>=15)) as WeatherDel,
        count(eval(NASDelay>=15)) as NASDel,
        count(eval(SecurityDelay>=15)) as SecurityDel,
        count(eval(LateAircraftDelay>=15)) as LateAircraftDel
        by Dest
| eval PCTDelayed=round(Delayed/Total*100,2)
| eval PCTCarrierDelay=round(CarrierDel/Delayed*100,2)
| eval PCTWeatherDelay=round(WeatherDel/Delayed*100,2)
| eval PCTNASDelay=round(NASDel/Delayed*100,2)
| eval PCTSecurityDelay=round(SecurityDel/Delayed*100,2)
| eval PCTLateAircraftDelay=round(LateAircraftDel/Delayed*100,2)
| sort - Total
| fields - Total, Delayed, CarrierDel, WeatherDel, NASDel, SecurityDel, LateAircraftDel
| head 5
```

As we are using the *ArrDel115* field as the base of this analysis, we must count the cause of delays that have 15 or more minutes. The results of these counts are then used to calculate the corresponding percentages. We sort the results based on the total number of flights by airport, as we are interested in knowing the delays on the busiest airports. Sorting by the percentage of delays, that is using sort -PCTDelayed, will list a number of small airports in the pacific coast with a few flights a day and prone to fog, as can be seen in Figure 10-24.

	Dest ↕	PCTCarrierDelay ↕	PCTDelayed ↕	PCTLateAircraftDelay ↕	PCTNASDelay ↕	PCTSecurityDelay ↕	PCTWeatherDelay ↕
1	CEC	11.36	65.67	81.82	9.09	0.00	2.27
2	MOD	20.00	64.71	89.09	1.82	0.00	0.00
3	RDD	21.57	58.62	72.55	7.84	0.00	0.00
4	EGE	72.22	56.25	44.44	5.56	0.00	0.00
5	CIC	15.79	46.34	92.11	5.26	0.00	0.00

Figure 10-24. Airports with highest percentage of delays

We go back to our original search, where we sort by the total number of flights. As we are only interested in the percentages, we eliminate the fields we used to calculate them. Finally, we just present the top 10 airports by using the *head* command. The table produced by this search can be seen in Figure 10-25. As we try to verify the results, we notice that the column that contains the *PCTDelayed* field is located in an odd place. Because this field presents the percentage of all delayed flights per airport we would like for it to be either the second column, right after the airport mnemonic code, or the last one.

	Dest ⇕	PCTCarrierDelay ⇕	PCTDelayed ⇕	PCTLateAircraftDelay ⇕	PCTNASDelay ⇕	PCTSecurityDelay ⇕	PCTWeatherDelay ⇕
1	ATL	25.82	12.01	41.14	32.35	0.13	1.77
2	ORD	25.04	18.47	43.45	34.65	0.09	1.60
3	DFW	26.95	22.76	48.03	26.65	0.20	1.97
4	DEN	28.70	12.24	40.19	26.87	0.22	1.70
5	LAX	27.21	18.07	41.00	31.73	0.12	1.80

Figure 10-25. *Cause of arrival delays by airport (tabular form)*

After exhaustive testing we found that fields created using the *eval* command are presented in tables and charts following ASCII order. ASCII is one way to represent characters internally in computers and the order can be described as some symbols, the numbers, some other symbols, upper case letters in alphabetical order, other symbols, lower case letters in alphabetical order and a final set of symbols. Fields created using the *stats* command are presented in the order they were created. This order can be clearly seen in Figure 10-23 where you see that *PCTOnTime* was calculated before *PCTDelayed*, but ends up in the last column of the table. To arrange the order of the columns in the results presented in Figure 10-25 as we want, we can add a character that comes before "P" in the ASCII table to the *PCTDelayed* field. This way, it will be placed right before the column of the *PCTCarrierDelay* field. Note that it will not be placed before the first column as that one is assigned to the field (or fields) used in the *by* argument, in this case the destination airport. As part of our experiments we found that we can use special characters to name the fields created using the *eval* command. The only down side is that when the fields have special characters they cannot be used for further calculations. With this knowledge, we rename the *PCTDelayed* field as *%Delayed* and now it shows up as the second column in the results table of this search as can be seen in Figure 10-26.

	Dest ⇕	%Delayed ⇕	PCTCarrierDelay ⇕	PCTLateAircraftDelay ⇕	PCTNASDelay ⇕	PCTSecurityDelay ⇕	PCTWeatherDelay ⇕
1	ATL	12.01	25.82	41.14	32.35	0.13	1.77
2	ORD	18.47	25.04	43.45	34.65	0.09	1.60
3	DFW	22.76	26.95	48.03	26.65	0.20	1.97
4	DEN	12.24	28.70	40.19	26.87	0.22	1.70
5	LAX	18.07	27.21	41.00	31.73	0.12	1.80

Figure 10-26. *Cause of arrival delays by airport (reordered columns)*

Now that we have the table columns in the order that we want, we can continue with the verification of the results. As the calculations we did are relatively simple, we are not so concerned about those individual results. However, it is of our interest to see if the addition of all the percentages for the airports equals 100. Las Vegas adds to 99.99, which is close enough to 100 to say the difference can be attributed to rounding errors. But Chicago O'Hare adds up to 104.83, which is more than can be excused by rounding errors. After researching the issue for this we concluded that the reason is that the fields used for causes of delays reflect delays on both origin and destination. The documentation at the TranStats web site does not go to this level of detail, so we cannot confirm our suspicion that if a flight had a delay on the departure and a delay on the arrival, the cause fields will add the minutes for both, resulting in the discrepancy we are observing.

We have mentioned before that when we have multiple columns of information in the results of searches, it is difficult to use charts to convey the results. However, in this search using a bar or column chart with the 100% stacked mode option makes for a great visualization as it presents the percentages of the causes for delays for each of the top 10 airports. We have to make one change to the search so that the results are accurate when using this type of chart; we need to eliminate the total percentage of delayed flights (*PCTDelayed*) so the easiest way is not calculating it anymore. The search ends up being:

```
* | stats count(ArrDel15) as Total,
          count(eval(ArrDel15=1)) as Delayed,
          count(eval(CarrierDelay>=15)) as CarrierDel,
          count(eval(WeatherDelay>=15)) as WeatherDel,
          count(eval(NASDelay>=15)) as NASDel,
          count(eval(SecurityDelay>=15)) as SecurityDel,
          count(eval(LateAircraftDelay>=15)) as LateAircraftDel
          by Dest
| eval PCTCarrierDelay=round(CarrierDel/Delayed*100,2)
| eval PCTWeatherDelay=round(WeatherDel/Delayed*100,2)
| eval PCTNASDelay=round(NASDel/Delayed*100,2)
| eval PCTSecurityDelay=round(SecurityDel/Delayed*100,2)
| eval PCTLateAircraftDelay=round(LateAircraftDel/Delayed*100,2)
| sort - Total
| fields - Total, Delayed, CarrierDel, WeatherDel, NASDel, SecurityDel, LateAircraftDel
| head
```

The stacked bar chart can be seen in Figure 10-27. Interestingly enough, the stacks are sorted in reverse order to the legend.

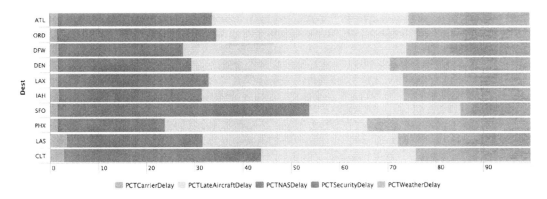

Figure 10-27. *Cause of arrival delays by airport (bar chart)*

Winter versus Summer Delays

Next, we want to compare the delays between January and July of 2012 for the top 10 airports, as we want to see if winter has any affect on the delays. We will do this by having two identical searches over two different periods of time and then joining the results. To run two or more searches Splunk offers a subsearch mechanism. Subsearches are contained within square brackets ([]) and are evaluated before the main search, this way the results of a subsearch can be used in the main search. Subsearches are limited to actions related to searches and not data transformation. In general, we only see subsearches used with the *join* and *append* commands.

Let us start with the search to calculate the delays. So far we have been using the time picker of the user interface, but that will not work as we have to select two different time periods. For this, we can specify the time ranges using the *earliest* and latest *attributes* of the search command. With this our main search is:

```
earliest=1/1/2012:0:0:0 latest=2/1/2012:0:0:0
| stats count(ArrDel15) as Total_Jan,
        count(eval(ArrDel15=1)) as Delayed_Jan
        by Dest
| eval PCTDelJan=round(Delayed_Jan/Total_Jan*100,2)
```

Now instead of using the * that indicates all the events, we restrict the choice of events to the time period between the date and time defined in *earliest* and *latest*. The "0:0:0" sequence after the date indicates the time as hour, minutes and seconds. Note that we defined the latest date and time to be February 1 at midnight because the *latest* argument is exclusive, meaning that it does not include the date and time specified. The date and time specified in *earliest* is inclusive.

▨ **Note**   Specifying a time range in a search or subsearch will override the time range selected in the time picker of the user interface.

The subsearch is the same as the main search, but we change the time period and the names of the calculated fields. To combine the results of both searches we use the *join* command, which is very similar to a SQL join used on relational databases. In the Splunk *join* command you can specify the exact fields to use for the join. If you do not specify any, it defaults to use all the fields that are common to both result sets. As our result sets are the same regarding their structure, we will not specify any fields. You can also specify the type of join to perform. There are two types of join, inner and outer (or left). Inner join is the default and only includes events that are matched on both result sets. An outer join combines all the events of both result sets, even if there are no matches, think of it as a merge. Once again, because our result sets are the same, we will use the inner join. The final version of the search is:

```
earliest=1/1/2012:0:0:0 latest=2/1/2012:0:0:0
| stats count(ArrDel15) as Total_Jan,
        count(eval(ArrDel15=1)) as Delayed_Jan
        by Dest
| eval PCTDelJan=round(Delayed_Jan/Total_Jan*100,2)
| join
  [
   search earliest=7/1/2012:0:0:0 latest=8/1/2012:0:0:0
   | stats count(ArrDel15) as ATotal_Jul,
           count(eval(ArrDel15=1)) as Delayed_Jul
           by Dest
   | eval PCTDelJul=round(Delayed_Jan/ATotal_Jul*100,2)
  ]
| sort - TotalJan
| fields - Delayed_Jan, Delayed_Jul
| head
```

We have indented the search to make it more readable as Splunk does not have the concept of indentation for search commands. The search command is the core of the search app in Splunk. As such, you do not even have to type it when used in the first clause; that is why we have only been typing the star (*) at the beginning of all our searches. However, after the first pipe you need to type it, as can be seen in the subsearch, otherwise you would get an error message stating that *earliest* is not a valid command. One thing to note is that we used a similar trick as on a previous search to order the table columns as we want to make it easier to analyze the results. As can be seen in Figure 10-28, we present the total number of flights and the delays as a percentage for both months. We added an A to the name of the *Total_Jul* field name to place it as the third column. It is kind of clumsy, but it gets the job done.

[subsearch]: Your timerange was substituted based on your search string				
Your timerange was substituted based on your search string				

10 results during September 2012

≡ ▦ .ıl ⤷ Export ✓ Options   10 per page ▾

Overlay: None ▾

	Dest ⇕	Total_Jan ⇕	ATotal_Jul ⇕	PCTDelJan ⇕	PCTDelJul ⇕
1	ATL	31031	34357	14.95	21.73
2	ORD	23849	26723	16.34	25.67
3	DFW	20714	23030	12.11	18.94
4	DEN	18797	21923	13.67	21.28
5	LAX	17600	20142	13.20	20.21
6	PHX	15213	15897	23.18	15.18
7	IAH	13456	16134	14.68	31.49
8	SFO	12981	15302	24.49	31.87
9	LAS	11197	12358	11.38	17.66
10	CLT	10944	11123	12.00	19.17

Figure 10-28. *Comparison of delays—January versus July 2012*

We left the month of September 2012 as the time range in the time picker of the user interface on purpose so that you can see the messages where the subsearch and the main search are overriding the time range. Interestingly enough, Splunk still states that there are "10 results during September 2012."

For this search, we chose to use the tabular form to display the results as we need the total number of flights in addition to the delays as a percentage of the total flights to do a proper analysis. Admittedly, a chart presenting only the percentages of delays would be enough to get a quick answer to our original question, if winter affects the delays, but when we looked at the results we noticed an anomaly. All airports have more delays during July except for Phoenix. Our first question is if the amount of flights was higher in January; thus, we need to gain access to the total number of flights. Of course, if we would not have found this anomaly, there would not have been the need to review the results in tabular form.

Creating and Using Macros

In the "Cause of Arrival Delays by Airport" search we repeated five times the same two calculations, counting the occurrences of the causes of delays that are more or equal than 15 minutes and calculating the percentages for each cause of delay. Splunk offers the ability to define macros that can be used in the search app. These macros are pieces of a search that can be reused in multiple places in a search and do not have to be a complete command. Optionally, macros can have arguments.

We can create two macros to handle those calculations, which should simplify a bit our lives and make the maintenance of the searches easier. For example, if we want to analyze delays of more than 60 minutes, we just have to change the macro once instead of the calculation five times.

We will call the first macro *countd*, which will have one argument: the field we want to count if it has a value of 15 or more minutes. The argument is surrounded by the dollar sign ($), which will be replaced with the value we call it:

```
count(eval($cause$>=15))
```

The second macro, *calcp*, calculates the percentage of a field and places the result in another one:

```
eval $result$=round($input_field$/Delayed*100,2)
```

Now the search using the macros looks like this:

```
* | stats count(ArrDel15) as Total,
        count(eval(ArrDel15=1)) as Delayed,
        `countd(CarrierDelay)` as CarrierDel,
        `countd(WeatherDelay)` as WeatherDel,
        `countd(NASDelay)` as NASDel,
        `countd(SecurityDelay)` as SecurityDel,
        `countd(LateAircraftDelay)` as LateAircraftDel
        by Dest
| `calcp(PCTCarrierDelay,CarrierDel)`
| `calcp(PCTWeatherDelay,WeatherDel)`
| `calcp(PCTNASDelay,NASDel)`
| `calcp(PCTSecurityDelay,SecurityDel)`
| `calcp(PCTLateAircraftDelay,LateAircraftDel)`
| sort - Total
| fields - Total, Delayed, CarrierDel, WeatherDel, NASDel, SecurityDel, LateAircraftDel
| head
```

You can see that to call a macro you just enclose the name and possible arguments between backticks (`). Defining the macro in Splunk is done in the Advanced Search screen of the Manager in the user interface. Figure 10-29 has a partial screenshot of the definition for the *calcp* macro.

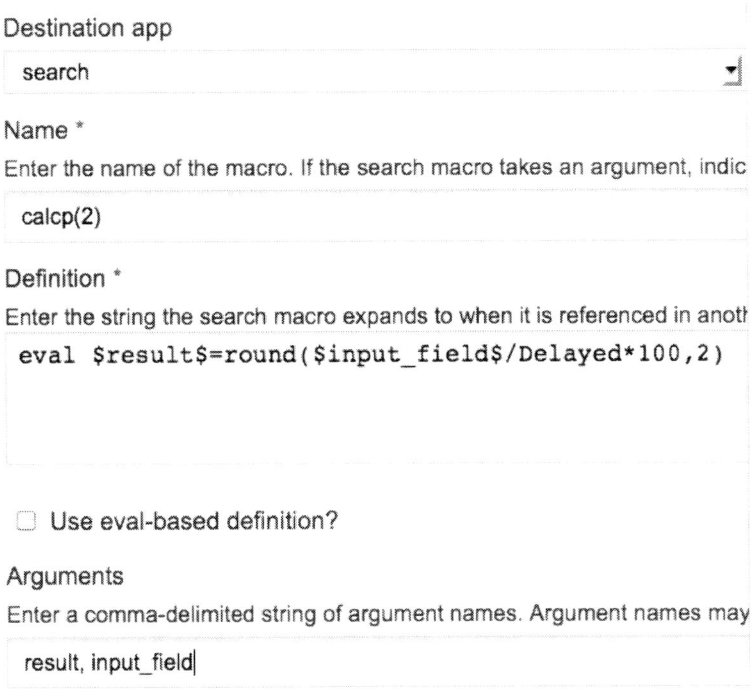

Destination app

```
search ▾
```

Name *

Enter the name of the macro. If the search macro takes an argument, indic

```
calcp(2)
```

Definition *

Enter the string the search macro expands to when it is referenced in anoth

```
eval $result$=round($input_field$/Delayed*100,2)
```

☐ Use eval-based definition?

Arguments

Enter a comma-delimited string of argument names. Argument names may

```
result, input_field|
```

Figure 10-29. *Defining a macro in Splunk*

In the screenshot you can see that we define the name of the macro but append it with "(2)". This means that the macro has two arguments, which are defined later in the screen. The definition contains the actual macro. Next you have the option to specify if the macro uses an eval-based definition, that is, if the macro is based on the eval command. Both our macros are straightforward expansions, text replacements; therefore, we do not check this item. Finally, you have the ability to define a validation expression for the arguments, which is a string that is an *eval* expression that either evaluates to a Boolean or a string. If the Boolean expression returns false or is null, the macro will return the string defined in the Validation Error message. If the validation expression is not a Boolean expression, it is expected to return a string or null. The string is used as the error string otherwise, when null, it is considered successful.

Report Acceleration

As we have been discussing in this chapter, most of the reports aggregate information to present summaries that are easily consumable for the intended audience. The amount of time that it takes to create a report is directly proportional to the amount of data to be processed. Calculating the average value of a specific field or something as simple as counting the number of events implies reading every single event, which in a large data set can take a long time. When you have reports that summarize large data sets on a regular basis this can be inefficient, especially if different users are running similar reports, thus incrementing the load on the system, which can quickly become a big problem.

The core of the problem is that every time you run a report it goes through every one of the required events every single time. One way to solve this repetitive problem is creating summaries of the data for that particular report on a regular basis. That way, when you run a report, it uses the summarized data, significantly reducing the processing time. For that chunk of data that has not been summarized yet, the normal processing applies, but because it is a substantially smaller amount of data, the processing time is usually a lot faster.

Splunk offers two methods to create data summaries for reports, report acceleration and summary indexing. Most of the reports that you have can probably be handled with the report acceleration feature, which is also the easiest of both methods. All you have to do is click on a checkbox and specify a time range when defining the search and Splunk will automagically take care of the rest. Not only that, but any search that is similar to the one you defined for your report will also benefit from the summary.

To show how report acceleration works we will use the Longest Flight by Airline search we defined earlier. When we were building this search we limited the time period to the most recent month of data, September 2012. To make an interesting example, we will state that this search will run on a monthly basis presenting the previous 12 months, that is, a rolling report of the last 12 months. Because TranStats takes a couple of months to upload the most recent month, we also define that the search starts 2 months before the current date and goes back 12 months from that point, that is, 14 months. In this case, earliest is defined as *-14mon* and latest as *-2mon*. Remember that when using months as a time modifier you have to use *mon*, *month*, or *months* instead of *m*, which stands for minutes. This seems to be a common mistake.

Without accelerating this report it takes about four minutes to go over almost five million events. In a world where waiting more than five seconds is unacceptable, four minutes is an eternity, so let us go through the necessary steps to accelerate this report. There are a couple of ways for doing this, when saving a report or by editing an existing report. We will show the latter, for which we go to the "Searches and reports" page of the Manager in the user interface. There we click on the name of our search, Longest Flight by Airline. In Figure 10-30 you can see that the dialog box contains all the information we have provided in the past. When we click on "Accelerate this search," a pull down menu shows up, presenting various choices for the summary range. We choose one year, as our report covers approximately the last 12 months. After that, all we do is click on the save button.

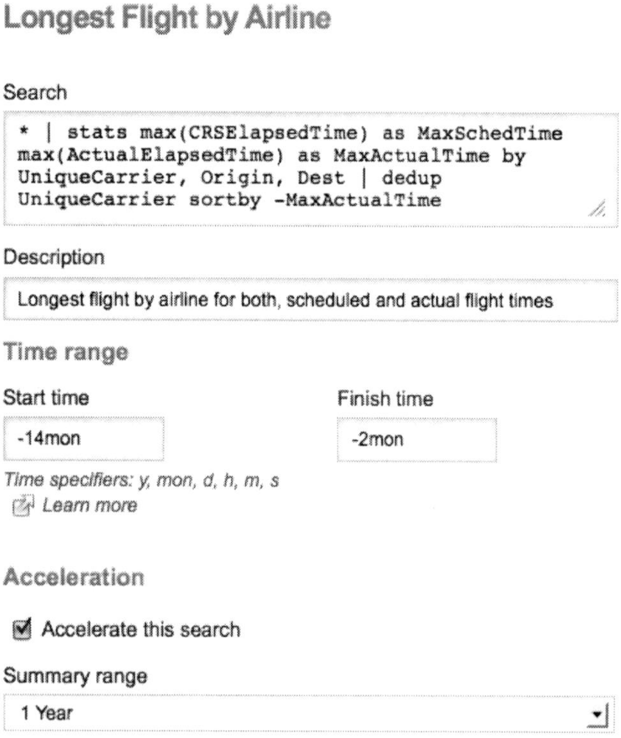

Longest Flight by Airline

Search

```
*  |  stats max(CRSElapsedTime) as MaxSchedTime
max(ActualElapsedTime) as MaxActualTime by
UniqueCarrier, Origin, Dest | dedup
UniqueCarrier sortby -MaxActualTime
```

Description

```
Longest flight by airline for both, scheduled and actual flight times
```

Time range

Start time

```
-14mon
```

Finish time

```
-2mon
```

Time specifiers: y, mon, d, h, m, s
📝 *Learn more*

Acceleration

☑ Accelerate this search

Summary range

```
1 Year                                          ▼
```

Figure 10-30. Accelerating a report

From this point on Splunk will take care of updating the summary on a regular basis in the background so that we will not even notice. The first time the summary is created immediately, but it can take a while to complete. To know if the summary is done you can go to the "Report Accelerations Summaries" page of the Manager in the user interface. In Figure 10-19 you will see under the title "Summary Status" that the summary for our report is complete. If it were in the process of updating the summary with the most recent information it would specify the amount of processing it has completed as a percentage. The "Summarization Load" is an important metric as it reflects the effort that Splunk has put into updating the summary. The bigger this number is, the more processing was involved. However, be aware that the calculation is based on the amount of time it takes to update the summary and how often it happens. For a report as ours, which will be updated every month the number will be very low. As a matter of fact, in Figure 10-31 you can see that the number is 0.00004 as it has not been updated for the last few weeks.

Report Acceleration Summaries

Showing 1-1 of 1 item   Results per page   25 ▾

Summary ID ⬍	Reports Using Summary	Summarization Load ⬍ ⓘ	Access Count ⬍	Summary Status ⬍
0ec1ae9ad188ea10	Longest Flight by Airline.	0.0004	10 Last Access: < 1 min ago	Pending Updated: 34d 7h 8m ago

Figure 10-31. *Report Acceleration Summary*

The other important metric is the "Access Count," which also shows the last time the report was run. If you have a report with a high load and it is seldomly used or it has not been used in a long time, you should seriously consider deleting this summary. In addition to *Complete*, the summarization status can also be *Pending, Suspended* and *Not enough data to summarize*. Pending means that it is close to review if an update is needed, whereas suspended means that Splunk has determined that the search is not worth summarizing because the summary is too large. This usually happens with searches that have a high data cardinality, that is, they produce a lot of events, almost as many as there are in the original data set.

If you are interested in more details of the summary, you can click on the strange number under the "Summary ID" column. Figure 10-32 shows the detailed information of a summary.

Summary: 0ec1ae9ad188ea10

Summary Status

Complete   Updated: 34d 7h 11m ago

Actions

Verify   Update   Rebuild   Delete

Reports Using This Summary

Search name	Owner	App
Longest Flight by Airline.	admin	search

Details ⬀ Learn more.

Summarization Load	0.0004
Access Count	10 Last Access: 3m ago
Size on Disk	44.75MB
Summary Range	366 days
Timespans	1d, 1mon
Buckets	5
Chunks	358

Figure 10-32. *Summary status*

Within the details, it is worth mentioning the size on disk. This feature is so powerful that it is easy to forget that it has a cost, that of regularly running the search and that the summary uses additional space on disk. Considering that the flight data occupies about 17GBs, 45MBs for this summary is almost negligible. Of course, this will grow over time and we should be mindful of the impact it can have on the longer term.

In Figure 10-32, you can also see that there are various actions available. Because data, fields, and the way that we manipulate them tend to change, sometimes without our knowledge, we can use the verify action to make sure that the summary is still producing correct results and that the data is still consistent. The other actions, update, rebuild, and delete, are standard maintenance issue.

The question now is, how fast is the search that we accelerated? Really fast, clocking at about three seconds. That is a couple of orders of magnitude faster, which in this case is well worth it. Again, remember that you cannot defy the laws of physics; you have to pay the price somewhere. In this case, we increase disk usage and distribute the processing over time in the background. In the summarization load and usage we have two good metrics that let us know if this powerful and very simple feature is worth it for every report.

Of course, not all searches qualify for report acceleration as there are a few limitations. The first one being that the searches must use reporting commands, such as *chart, timechart, stats, top, rare,* and others. Additionally, the commands feeding the reporting commands must be streaming commands, such as *search, eval, fields, regex, lookup, multikv,* and others. The second limitation is that the data to be summarized has to have at least 100,000 recent events.

The way the summarization works is that Splunk will regularly run the search in the background and store the summary information so that it is available the next time it is needed. Splunk will figure out the frequency it needs to run the background search, but we can say that it is dependent on the time range you define.

All of the searches we have done in this chapter qualify for report acceleration. You might have thought that the searches for the longest and shortest flights would not qualify because they use the *dedup* command, which is not a streaming command, but it is used after the reporting commands, not before; thus, it meets the first requirement. Regarding the recent event count, we know that every monthly update brings well over 400,000 events, so that is not an issue.

The second method for creating data summaries is summary indexing. It works basically the same as report acceleration, but internally Splunk handles things differently. Setting it up and using it is a more involved process. As most of the reports can be done with report acceleration, summary indexing is relegated for those cases where the search includes nonstreamable commands before a reporting command. We will not be covering summary indexing in this book. If for any reason you may need to use it, please refer to the Splunk documentation.

Accelerating Statistics

As we saw at the beginning of this chapter, running a simple search over all the flight data set can take as much three hours. So far we have used the strategy of testing our searches on the most recent month of flight data. If the search spanned over a month we tested with a couple of months and if the search spanned over years we tested with a couple years. Inevitably, the moment will come that your searches are ready to be used in prime time and some of them will cover all of the data; in our example, that is about 26 years of flight data. Having to wait three hours to get the results of a single search might be bearable, but when you have a dozen of searches and you have to run them on a regular basis, this is totally unacceptable.

Whereas report acceleration focuses on creating data summaries for specific searches or reports, Splunk has a mechanism referred to as tsidx that allows for the creation and use of general summaries of indexed data. Using tsidx is a two-step process. The first one is creating a general summary of the data using the *tscollect* command. Then you can use the *tstats* command to perform the statistical queries of your interest with an incredibly fast performance.

The *tscollect* command is very simple; you just have to define the name of the summary file where Splunk will store the general summary:

```
* | tscollect namespace=summary_fd
```

As easy as it is, creating the general summary for all the 147 million events of the flight data ran for over 8 hours and it created a file about half the size of the flight data. Be aware that you cannot use the *tscollect* command to replace information within a summary file; however, you can append information to it. For example, once we get the flight data for October 2012 loaded into the main index, we can append it to *summary_fd* like this:

```
earliest=10/1/2012:0:0:0 latest=11/1/2012:0:0:0
| tscollect namespace=summary_fd
```

The *tstats* command has a different syntax than the *stats* command. Let us illustrate how it works with an example using one of the searches we defined earlier in this chapter, the average flight time per airline, which using the *stats* command is:

```
* | stats avg(CRSElapsedTime) as AverageFlightTime
        by UniqueCarrier
| eval AverageFlightTime=round(AverageFlightTime)
| sort -AverageFlightTime
```

We will now replace the *stats* command with the equivalent *tstats* command:

```
| tstats avg(CRSElapsedTime) as AverageFlightTime
        from summary_fd
        groupby UniqueCarrier
```

As you can see, we can still do calculations on fields using functions and create a new field to contain the results. Next we have to specify the name of the file that contains the general summary and finally the *by* argument is now *groupby*. Notice that we no longer have a star (*) at the beginning of the search. This is because you are not doing a search and passing the events on to the *tstats* command. We still have the pipe before the command because we have to tell Splunk that this is an actual command and not an attribute of the search command. The rest of the search, the *eval* and *sort*, remains the same.

How fast is it? When we run this search on the month of September 2012 using good old *stats* it takes about 15 seconds. Using *tstats* over all the 26 years of flight data, it takes about 30 seconds. It seems that investing more than eight hours in creating the general summary and using the additional 8.6GB of disk space are well worth it.

One of the down sides of *tstats* is that not all statistical functions are available. At the moment of this writing the functions are limited to *count, sum, sumsq, distinct, avg,* and *stdev*. However, Splunk offers a way to bridge the *tstats* command with the *stats* command to gain access to those functions that are not available. It is a bit quirky, but it works. Let us use the search we did to count the airlines by year:

```
* | stats dc(UniqueCarrier) by Year
```

The equivalent search using *tstats* is the following:

```
| tstats prestats=t
        dc(UniqueCarrier)
        from summary_fd
        groupby Year
| stats dc(UniqueCarrier)
        by Year
```

Just to avoid any confusion, the *distinct* function available in *tstats* lists the distinct or unique values of a field just like the *values* function, which is different than the *distinct_count,* or *dc* function that calculates a count of the distinct values of a field. The *prestats* argument set to t (for true) indicates that the following function is not supported by *tstats* and it should create a bridge to the *stats* command that follows immediately and has the same function. As we said, it's a bit bizarre, but at least we do have a mechanism to overcome these limitations of *tstats*. The performance improvement for this search is mind-blowing; the original search over the 147,122,177 events using *stats* took close to three hours. With *tstats* it took 15 seconds!

The *groupby* clause in *tstats* can handle multiple fields. We will simplify a search we used earlier to find the longest flights by airline to illustrate this. The original search was:

```
* | stats max(CRSElapsedTime) as MaxSchedTime
         max(ActualElapsedTime) as MaxActualTime
         by UniqueCarrier, Origin, Dest
| dedup UniqueCarrier
         sortby -MaxActualTime
```

Changing it to use *tstats* is pretty straightforward, very much like the other examples:

```
| tstats prestats=t
         max(CRSElapsedTime),
         max(ActualElapsedTime)
         from summary_fd
         groupby UniqueCarrier, Origin, Dest
| stats max(CRSElapsedTime) as MaxSchedTime,
         max(ActualElapsedTime) as MaxActualTime
         by UniqueCarrier, Origin, Dest
|   dedup UniqueCarrier
         sortby -MaxSchedTime
```

Because the *max* function is not a supported by *tstats,* we bridge it over to *stats* by setting the *prestats* argument to true. As *stats* is the command that will actually be calculating the maximum flight times it's there where we assign the results to a new field. We could have typed it in the *tstats* command, but it would not have any effect. For didactical purposes, we have changed the sort field to the maximum scheduled flight time. Running the normal search on the flight data of September 2012 takes about three seconds. The accelerated search over all the 26 years takes a little over two minutes. In Figure 10-33 you can see results of the search with the modified sort.

	UniqueCarrier ⇕	Origin ⇕	Dest ⇕	MaxSchedTime ⇕	MaxActualTime ⇕
1	UA	GUM	HNL	1865	477
2	AA	DFW	HNL	1613	727
3	CO	HNL	GUM	1565	693
4	EA	ATL	LAS	1560	346
5	TW	DTW	MKE	1495	181
6	B6	JFK	TUS	1484	521
7	NW	CMH	MEM	1440	284
8	WN	DET	MDW	1440	1440
9	MQ	ORD	GRB	1435	260
10	OO	BZN	DEN	1375	287

Figure 10-33. *Longest flights by airline, all years*

The reason we did this is to show you the strange numbers we get on almost all the rows under the maximum scheduled time column. These are obviously wrong as 1,865 minutes—about 31 hours—is totally unrealistic for a commercial flight. This is confirmed by the fact that the maximum actual flight time for that same flight over the last 26 years is 477 minutes, or about 8 hours, which is much more realistic number for a flight between Guam and Honolulu. We need to find more information about this. First we want to know if this is a regular error or it only happens once. For that we do the following simple search, which produces the output found in Figure 10-34.

```
UniqueCarrier=UA CRSElapsedTime=1865
```

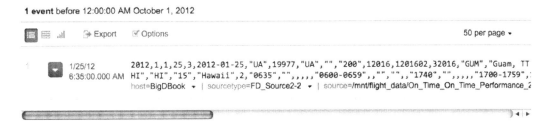

Figure 10-34. All United flights with scheduled flight time of 1,865 minutes

This search returns only one event, but we still do not know if it is an isolated error. For this we search for any flight in that route that has a scheduled time bigger than the biggest actual flight time over the 26 years:

```
UniqueCarrier=UA Origin=GUM Dest=HNL
| where CRSElapsedTime>477
```

This search only returns the same event as the previous one, so this is an isolated incident. We suspect there was some data entry problem for United Flight 200 on January 25, 2012. This last search allows us to explain some performance tips in searches. Splunk is quite slow when it has to read a large amount of events from raw data, therefore the reason for *tsidx* to exist. However, Splunk is extremely fast finding specific items in events, even more so when they are key value pairs. The above search took about one second for all the 147 million events. The first section of the search zooms in directly into the required events by looking directly into the index table Splunk keeps internally instead of reading every event. The second clause applies the conditional, which tends to be a more expensive operation but only on those events that resulted from the first search. Alternatively, we could have formulated the search as:

```
* | where UniqueCarrier=UA AND Origin=GUM AND Dest=HNL AND CRSElapsedTime>477
```

This would have evaluated the conditional for every single raw event and it probably would have taken longer than three hours. The difference is matching 336 events that contain United, Guam and Honolulu, and then evaluate if the scheduled time is greater than 477 minutes against evaluating if the airline is United and the origin Guam and the destination Honolulu and the scheduled time greater than 477 for every single one of the 147+ million events.

For completeness, we did a quick check with the anomaly for American Airlines and found two events that have 1,613 and 1,517 minutes for scheduled flight time, one in 1993 and another one in 1991. Obviously, these are isolated incidents, but they could potentially affect our analysis by incorrectly skewing results.

One way to solve this problem is by erasing those events using the *delete* command. This command does not actually delete events; it just marks them so that they are not used in searches. Of course, this will not affect a tsidx summary that was created before the events were deleted. An example of one way that you could delete specific events presenting the anomalous scheduled flight times is:

```
UniqueCarrier=UA CRSElapsedTime=1865 | delete
```

You must be extremely careful using this command, as far as we know there is no way to undelete events. Splunk is quite sensitive about this command and by default no user, not even the admin has the capability to delete events. They suggest you create a special user for when you will delete events.

The final example we have shows how to deal with multiple functions within a single *stats* command that assigns the results to new fields, and how to handle conditionals. For this, we will use the Arrival Delays by Airline search we built earlier in this chapter:

```
* | stats count(ArrDel15) as Total,
        count(eval(ArrDel15=0)) as OnTime,
        count(eval(ArrDel15=1)) as Delayed
        by UniqueCarrier
| eval PCTOnTime=round(OnTime/Total*100,2)
| eval PCTDelayed=round(Delayed/Total*100,2)
| sort - PCTDelayed
```

The first issue that we see in this search is that it uses the *eval* function within the *count* function. Whereas *count* is supported by the *tstats* command, *eval* is not. The way around this is using the *where* clause of the *tstats* command, but the problem here is that there can be only one clause, so the logic in the *stats* command will have to be broken down in many parts to make it work with *tstats*. We start by obtaining the count of delayed flights:

```
| tstats count as Delayed
        from summary_fd
        where ArrDel15=1
        groupby UniqueCarrier
```

The *where* argument is used in an equivalent manner to the way the *eval* function and the above search ends up counting all the delayed flights by airline. Be aware that you must follow the order of the *from, where* and *groupby* clauses for the *tstats* command. You will get very misleading messages when you do not do so and will spent a lot of time trying to understand what is happening. Next we need another search to find either the total number of flights by airline or the on-time flights:

```
| tstats count as Total
        from summary_fd
        groupby UniqueCarrier
```

This search is quite simple and all we have left is joining both tstat commands, calculate the percentages and sort appropriately. The final search looks like this:

```
| tstats count as Delayed
        from summary_fd
        where ArrDel15=1
        groupby UniqueCarrier
| join
  [
  | tstats count as Total
          from summary_fd
          groupby UniqueCarrier
  ]
| eval PCTDelayed=round(Delayed/Total*100,2)
| eval PCTOnTime=round((Total-Delayed)/Total*100,2)
| sort - PCTDelayed
| fields - Delayed, Total
```

Essentially we have completely reformulated the search to work around the limitations of the *tstats* command and achieved the same results. Whereas in the original search we counted in one single command the total, on-time and delayed flights, in the new search we broke the counts of the total and delayed flights in two separate *tstats* commands, which were then combined using the *join* command, as we did in the Comparison of Delays search earlier in this chapter. The on-time percentage is based on the subtraction of the total and delayed flights counts. The only functional difference with the original search is that we eliminated the count results using the *fields* command and left only the percentages. Using a 100% stacked bar chart provides a good visualization of the aggregated arrival delays over the last 26 years, which can be seen in Figure 10-35.

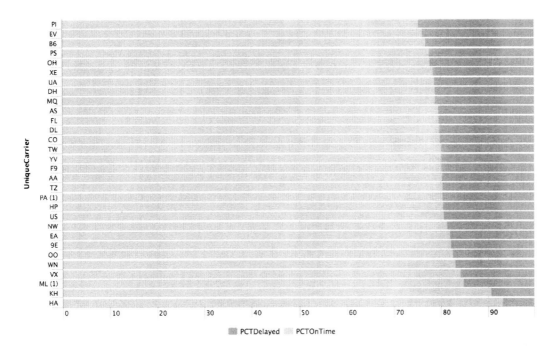

Figure 10-35. *Arrival delays by airline, all years*

As expected with the tsidx mechanism, the performance difference is huge. The original search using only the month of September 2012 ran in 10 seconds, whereas using the reformulated search it took 15 seconds to go over the 147+ million events.

Summary

In this chapter, we have gone through a typical data analysis session using the Airline On-Timer Performance data, during which we learned how to use some of the commands available in Splunk. We also discussed various visualization options, which are dependent on the results of the searches and the information that is being conveyed to the consumers of the analysis. Finally, we went over some performance tips on searches and reports, which can make amazingly big differences by dramatically cutting the response times of searches.

CHAPTER 11

■ ■ ■

Analyzing a Specific Flight Over the Years

Using the analysis of a specific flight over the years as an example, we continue to explore Splunk commands. This time we focus on lookup tables, both static CSV based and using a database. It can be argued that field lookup tables are inherited from relational databases and their constant need to normalize data. As such, much of the data that we deal with on a daily basis is coded and those codes are explained in separate tables, which are called lookup tables. In Splunk you can create and upload lookup tables that can be later used in searches to expand the aforementioned codes.

Airline Names

In Chapter 10 we mentioned that we would be using lookup tables to present the name of the airlines instead of those acronyms we have been using so far, which are all we have available in the three fields that provide us airline information: *UniqueCarrier, AirlineID,* and *Carrier*. We will start by creating a simple lookup table in Splunk to replace these codes with the actual airline names.

When you go to the download page on the TranStats web site you will see that some of the fields described have a link to the right for the corresponding lookup table. This can be seen in Figure 11-1. When you click on any of those links it will download a CSV file that contains the detailed information associated with that particular field. For this example we downloaded the lookup table for the *UniqueCarrier* field, which is the one we have been using for the searches in Chapter 10.

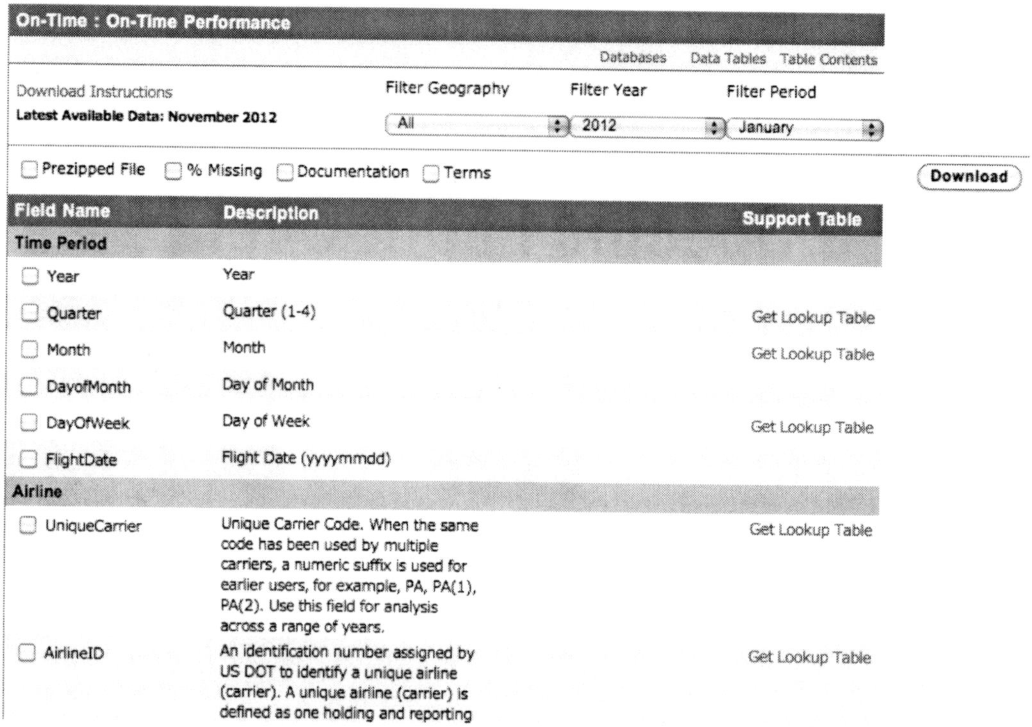

Figure 11-1. On-Time performance download page

The first thing that we notice when we download the lookup table is that the file name has a strange character at the end of the csv extension, a double quote (") in MacOS and a dash (-) in Windows. Removing this extra character makes it usable with a spreadsheet application. When we open it up, it comes as a surprise that there are a lot more than the 30 airlines we have been handling and many of the names are ones that we have not heard of before. As a matter of fact, in the file we downloaded at this writing, there are 1,543 airlines. After a quick review we notice that most of them are foreign airlines or airlines that no longer exist. In Table 11-1 we show the first five lines of this file.

Table 11-1. Sample of the UniqueCarrier field lookup file

Code	Description
02Q	Titan Airways
04Q	Tradewind Aviation
05Q	Comlux Aviation, AG
06Q	Master Top Linhas Aereas Ltd.
07Q	Flair Airlines Ltd.

To make a lookup table available for searches within Splunk there are a couple of steps that have to be taken. The first one is to upload the file that contains the lookup table. This can be done from the user interface by selecting "Add new" in the "Lookup table files" option of the "Lookups" menu in the Manager. Figure 11-2 contains a screen shot with the dialog to upload a lookup file.

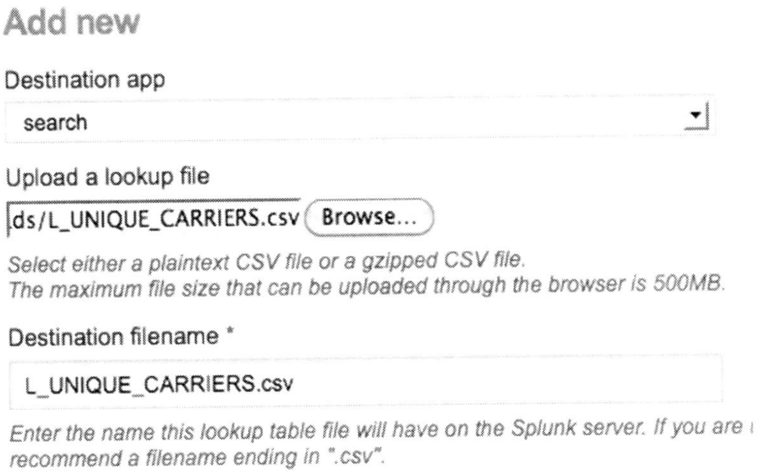

Add new

Destination app

[search ▾]

Upload a lookup file

[ds/L_UNIQUE_CARRIERS.csv] (Browse...)

Select either a plaintext CSV file or a gzipped CSV file.
The maximum file size that can be uploaded through the browser is 500MB.

Destination filename *

[L_UNIQUE_CARRIERS.csv]

Enter the name this lookup table file will have on the Splunk server. If you are
recommend a filename ending in ".csv".

Figure 11-2. *Uploading a lookup file*

For this example, in the upload dialog leave the "Destination app" as search, just be aware that not defining correctly where the lookup table will be used, that is, in which app, can cause many headaches when debugging related issues. Splunk will take plain text CSV files or compressed CSV files (using gzip only). There is a limit to the size of the file to be uploaded via the user interface of 500 MB. Bigger files can be uploaded using the configuration files directly. As the file will be stored within the Splunk area, we need to provide a destination filename. Best practices call for using the same name as the original file, including the extension. After you click on the "Save" button, the file is uploaded into Splunk.

The second step is to define the field lookup. This is done under the same "Lookups" menu with the "Lookup definitions" option. The dialog screen can be seen in Figure 11-3. We need to give the field lookup a name so we can refer to it in the search; we typically use the field name preceded by "L_". There are two types of lookups: file based, such as the one we are defining, and external, which can use a Python program to match the fields. We will not cover external lookups in this book, as that is an advanced topic.

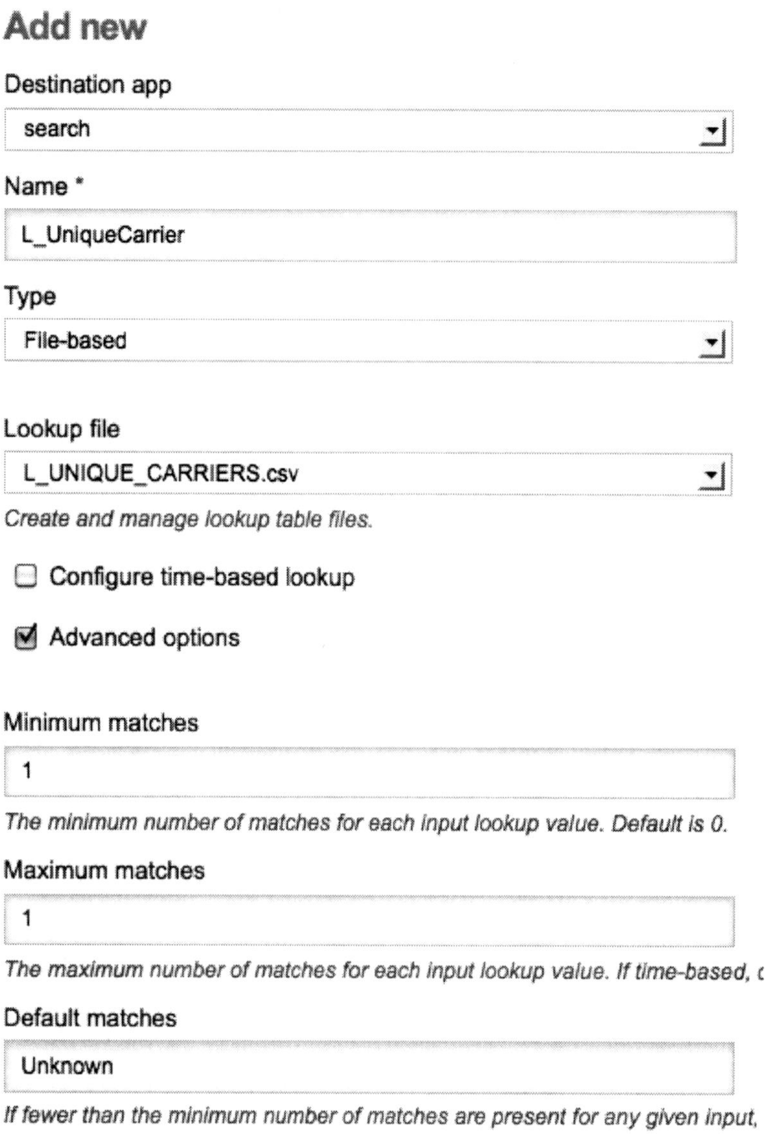

Add new

Destination app

| search | ▾ |

Name *

| L_UniqueCarrier |

Type

| File-based | ▾ |

Lookup file

| L_UNIQUE_CARRIERS.csv | ▾ |

Create and manage lookup table files.

☐ Configure time-based lookup

☑ Advanced options

Minimum matches

| 1 |

The minimum number of matches for each input lookup value. Default is 0.

Maximum matches

| 1 |

The maximum number of matches for each input lookup value. If time-based, c

Default matches

| Unknown |

If fewer than the minimum number of matches are present for any given input,

Figure 11-3. *Defining a field lookup*

The next pull down menu has the names of the lookup files that have been uploaded so far. Just select the appropriate one. Leave unchecked the "Configure time-based lookup"; this option allows the use of lookups based on timestamps, which will not be reviewed in this book. Check "Advanced options," which we will use to specify the minimum and maximum number of matches when using the lookup table. We want the maximum to be 1, so that we get only one value. Technically speaking this is not needed with this table because there is supposed to be only one value for each code, but we have already seen a few errors in the flight data. In the case there are no matches, we want the string "Unknown" to be used; therefore, we specify a minimum of one match, as the default is zero.

Now that we have the lookup completely defined, we can start using it. The following is the most basic form of this command:

```
lookup <lookup-definition> lookup-field
```

You must be aware that the lookup field name must match the name of the desired column on the header line of the lookup table. Let's use an example to illustrate this by counting the number of flights by airline during the most recent month:

```
* | stats count by UniqueCarrier
| lookup L_UniqueCarrier UniqueCarrier
```

This will not work because there is no field name in the header line of the lookup table called *UniqueCarrier*. As you can see in Table 11-1 the field names are *Code* and *Description*. We have two options; on one we change the header line in the lookup table, whereas on the other one we rename the *UniqueCarrier* field. The first option is high maintenance, as we would have to edit the file every time we get a new version. The second option is easier:

```
* | stats count by UniqueCarrier as Code
| lookup L_UniqueCarrier Code
```

This one will not work either because it is the count that we are naming as *Code*; we are not renaming the field. The correct way to do it is:

```
* | stats count by UniqueCarrier
| lookup L_UniqueCarrier Code as UniqueCarrier
```

Once we verify that this search works correctly we can apply it to all the flight data using the *tstats* command to save us time:

```
| tstats count from summary_fd groupby UniqueCarrier
| lookup L_UniqueCarrier Code as UniqueCarrier
```

Partial results from this search can be seen in Figure 11-4. Without the lookup command the results would have only two columns, the unique carrier code and the count. With the lookup we add the description column. Should the lookup table have more columns, all of them would have shown up in the results. You can specify the fields that should show up with the *output* argument of the *lookup* command and list right after the fields you want to appear.

UniqueCarrier ⇕	count ⇕	Description ⇕
1	9E	1045396	Pinnacle Airlines Inc.
2	AA	17010769	American Airlines Inc.
3	AS	3407700	Alaska Airlines Inc.
4	B6	1594795	JetBlue Airways
5	CO	8888536	Continental Air Lines Inc.
6	DH	693047	Independence Air
7	DL	18986659	Delta Air Lines Inc.
8	EA	919785	Eastern Air Lines Inc.
9	EV	3199446	Atlantic Southeast Airlines
10	F9	651342	Frontier Airlines Inc.
11	FL	2182817	AirTran Airways Corporation
12	HA	537423	Hawaiian Airlines Inc.
13	HP	3636682	America West Airlines Inc. (Merged with US Airways 9/05. Stopped reporting 10/07.)
14	KH	154381	Aloha Air Cargo
15	ML (1)	70622	Midway Airlines Inc. (1)
16	MQ	5634687	American Eagle Airlines Inc.
17	NW	10585760	Northwest Airlines Inc.
18	OH	1765828	Comair Inc.
19	OO	5289490	SkyWest Airlines Inc.
20	PA (1)	316167	Pan American World Airways (1)

Figure 11-4. *Flight count by airline using the lookup table*

The report has a couple of annoying items, such as the merger information for America West Airlines and US Airways, and the "(1)" after the names of Midway Airlines and Pan American World Airways. As you might remember the "(1)" appended after the unique carrier code signifies that another airline is currently using that code. We fell that we should clean up this report by eliminating the merger information, and since we are at it we could do it also for the "(1)". As we are in cleaning mode we should use comma separators for 1,000s on the final count of flights, use the proper titles and eliminate the code column. The final search for our report is:

```
| tstats count from summary_fd groupby UniqueCarrier
| lookup L_UniqueCarrier Code as UniqueCarrier
| eval Airline=replace(Description, "^(.+)\(.+\)", "\1")
| eval Flights=tostring(count, "commas")
| fields - Code, count, Description
| sort Airline
```

The *replace* function of the *eval* command applies the regular expression described on the second argument to the field in the first argument. The regular expression has two parts; on the first one it takes one or more characters from the beginning to the first right parenthesis and keeps that match for future use, as indicated by enclosing that part of the expression in between parenthesis. The second part matches one or more characters surrounded by parenthesis as indicated by prefixing the parenthesis with a backslash (\). If a string in the *Description* field matches this regular expression, then it is replaced by the regular expression on the third argument, which in this case states it should use the first part of the match, which was kept for future use.

Adding commas to the count field is done using the *tostring* function of the *eval* command. You have to be aware that a sort based on the *Flights* field will not be numeric any more as now it is a string. If needed, the *sort* command has a function that allows us to treat a string field as a number. A partial output of the results from this search can be seen in Figure 11-5, where the merger notice of America West Airlines no longer shows up.

Airline ‡	Flights ‡
1 ATA Airlines d/b/a ATA	208,420
2 AirTran Airways Corporation	2,182,817
3 Alaska Airlines Inc.	3,407,700
4 Aloha Air Cargo	154,381
5 America West Airlines Inc.	3,636,682
6 American Airlines Inc.	17,010,769
7 American Eagle Airlines Inc.	5,634,687
8 Atlantic Southeast Airlines	3,199,446
9 Comair Inc.	1,765,828
10 Continental Air Lines Inc.	8,888,536
11 Delta Air Lines Inc.	18,986,659
12 Eastern Air Lines Inc.	919,785

Figure 11-5. *Improved report of flight count by airline*

A final word regarding the use of the lookup command. Be aware that a search will be done on the lookup table for each event or partial result of search, so there is a huge difference between:

```
* | lookup L_UniqueCarrier Code as UniqueCarrier
| stats by Airline
```

and this search, which will produce the same results:

```
* | stats by UniqueCarrier
| lookup L_UniqueCarrier Code as UniqueCarrier
```

The first search will go to the lookup table for each and every one of the events, whereas the second one will only go as many times as unique carriers result from the search, which in the flight data is 30 airlines against 147 million total events in the first search.

One optimization that Splunk does on itself with lookup table files is that if they are larger than 10 MB, it automatically indexes them so that the lookup searches are faster. This size is the default and can be modified in the configuration files.

Automating Field Lookups

In some cases we would like for the lookup to happen automatically, without us having to specify the lookup command. Splunk offers the ability to define automatic field lookups, which is really simple as can be seen in Figure 11-6 by going to the "Automatic lookups" option of the "Lookups" menu in the Manager screen in the user interface.

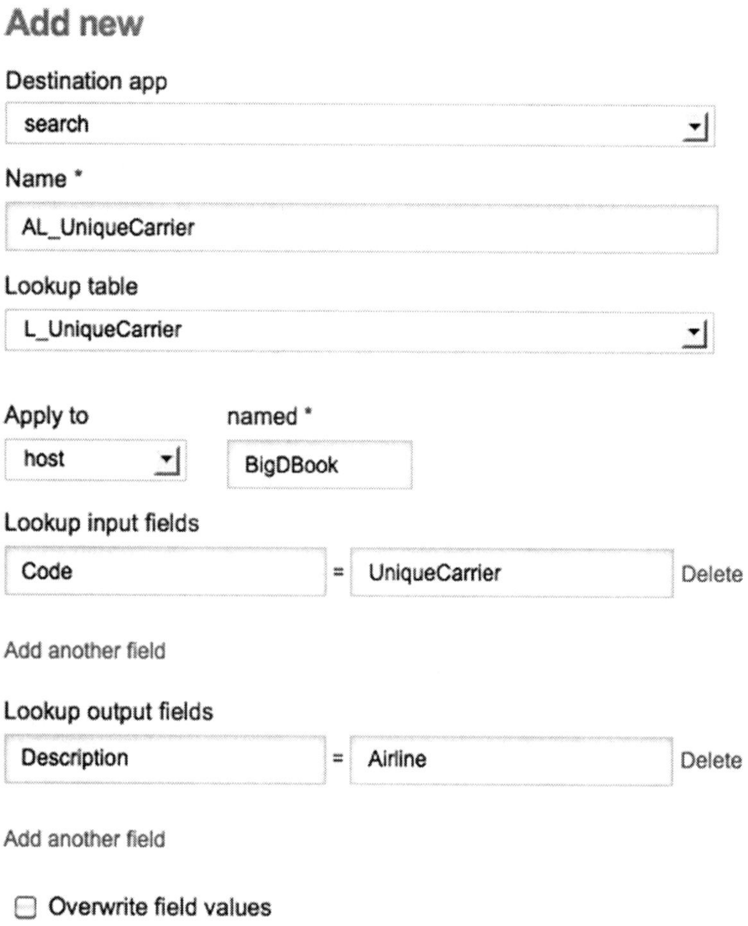

Add new

Destination app

| search | ▾ |

Name *

| AL_UniqueCarrier |

Lookup table

| L_UniqueCarrier | ▾ |

Apply to   named *

| host ▾ | | BigDBook |

Lookup input fields

| Code | = | UniqueCarrier | Delete |

Add another field

Lookup output fields

| Description | = | Airline | Delete |

Add another field

☐ Overwrite field values

Figure 11-6. *Defining an automatic lookup*

As a naming convention, we like to preface the names of automatic lookups with "AL_", so this one is called *AL_UniqueCarrier*, based on the *L_UniqueCarrier* lookup definition. The only thing with automatic lookups is that you have to tie them to events based on either a source, a source type or a host. For this example we will use the use host, as all the flight data was indexed on a single host called *BigDBook*. The next section is defining the lookup input fields, that is, the fields that will be used to find a match in the lookup table. Our example is based on one field, *Code* in the table and *UniqueCarrier* in the flight data. Note that you have to place the CSV field name on the left and the field you want to match on the right, otherwise the lookup will fail very quietly. The same goes for the fields you want to add to the events returned from the lookup table. Finally, leave the "Overwrite field values" unchecked, as this is used only when you want to overwrite any fields in your events with the corresponding values you matched in the lookup table. In this particular case we are just adding *Description* as a new field to the events and rename it *Airline* in the definition. From now on we can use the *Airline* field directly in any search without having to specify a lookup command. For example:

```
* | stats count by Airline
```

will produce the same results as:

```
* | stats count by UniqueCarrier
| lookup L_UniqueCarrier Code as UniqueCarrier
| rename Description as Airline
| fields - Code, UniqueCarrier
```

Interestingly enough, by creating an automatic lookup, we can do reverse searches. By this we mean you can use a resulting field of a lookup table to do a search. For example:

```
Airline="Virgin America"
| stats count
```

The *Airline* field is available as a result of the automatic lookup. Note that because we specified that the *Description* field is automatically mapped to *Airline*, it is no longer available. Any search that uses *Description* will fail as well as those that use the *Code* field because in the automatic lookup definition we mapped *Code* to *UniqueCarrier*. The *UniqueCarrier* field is still available as it is part of the original events.

Creating Lookup Tables from Searches

So far we have been using CSV files to upload data into Splunk or to use as lookup tables. Splunk offers the ability to create CSV based lookup table files. We will optimize the *L_UniqueCarrier.csv* lookup table we used earlier in this chapter to illustrate how to create a new lookup table file and reuse it as a new in our searches.

The lookup table for the *UniqueCarrier* field we originally downloaded from the TranStats web site has 1,543 airlines. We know that we refer to only 30 of them throughout all the flight data. Additionally, there are a couple of items that we changed in the name of the airlines when we cleaned it up, such as the merger notes for two of the airlines. We will redo this cleanup and eliminate all the airlines not used in the 26 years worth of flight data with this search:

```
| tstats count from summary_fd groupby UniqueCarrier
| lookup L_UniqueCarrier Code as UniqueCarrier
| eval Airline=replace(Description, "^(.+)\(.+\)", "\1")
| fields - count Description
| outputlookup L_UniqueCarrier_short.csv
```

The first four clauses of this search are already known. The new command is *outputlookup*, which creates a CSV file and also creates a lookup table file definition using the name that was given as part of the command. Note that you can also specify a *.csv.gz* extension, which will automatically compress the CSV file. The *outputlookup* command has a number of arguments, of which the only one we will mention is *append*. This argument can be used to append data to an existing lookup table file, as long as it is not a compressed file. Only fields that exist on the header line of the file will be used.

United Flight 871

Now that we know how to create and use lookup tables, we will complicate things a little by trying to answer a query. In the process of doing so we will be using database lookup tables. The query we have is to present the history of United Airlines flight 871, by showing the segments or city pairs (origin and destination) this flight number has had over the years as well as which model of aircraft was used for those segments.

This query will require that we use a couple of lookup tables because the information in the flight data events only includes the registration number of the airplane (the *TailNum* field). To obtain the model of the airplane we will have to refer to the aircraft registry database available at the Federal Aviation Administration (FAA) web site at: `http://www.faa.gov/licenses_certificates/aircraft_certification/aircraft_registry/releasable_aircraft_download/`. On this web site we learn that this database is updated on a weekly basis, its compressed size at the time of this writing is 45MB, and it is composed of various files.

After downloading and expanding the file we quickly review the *ARData.pdf* file, where we notice that the aircraft registration master file does not contain the model name of the airplane but rather a code, which consists of three parts, a manufacturer code, the model code and a series code all grouped up into a single field called *MFR_MDL_CODE*. Further exploration takes us to find that the aircraft reference file is the one that has the relationship between the manufacturer, model, code field of the master file and the manufacturer name and model name, which is what we want.

The master file contains over 352,000 aircraft and is about 207MB when not compressed, whereas the reference file has over 76,000 records and is 6.7MB in size. As explained earlier in this chapter, uploading the master file as a lookup table in Splunk has to be done by using the configuration files as it exceeds the maximum size of 100MB of the user interface. Additionally, the lookup file would be automatically indexed by Splunk as it exceeds the 10MB default size. However, for didactical purposes we will upload both tables to a MySQL database and use it by defining database lookup tables using the Splunk DB Connect app.

Within the existing MySQL instance we used in Chapter 9, we created a different database called *AircraftRegistration*, which contains both tables, *master* and *reference*. Having done that we create in Splunk a new "External Database" in the same fashion as we did in Chapter 9 to create the FD_DB database, but we will call this one AR_DB. Now that we have defined the new external database, we can proceed to create the two database lookups we need. This can be done by selecting the "Database lookups" option in the "Lookups" menu. The dialog to create the database lookup for the master table can be seen in Figure 11-7.

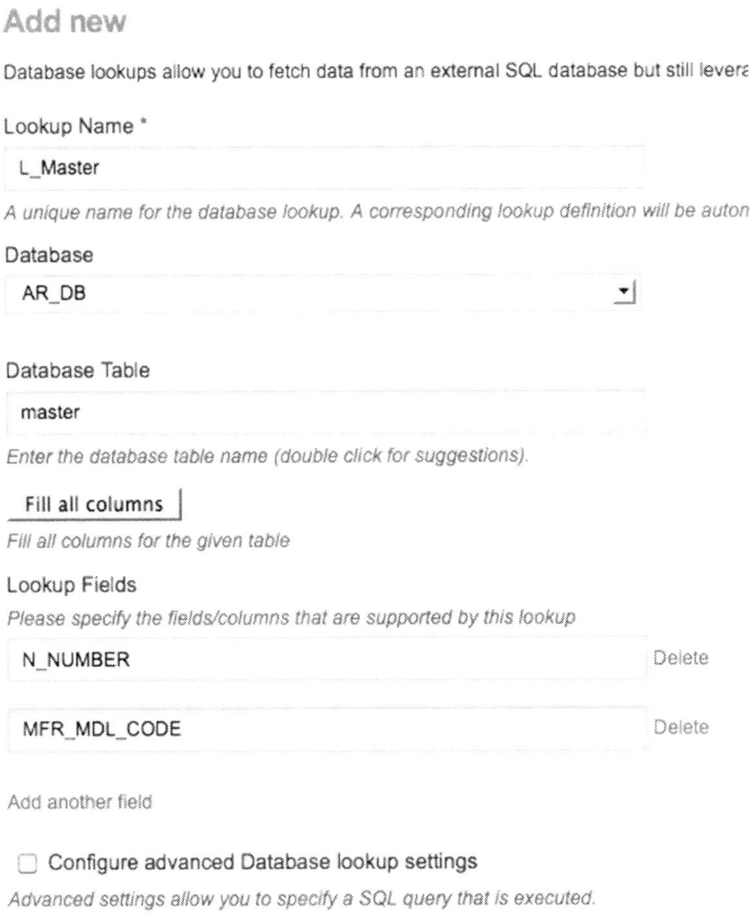

Add new

Database lookups allow you to fetch data from an external SQL database but still levera

Lookup Name *

```
L_Master
```

A unique name for the database lookup. A corresponding lookup definition will be auton

Database

```
AR_DB                                              ▾
```

Database Table

```
master
```

Enter the database table name (double click for suggestions).

| Fill all columns |

Fill all columns for the given table

Lookup Fields

Please specify the fields/columns that are supported by this lookup

```
N_NUMBER                                           Delete
```

```
MFR_MDL_CODE                                       Delete
```

Add another field

☐ Configure advanced Database lookup settings

Advanced settings allow you to specify a SQL query that is executed.

Figure 11-7. *Create a database lookup*

Slightly modifying our naming convention for lookup tables, we call this lookup table *L_Master*. After specifying the database and table for the lookup, we define which fields that are going to be used. The master table has 34 fields, but we are only interested in *N_NUMBER*, which maps to *TailNum* from the flight data, and *MFR_MDL_CODE*, which provides the relationship with the manufacturer and model names in the reference table. In a similar fashion we define the *L_Reference* lookup table, with only three of the 11 fields: *CODE*, which maps to the *MFR_MDL_CODE* field of the master table, *MFR*, the name of the manufacturer and *MODEL*, which is the name of the model of the airplane.

Now that we have the lookup tables up and running we can start formulating the search we need to answer the query. The first step of our exploration is understanding how many United flight 871 are over time. The search is pretty simple and the output can be seen on Figure 11-8.

```
UniqueCarrier=UA FlightNum=871
| stats count by Year
```

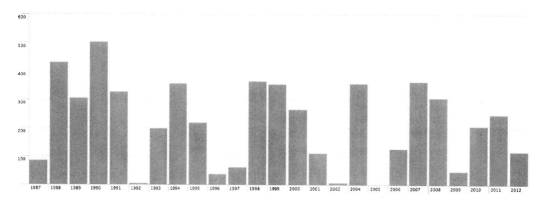

Figure 11-8. *Count of United 871 over time*

As you can see in Figure 11-8, the use of that flight number has not been very constant over the years. Another important piece of information found in the results of the previous search is that over all the 26 years there have been 5,402 flights from United with that number. By hovering over the 2012 column we learn that there have been 116 flights. Rather than test our searches with the most recent month, we will use only the 10 most recent flights. We do this because if there are any issues with our searches it will be a lot easier to figure out the problem with only 10 events than 116. Building the search step by step, the next logical thing to do is the first lookup:

```
UniqueCarrier=UA FlightNum=871
| head 10
| lookup L_Master N_NUMBER as TailNum
```

However, when looking in the left side bar that shows the fields, the *MFR_MDL_CODE* does not show up. We do notice the *Airline* field there, so the automatic lookup we defined earier is working correctly. To make sure that we are not matching the tail number in the lookup table we modify the search as follows:

```
UniqueCarrier=UA FlightNum=871
| head 10
| lookup L_Master N_NUMBER as TailNum
| stats count(MFR_MDL_CODE)
```

The result for the count is zero: there are no matches. We look at the first event that is returned by the query, with a tail number of N441UA, and proceed to look for it in the lookup table using a spreadsheet. We quickly find it, and also the reason the lookup was not working. As it turns out, whereas the tail number of the flight data includes the N at the beginning of the registration, the master file does not, so the solution is to trim the N before we do the lookup. We do this using the *ltrim* function of the *eval* command, which trims the specified string from the left side of the field:

```
UniqueCarrier=UA FlightNum=871
| head 10
| eval N_NUMBER=ltrim(TailNum, "N")
| lookup L_Master N_NUMBER
```

As we are modifying this search, we also take advantage and use the *eval* command to assign the trimmed tail number to the *N_NUMBER* field so it can be directly used by the lookup, thus we can drop the as `TailNum` from the lookup command. In Figure 11-9, you can see that the *MFR_MDL_CODE* field now exists in the left side bar and when we click on it, the popup window shows it has two values.

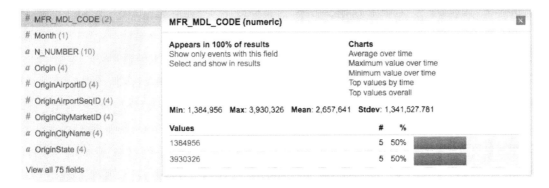

Figure 11-9. *Fields side bar and the* MFR_MDL_CODE *field detail*

We now verify that those values exist in the second lookup table. 1384956 corresponds to a Boeing 757 and 3930326 to an Airbus A320, both of which make sense for a commercial airline like United. Additionally, we manually verify that those 10 events actually match the count shown in Figure 11-9. Another reason we chose to just use 10 events instead of the most recent month is that it is a lot easier to verify 10 events and all the associated relations in lookup tables. Now that we have the values we can add the second lookup to obtain the model name as well as the manufacturer's name.

```
UniqueCarrier=UA FlightNum=871
| head 10
| eval N_NUMBER=ltrim(TailNum, "N")
| lookup L_Master N_NUMBER
| lookup L_Reference CODE as MFR_MDL_CODE
```

Once again, we verify that the expected new fields resulting from the second lookup show in the left side bar, and both *MFR* and *MODEL* do so presenting the expected values we had already discerned, so things are working correctly. Now we have all the information we need so let's go ahead and try to answer the query:

```
UniqueCarrier=UA FlightNum=871
| head 10
| eval N_NUMBER=ltrim(TailNum, "N")
| lookup L_Master N_NUMBER
| lookup L_Reference CODE as MFR_MDL_CODE
| stats values(DestCityName) as Destination by Year OriginCityName MFR MODEL
```

Using the output of this search, which can be found in Figure 11-10 we verify that the most recent 10 events actually match the origins and destinations. As everything matches, we are ready to drop the | head 10 section of the search and run it for all the flight data. When we do this and examine the results we find that in 2004 the airplane used for the Chicago, IL, to Des Moines, IA, segment is a P2004 Bravo, which is a single engine airplane with 2 seats made by a company called Costruzioni Aeronautiche Tecna. This obviously is a mistake in the records and we decide to ignore it. The other strange thing we found in these results is that they start on 1995, as can be seen in Figure 11-11. We know that there are United flights with this number since 1987, the beginning of the flight data, which we learned when we started exploring the flight data for this query and can be seen in Figure 11-8. After some research we found that the events before 1994 do not contain the tail number of the aircraft used on those flights.

Year ⇕	OriginCityName ⇕	MFR ⇕	MODEL ⇕	Destination ⇕
1 2012	Chicago, IL	AIRBUS INDUSTRIE	A320-232	Portland, OR
2 2012	Houston, TX	BOEING	757-222	Portland, OR
3 2012	Philadelphia, PA	BOEING	757-222	Chicago, IL
4 2012	Portland, OR	AIRBUS INDUSTRIE	A320-232	San Francisco, CA

Figure 11-10. *United 871 city pairs with airplane information*

The database lookup tables do not offer an option like the CSV file based lookup tables, where you can define a string to return in the case the minimum match has not been reached (see Figure 11-3).

Unfortunately, visualizing these results is a real challenge. Other than the table that we show in Figure 11-10, we cannot come up with anything decent, not even if we break down the table into sub-groups of columns and try using charts as we did in Chapter 10 (Figure 10-9) with the dashboard sample for the top five airports by airline. The core of the challenge is finding a way to represent the segments or city pairs (origin and destination). The ideal way would be to use a map and draw a line between the city pairs, but we were not able to readily find an application that would provide us with this functionality. However, as we were looking for it, we stumbled on a Network Graph, which is part of Google Fusion Tables at http://tables.googlelabs.com. Here you can upload a CSV file with you data and create a number of visualizations, most of them already available in Splunk.

To export the results of a search in Splunk is extremely simple. Just look for the Export link, which is next to the output mode selection. This can be seen in Figure 11-11 circled in red. When you click on this link, a dialog box will ask you to name the output file and select the format. The choices are CSV, XML and JSON.

Figure 11-11. *Location of the export results link*

We created a CSV file with the results of the United 871 search covering all the flight data, and uploaded it to Google Fusion Tables. We chose to create the Network Graph using the *OriginCityName* and *Destination* columns. The interesting graphic can be seen in Figure 11-12.

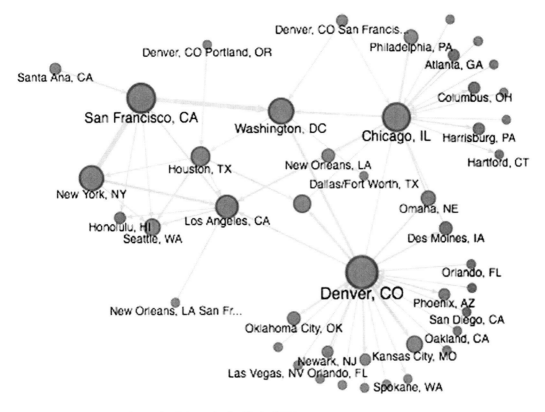

Figure 11-12. *Network graph of city pairs for United 871*

Summary

In this chapter you learned how to create and define lookup tables. We did this for both static files and database tables. We did a couple of searches that used a simple and a rather complex lookup to illustrate its usage. We also discussed how to create lookup tables based on the results of a search, as well as how to export the results of a search as a CSV, XML, or JSON format file.

CHAPTER 12

Analyzing Tweets

In this chapter we will review how to process and analyze tweets. We will go over the Twitter object format and the fields it contains. You will also learn about JSON, the data interchange format used by Twitter to share tweets and other information, as well as how to load data with this format into Splunk, including timestamp considerations. New commands will be used to create a number of searches for handling tweets in a historic way and on a real-time basis.

Twitter is a microblogging service where users can post or read text-based messages that can have a maximum of 140 characters. Each one of these messages is known as a tweet. Twitter is extremely successful and many people and companies use it. Statistics from various sources on the Internet show that for the year 2012 there were about 175 million tweets a day from over 465 million registered accounts, with 1 million accounts being added every day. The top three countries using Twitter are the United States, Brazil, and Japan. The busiest moments of Twitter have seen as many as 25,000 tweets per second.

Users may subscribe to read tweets of other users. This is known as following, and subscribers are called followers. By the end of 2012 the three most popular Twitter users by number of followers were:

- Lady Gaga with 32,987,239

- Justin Bieber with 32,874,390

- Katy Perry with 31,134,772

Tweets are public by default, that is, anybody can see them, but users have the option to send tweets to their followers only and approve who follows them. There is also the ability to send direct messages, which are private messages from one Twitter user to another. These are restricted in the sense that a user can only send a private message to a follower.

There are a few additional concepts around tweets, such as the @reply. When a tweet begins with @username, it means that it is a reply to a specific tweet of the user that sent it. @replies are treated as normal tweets, and as such they are sent to the followers of the user. Because of the 140-character limitation on a tweet, a special lingo has developed over the years. Although we will not delve into this lingo, we should point out that some things such as the @username not only are used anywhere in a tweet but also are becoming quite popular outside of Twitter.

In the lingo you will find hashtags, which also grew as a rather informal way of organizing or categorizing tweets. Users place a # in front of a word they consider represents a category or subject of the content of the tweet. Although this is an interesting way of categorizing content, because of its casual nature it is not always exact. For example, during the presidential elections of 2012 in the United States there were many variations of the same intended category: #Elections, #elections, #Elections2012, elections2012, #Elections12, in addition to some variations with misspellings of the word election.

When you send a tweet from somebody else, it is called retweeting. When you use the Twitter retweet feature, it will automatically add the letters RT and the user name of the author at the beginning of the tweet. Sometimes people will add themselves RT, MT, or MRT to a tweet to indicate they are retweeting or have modified a tweet before retweeting it, but this is not an official Twitter feature and those tweets are missing additional information that Twitter includes on retweets.

When a hashtag increasingly shows up in tweets it is said to be trending, or becoming popular. Trends tend to be short-lived and there are many web sites that offer their views of trending items based on their own algorithms. Twitter itself offers personalized trends, which match the topics in which you have expressed interest. Their algorithm focuses on items of immediate popularity instead of those that have been popular for a while or on a daily basis. Later in this chapter we will offer our own rather simple version of trends.

When a URL is included as part of the content of a tweet, Twitter will automatically check it against a list of potentially dangerous web sites. Whether or not the URL is pointing to a suspicious web site, Twitter will also convert it to a new URL using what they call the t.co service, which is a URL shortener. This way any URL will always be 20 characters long, even if the link is shorter than that. Some Apps will adjust the character count of the message to reflect this, whereas others will not.

Each user has a profile, and they can define their location, time zone, name and screen name, and also provide a picture of themselves, along with a number of other attributes. Each public tweet published by a user is sent to their followers and automatically posted on their profile page, making it searchable in the Twitter web site as well.

In addition to the web site, Twitter offers Apps for various mobile phones and tablets. There are also a large number of third-party Apps or clients that allow you to interact with Twitter. These interactions or integrations are done using the Twitter APIs, which provide simple interfaces to most of the functionality. Additionally, Twitter offers a set of APIs dedicated to the streams that contain the tweets. From all of these APIs we will be using the one that allows us to access the stream called the 1 percent sample, which is a real-time random sample of about 1 percent of the total tweets. The stream containing all the tweets, which is not publicly available, is known as the Twitter firehose.

Tapping the Sample Stream

Using a program we call get1pct.py we collected 24 hours worth of tweets, which we will use later in this chapter to do some basic analysis before we move on to use a Splunk App that allows us to analyze the stream on a real-time basis. You can find the program in this book's download package. To use the program, just call it as follows and it will capture tweets into a file called SampleTweets.out until you stop it:

```
get1pct.py -n twitter_username -p password
```

The Twitter streams use JSON as the standard format. JSON stands for JavaScript Object Notation, and is a text-based standard for data interchange. It provides for defining a multilevel structure by using curly brackets "{}" and multiple values for a field by using square brackets "[]". Ultimately, a tweet using JSON is a list of key values organized in multiple levels. You can see a typical tweet in JSON format in Figure 12-1, which is (inevitably) quite unreadable.

{"contributors":null,"truncated":false,"text":"#Jozan TERBAEK!","in_reply_to_status_id":null,"id":287211786338447361,"source":"<a href=\"http://blackberry.com/twitter\" rel=\"nofollow\">Twitter for BlackBerry\u00ae</a>","retweeted":false,"coordinates":null,"entities":{"user_mentions":[],"hashtags":[{"indices":[0,6],"text":"Jozan"}],"urls":[]},"in_reply_to_screen_name":null,"id_str":"287211786338447361","retweet_count":0,"in_reply_to_user_id":null,"favorited":false,"__time":"Fri Jan 04 15:00:00 +0000 2013","user":{"follow_request_sent":null,"profile_use_background_image":true,"default_profile_image":false,"id":388260118,"verified":false,"profile_image_url_https":"https://si0.twimg.com/profile_images/3032686002/e455b46d716ffe853e208283856d031a_normal.jpeg","profile_sidebar_fill_color":"FFFFFF","followers_count":494,"profile_sidebar_border_color":"FFFFFF","id_str":"388260118","profile_background_color":"FFFFFF","listed_count":2,"profile_background_image_url_https":"https://si0.twimg.com/profile_background_images/725499241/f751417e0808339 5c396723523816409.jpeg","utc_offset":-32400,"statuses_count":27938,"description":"Dear @justinbieber, thanks for being my inspiration,my idol. I will never stop supporting you and loving you. You will always still in my heart forever \u2661 16/7 \u2665","friends_count":163,"location":"All Around The World","profile_link_color":"000000","profile_image_url":"http://a0.twimg.com/profile_images/3032686002/e455b46d716ffe853e208283856d031a_normal.jpeg","following":null,"geo_enabled":false,"profile_banner_url":"https://si0.twimg.com/profile_banners/388260118/1356699609","profile_background_image_url":"http://a0.twimg.com/profile_background_images/725499241/f751417e08083395c396723523816409.jpeg","name":"Justin Bieber \u2661","lang":"en","profile_background_tile":true,"favourites_count":7296,"screen_name":"Aqilaaaah","notifications":null,"url":null,"created_at":"Mon Oct 10 13:16:27 +0000 2011","contributors_enabled":false,"time_zone":"Alaska","protected":false,"default_profile":false,"is_translator":false},"geo":null,"in_reply_to_user_id_str":null,"lang":"nl","created_at":"Fri Jan 04 15:00:00 +0000 2013","in_reply_to_status_id_str":null,"place":null}

Figure 12-1. *A tweet in JSON format*

To explain the basic fields we can expect in a typical tweet we will be formatting them by placing one field (key value pair) per line and indenting the various levels. You need to be aware that the fields do not necessarily come in the same order for every tweet and not all fields appear in all tweets. In other words, if a field does not appear it can be considered to have a null value or to be an empty set. The good news is that Splunk handles these conditions transparently when indexing the data. After formatting the tweet it becomes more legible:

```
{
  "contributors":null,
  "truncated":false,
  "text":"Searchtemplate on a form for a dashboard and eval http://t.co/Cc4gIIDe",
  "in_reply_to_status_id":null,
  "id":287365037834788865,
  "source":"<a href=\"http://ifttt.com\" rel=\"nofollow\">IFTTT</a>",
  "retweeted":false,
  "coordinates":null,
  "entities":{[+}},
  "in_reply_to_screen_name":null,
  "id_str":"287365037834788865",
  "retweet_count":0,
  "in_reply_to_user_id":null,
  "favorited":false,
  "__time":"Sat Jan 05 01:08:58 +0000 2013",
  "user":{[+]},
  "geo":null,
  "in_reply_to_user_id_str":null,
  "possibly_sensitive":false,
  "lang":"en",
  "created_at":"Sat Jan 05 01:08:58 +0000 2013",
  "in_reply_to_status_id_str":null,
  "place":null
}
```

Following the order of the fields in this example, we will comment on those that we consider of interest at this point; a few others will be discussed later in this chapter. If you want a detailed description of each field you can find it on the Twitter web site at http://dev.twitter.com/docs/platform-objects/tweets. The first field of interest is text, which contains the actual tweet. Next is the source field; this one tells us which App or client was used to create the tweet. For the entities field we use the Splunk notation {[+]}, which is not a JSON standard, to signify that this field has additional subfields we will examine later. The __time field does not exist in the tweet; it was created by the Python program and will be discussed in detail when we get to the section that goes over loading the data and timestamp issues. The following field, user, is the one that contains all the information regarding the user that published the tweet in the form of subfields; again we use the {[+]} notation to indicate this. After this, we have the lang field, which contains a two-letter code for the language following the ISO-639-1 standard. Although this field is not explained in the Twitter documentation at the time of this writing, we suspect that it indicates a best guess to the actual language of the tweet and will look at this in detail later in this chapter. The final field of interest on this first pass is created_at, which states the time the tweet was created. All times in this field are based on the UTC/GMT time zone, which is noted by the +0000 string.

Expanding the user field, the typical subfields look like this after being formatted:

```
"user":{
  "follow_request_sent":null,
  "profile_use_background_image":true,
  "default_profile_image":false,
```

```
    "id":121214676,
    "verified":false,
    "profile_image_url_https":
    "https://si0.twimg.com/profile_images/825701927/answers-avatar_normal.png",
    "profile_sidebar_fill_color":"DDEEF6",
    "profile_text_color":"333333",
    "followers_count":656,
    "profile_sidebar_border_color":"C0DEED",
    "id_str":"121214676",
    "profile_background_color":"C0DEED",
    "listed_count":28,
    "profile_background_image_url_https":"https://si0.twimg.com/images/themes/theme1/bg.png",
    "utc_offset":-28800,
    "statuses_count":12923,
    "description":null,
    "friends_count":9,
    "location":"San Francisco, CA",
    "profile_link_color":"0084B4",
    "profile_image_url":
    "http://a0.twimg.com/profile_images/825701927/answers-avatar_normal.png",
    "following":null,
    "geo_enabled":true,
    "profile_background_image_url":"http://a0.twimg.com/images/themes/theme1/bg.png",
    "name":"Splunk Answers",
    "lang":"en",
    "profile_background_tile":false,
    "favourites_count":0,
    "screen_name":"splunkanswers",
    "notifications":null,
    "url":"http://answers.splunk.com/",
    "created_at":"Mon Mar 08 20:09:54 +0000 2010",
    "contributors_enabled":false,
    "time_zone":"Pacific Time (US & Canada)",
    "protected":false,
    "default_profile":true,
    "is_translator":false
},
```

Again, following the order of this example, we explain the most relevant fields for our purpose. The first one is followers_count, the number of Twitter users that are subscribed to receive tweets from this user. The next field is utc_offset, the difference between the UTC/GMT time zone and the user's declared time zone in number of seconds, which is useful for mathematical operations based on the Unix time. Then comes statuses_count, the number of tweets the user has published since the start of the account. The next one is friends_count, the number of users that this particular user is following. Immediately after we find location, the declared geographical location of the user.

The next two fields of interest are the name and the screen name, where the latter is unique and is used by Twitter as the username. The former is not necessarily the real name. In between these two fields you will find lang, which is the code for the declared language of the user. This does not mean that the tweets are in this language. (See the note later in this section.) The created_at field contains the timestamp when the user profile was created or last updated. Do not confuse this field with the one on the higher level with the same name, as that one contains the timestamp of the creation of the tweet. The final field contains the name of the declared time zone of the user.

■ **Caution**   It is very important to note some fields are declared by the user in their App or client and are not always correct. The contents of `location` and `lang`, and to a lesser degree `time_zone`, and `utc_offset`, may or may not have anything to do with the content of their tweets. Our experience has shown that the contents of these fields tend to be unreliable. Keep this in mind when you are doing your analysis.

When a tweet is a retweet using Twitter's native functionality, the tweet will contain a high-level field called `retweeted_status`. This field has subfields with all the associated information of the original user, including the original tweet. When dealing with the text of retweets it is better to use the original tweet when it is close to the 140-character limit. This is because the retweet will chop off the end of the original tweet in favor of the adding the screen name and the letters RT at the beginning of the text of the retweet. Please note that the high-level `retweeted` field does not necessarily indicate that the current tweet was retweeted.

In the sample stream we also find some events that are not tweets but instructions to delete tweets. Twitter does not allow users to modify tweets that have already been published; but it does allow them to delete tweets or retweets they posted. Although we have no use for these records, as they might refer to tweets that are not included in the 1 percent sample, we decided to leave them as is in the file we will index into Splunk. These delete events look like this once formatted:

```
{
  "delete":{
    "status":{
      "user_id_str":"400166241",
      "user_id":400166241,
      "id":139980458426499072,
      "id_str":"139980458426499072"
    }
  }
}
```

Loading the Tweets into Splunk

We modified the file generated by the `get1pct.py` program to have exactly 24 hours worth of data. Our sample data goes from Friday, January 4th, at 15:00:00 to Saturday, January 5th, until 14:59:59. The file contains 4,815,166 events and is about 12 GB in size. Because of restrictions in Twitter's Terms of Service, the actual tweets cannot be distributed. We suggest that you run the Python program and collect a sample with a size to fit the indexing restrictions of your Splunk license.

The first challenge we have to deal with is defining the correct field for the Splunk timestamp. The issue is that in Splunk we can only use regular expressions to define the field or fields that contain the timestamp. There are two fields named `created_at`, one at a high level, for the actual tweet, which is the timestamp we want, and another one within the `user` object. Whereas the fields are clearly different for a human or a JSON interpreter, they are the same for a regular expression, as there are no guarantees on the order of the fields and that all the fields are present. We cannot use the same approach that we used in Chapter 9, where the timestamp was spread over two fields, but we knew exactly which fields we wanted based on counting commas. This is the reason why the Python program creates a new field called `__time` based on the high-level `created_at` field. Please note that this new field has two underscores before its name, so it is not confused with the internal Splunk field `_time`, with only one underscore. With this new field it becomes really easy to define the timestamp in Splunk.

Out of abundant caution we decide to create a small test file with just the first 100 tweets of the sample data. With this we follow the same steps we used in Chapter 9 under the "Pre-Processing the Flight Data" heading to define a new source type, which specifies a timestamp location for the tweets:

- Define a new input file and preview the data before indexing

- Define a new source type

- Select "adjust timestamp and event break settings"

- Select "Every line is one event" and click on Apply

- Select the Timestamps tab

- Under Location select "Timestamp is always prefaced by a pattern" and type the following: "__time":" (remember, there are 2 underscore characters before the word time)

- Under Format select Specify timestamp format and type the following: %a %b %d %T %Z %Y, where:

    - %a is the name of the day of the week, either abbreviated or full name

    - %b is the name of the month, either abbreviated or full name

    - %d is the day of the month; leading zeroes permitted, but not required

    - %T is the time as %H:%M:%S

    - %Z is the time zone or name or abbreviation

    - %Y is the year including the century

- Select "Specify time zone" and choose from the pull down menu "(UTC) Coordinated Universal Time," as we know that all of the timestamps are in this time zone.

- Click on Apply. Now you should see the timestamp highlighted on each event or tweet and the correct timestamp on the second column. You will also see warnings for the delete events because they do not have a timestamp. Splunk will use the timestamp of the previous event for these ones.

- Click on "Continue" and specify the name of the new source type. We saved it as "twitter."

- You can now start indexing the test data

Once we have these 100 events indexed into Splunk we can do a quick verification and confirm the loading operation worked correctly. In Figure 12-2 you can see a partial screen shot of the output when listing all the events. There are a number of things to note in this figure. In the first place, even though you cannot see it, at the top of the results it states that there are 100 events over all time, therefore confirming it has indexed all of the events we provided. The second item to notice is that the left-side bar presents a number of fields, which are similar to those we just reviewed, thus Splunk is understanding the JSON format. Note that the way Splunk represents a multilevel field is in the form of a path using a dot (.) to separate the level. For example, the created_at subfield for the user field is referenced as user.created_at and this is the way that we will be referring to fields when we create searches using data that comes in JSON format.

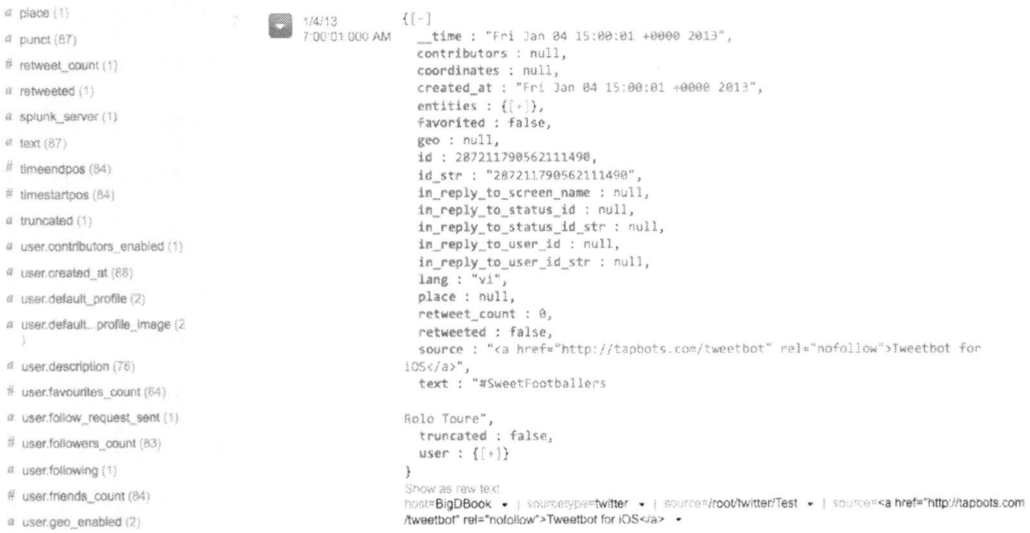

Figure 12-2. *JSON events listed in the user interface*

The event itself presents the fields in alphabetical order with the correct Splunk timestamp, which matches the
__time field. Additionally, the first event is the last one in our test file. As you may remember, Splunk presents the
events starting from the most recent to the oldest, so this is correct as our test file goes from earliest to most recent.
You will also notice that the events start with {[-]. If you click on the minus sign, Splunk will collapse the information
presented at that level, the opposite effect to when you click on the plus sign in {[+]}, which expands to present all the
subfields of the chosen field. Notice that the entities, retweeted_status, and user fields offer the option to expand
to view their subfields. At the bottom of the listed events there is a link to show the event as raw text. When you click
on this link, it will present the event in the same way as shown in Figure 12-1.

Another thing to notice is that fields in the left-side bar such as id and user.created_at have a count of 88. This is
because of the 100 events, 88 are tweets and 12 are delete events. Finally, the source field on the side bar is presenting
a count of 31, which looks wrong as there is only one source file. When you look at the end of the listed events you will
notice that there are two "source" items, one that correctly presents the name of the file that contains the test events
we just indexed and another one with the contents of the source field from the tweet. Admittedly this is confusing, as
Splunk is dealing with two fields that have the same name but is able to distinguish some level of difference between
them. We will review this issue in detail when we introduce the spath command later in this chapter.

Other than the confusing situation with the source field, the indexing of the test file with the tweets seems to have
been successful and all is in order. With this we move to delete the 100 events we just loaded and index the full file
with the almost five million tweets.

A Day in Twitter

Using the file with the almost five million events we just indexed we will do a historic analysis of the tweets. To get
familiar with the field notation for JSON let's start by counting the number of delete events within the file we collected:

```
* | stats count(delete.status.id)
```

It is tempting to just search for the word "delete", but if the word is part of any field of a tweet we would get an incorrect count. That is why we chose to count a field that is unique within the delete structure. Any field will do; we chose the `id` subfield, so that when the field exists in the event we are certain it is a delete event. The total count is of 493,382, which is a little over 10 percent of the stream; that is a lot of people changing their minds.

Next we would like to know the number of tweets by hour of the day. Independently of the format of the data, this search is very simple because all we have to do is count events:

```
* | stats count by date_hour
```

The `date_hour` field is a default field available in Splunk as part of a group of fields called date time, which provide a better granularity to event timestamps. These also include `date_minute`, `date_month`, `date_second`, `date_wday`, `date_year` and `date_zone`. The output of this command in the form of a chart can be seen in Figure 12-3. Note that the y-axis starts at 140,000 and not at 0.

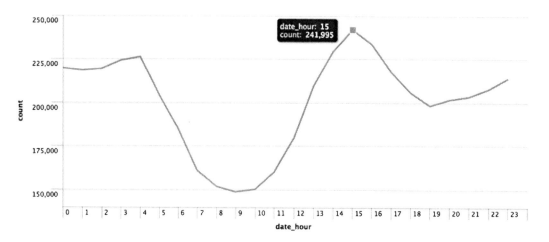

Figure 12-3. *Number of tweets by hour of the day*

You must remember that Twitter is a worldwide phenomenon, so when we count tweets they are happening all around the world. As mentioned earlier, the timestamp is based on the UTC time zone. The hour block with the most tweets is 3 PM with 241,995. To put that in context, it is 3 PM in London, 10 AM in New York, 7 AM in San Francisco, and midnight in Tokyo, just starting the next day. Remember also that our sample data starts at 15:00 of Friday, January 4th, so the chart should really start at 15 and not at 0 to provide a realistic view. Unfortunately, the charting options of Splunk do not offer the ability to decide on which value to start an axis.

On a slightly different search, we look at the number of tweets by time zone using both fields available, `time_zone` and `utc_offset`. The fields are part of the user object and, as mentioned earlier, they are user defined, or use the default value of the App they are using. If correct, it usually refers to the main geographical location of the user. We have found that many people get confused when dealing with both of these fields. A time zone can have many names—after all, they extend all the way from the North Pole to the South Pole—so a top five of the time_zone field is about the top names, whereas the utc_offset is about the actual time zone. Table 12-1 shows the top five tweets by either field.

Table 12-1. *Tweets per time zone*

Position	time_zone	Percentage	utc_offset	Percentage
1	Null	27.64%	Null	27.64%
2	Central (US & Canada)	6.48%	-5 hours	9.21%
3	Pacific (US & Canada)	6.38%	-6 hours	7.31%
4	Eastern (US & Canada)	5.20%	+9 hours	7.30%
5	Tokyo	4.76%	-3 hours	6.45%

Surprisingly the most popular time zone for both fields is null with over 27 percent. The time zone name with the most tweets is the central United States; however, the actual time zone with the most tweets is –5 hours from UTC, which corresponds to the following names: Eastern Time (US and Canada), Indiana (Eastern), Bogota, Lima, and Quito. To obtain a list of the time zone names by UTC offset, just use this search:

```
* | stats values(user.time_zone) by user.utc_offset
```

You will find that some time zones have a large number of names associated with them; for example, +1 hour has 20 names, basically the names of the capitals of the European countries. Please note that even though most offsets are based on whole hours, there are a few that are offset by 30 or 45 minutes, most notably India with 5 hours and 30 minutes from UTC and four time zone names: Chennai, Kolkota, Mumbai, and New Delhi.

Now let us have a look at the language of the tweets. The situation is rather unclear, as we are dealing with two fields, one at the high level that is not documented by Twitter at this writing (but we suspect is a best guess at the language of the tweet), and another language field that is part of the user structure. The latter is the declared language of the App the user has, or it can also be the default of that App. We ran a search for the first 20 events where these two fields have different contents and there are eight events that fit this profile. If we count that there are two delete events in these 20, the ratio is almost 50 percent. Looking carefully at the language of the actual tweet (the text field) of these eight events, five had the right language, two were wrong, and one did not have enough text to guess a language at all, as all it contained was "@username (Y)."

Using the following search to find how many events have a different language on the text as compared to the one defined in the user profile we get 1, 271, 427 events distributed over 801 different combinations of languages. That is almost 30 percent of the total tweets. Interestingly enough, about 21 percent of these tweets have English defined as the user language, but the tweet language is flagged as Indonesian.

```
* | rename user.lang as tweet_user_lang
| where tweet_user_lang != lang
| stats count by tweet_user_lang, lang
```

▓ **Caution** Be careful when using the eval and where commands when working with JSON-based events. They will interpret the dot (.) as the concatenation operator. To avoid this, you can either enclose the JSON field path in single quotes (') or rename the field. We consider renaming the field a best practice.

We also noticed that from these results there are about 81,000 events that have the code "und" for the tweet language, which we assume means undetermined. Trying to guess the language of some text, especially when it is as short as a tweet, is a difficult proposition, especially nowadays when mixing languages is becoming more common. Combinations of languages with interesting names such as Spanglish (Spanish and English), Portuñol (Portuguese and Spanish),

Frarabic (French and Arabic), Hinglish (Hindi and English), and others make it even more challenging. At this point all we can say is that you should handle the analysis of tweet languages with extreme care.

Another field with questionable content is location. This field is provided by the user and we have seen it go from the name of a city to "in the heart of @username" and anything in between and beyond. Location is not really a field that can be used for any serious analytics. Some of the Apps for Twitter offer the ability to geotag their tweets, that is, attach the location of the tweet in the form of latitude and longitude. The field for this is coordinates. There is another field called place, which allows for an address and a bounding box (a polygon of geocoordinates). At this time these fields are not widely used. We did a quick search and found that about 1.5 percent of all the tweets in our sample had geolocation information. We cover how to work with geolocations in Chapter 13.

There are three fields that provide interesting information and are rather straightforward: followers_count, friends_count, and statuses_count. These are part of the user object and are calculated by Twitter for each tweet. Searches for the user with the most followers, or following the most users, or the one with most tweets are extremely simple. The only thing you need to be very aware of is that these metrics are extremely dynamic, so you must use the most recent tweet to obtain the most accurate results. As usual, we suggest that you be very careful when formulating the searches. There is a big difference between the user with the most tweets in the period of time that is being analyzed and the user with the most tweets over time. In the first case, you can use the following search:

```
* | top user.screen_name
```

In the second case, you could use a search like:

```
* | stats max(user.statuses_count) as max_count by user.screen_name
| sort - max_count
```

In the first search, we just count the events by user for the data we have. In the second one, we look for the biggest count of tweets for each user, as this number changes with each tweet. For the first search we get favstar_bot, a robot, as the user with the most tweets at 31, whereas for the second search we get Yougakudan_00, another robot in Japan, with well over 36 million tweets.

Searches to count the number of followers and friends are similarly simple. When we do the search by the number of followers, none of the top three users we mentioned at the beginning of the chapter show up in the results. If a user has not tweeted during the period of our sample, that user will not show in the results, and none of those top three tweeted during that period. The top user we obtain is Kim Kardashian, with 17,084,687 followers.

Most Popular Words

For our next search we want to find the most popular words in a tweet. We will use a couple of new commands to achieve this. The main idea is to break down the text field into a multivalue field, where each value is an individual word. We consider a word to be anything enclosed by two spaces. Once we have done this we can go ahead and count the words. The first command, makemv, takes a field and converts it into a multivalue field using a space as the default delimiter. As Splunk sees everything as events, we will use the mvexpand to convert each value of the multi-value field into separate events. We tested the search first with only one tweet, so that it would be easy for us to count the words and make sure that it is working correctly. The next step was to do the search with the first 200 events to get an idea of what the results look like:

```
* | head 200
| makemv text
| mvexpand text
| top text
```

The output of this search can be seen in Figure 12-4. Not surprisingly, the most popular word for these first 200 tweets is RT, with over double the count of the word "to", which came in the second place. You can also see that most of the words are prepositions, nouns, or pronouns in both English and Spanish, which can generally be considered noninformational. We are looking for some more substance than that. One way to do this is by creating a list of words to exclude from the results. This is typically called a stop word list, and many can be found on the Internet. One thing that we have to consider is the language of the stop word list. We created a rather simple stop list with only 44 English words and will focus only on tweets in English.

	text ⇕	count ⇕	percent ⇕
1	RT	41	2.321631
2	to	18	1.019253
3	a	12	0.679502
4	...	10	0.566251
5	the	9	0.509626
6	is	9	0.509626
7	of	8	0.453001
8	de	8	0.453001
9	I	8	0.453001
10	que	7	0.396376

Figure 12-4. *Most popular words in the first 200 tweets*

To illustrate how to formulate a search that excludes words from a stop list, we will use a lookup table. You already saw how to use lookup tables in Chapter 11, but this time we will use them slightly differently. The first step is to create a simple file that has one word per line and remember to include a header line. For this example, we will use "Word" as the name of the field in the header line. The next step is to define a lookup table, which we call StopWords.csv. This can be found in the download package of this book. Note that we use the csv extension even though the file technically does not contain comma-separated values, only one column. There is no need to create a lookup definition as the inputlookup command will take either a filename or a tablename as an argument. This command loads the contents of the lookup table as search results. Putting it all together the search look like this:

```
lang=en
| makemv text
| mvexpand text
| search text NOT
  [
    | inputlookup StopWords.csv | rename Word as text
  ]
| top text
```

As you can see, the search is based on a subsearch, which loads all the stop words and renames the field to text. The stop words are then used in a search with the words we found in the previous search and combined with a Boolean "not", meaning that we do not want to count words with the top command that show up in both. Please note that the "not" has to be uppercase for Splunk to understand it as a Boolean operator. Another thing we have to mention is that you cannot rename the text field in the first part of the search because the new field by itself will not be available to the second search command.

As we are playing with words, we will play a little with the visualization of the results. We used IBM's Many Eyes web site, where you can load your data and experiment with many visualization types. After playing around with the results of the previous search we especially liked two different visualizations: a bubble chart, which we present in Figure 12-5, and a word cloud, seen in Figure 12-6. This fun web site can be found at the rather strange URL http://www-958.ibm.com. Just be aware that whatever data you upload will be made public.

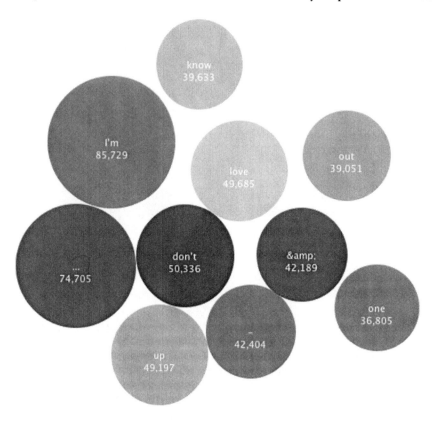

Figure 12-5. *Bubble chart of the top 10 words in English tweets*

Figure 12-6. *Word cloud of the top 50 words in English tweets*

Obviously the stop word list we used can be further improved as the top 10 words list still includes many noninformational words. The bubble chart works fine for the top 10 values, but it produces a rather skimpy word cloud, so we decided to use the top 50, which makes for a more complete word cloud. With this we conclude the section where we examine the Twitter information from a historic perspective. Next we move on to analyze real-time trends.

Real-Time Twitter Trends

In Splunkbase, Splunk's App library, you will find an App that provides a scripted input for Splunk that automatically extracts tweets from Twitter's 1 percent sample stream and has a flashy dashboard:

`http://splunk-base.splunk.com/apps/56296/twitter-for-splunk`

The App was originally written by David Foster of Splunk. It is available for free and is covered by the Apache 2 license. You must have a valid Twitter username to be able to use this App; also, as we saw earlier, the sample Twitter stream generates about 12 GB worth of data a day, so make sure your Splunk license allows to index that amount of data. To install the App from within Splunk follow these steps:

- Go to the Manager, choose Apps, and then click on "Find more apps online"

- Search for the Twitter App; the search box is on the upper right part of the screen, just type "twitter"

- Locate "Twitter for Splunk" and click on "Install free"

- The App will be downloaded and it will ask you to set it up. All you have to do is provide your Twitter account name and password. If they are correct, the App will start running and after collecting data for a couple of minutes it will present a dashboard with six panels like the one shown in Figure 12-7:

  - Top Hashtags—Last 15 minutes

  - Top Mentions—Last 15 minutes

  - Tweet Time Zones—Last 15 minutes

  - Top User Agents—Last 24 hours

  - Tweet Stream (All Users)—Real time, last 30 seconds

  - Tweet Stream (First Time Users)—Real time, last minute

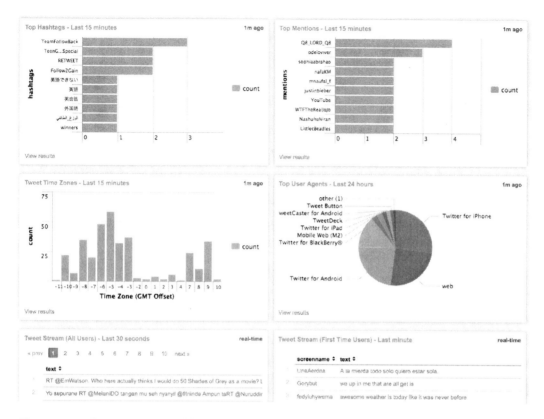

Figure 12-7. *The Twitter App dashboard*

We will review the searches that make up each one of these panels, but first a quick comment on the way the data is collected. The App includes a Python program that collects the data directly from the Twitter 1 percent sample stream and indexes it into Splunk in an index called "twitter". This is called a scripted input, where instead of having Splunk index a file (one shot) or monitor a directory or a TCP/UDP port, it obtains the data from a program. If you are interested in reviewing this program, you can find its location by going to Manager ➤ Data Inputs ➤ Script. There are two versions of the program, one for Unix and another for Windows. We will not review either program as that is out of the scope of this book. On a related note, the program we used to collect the 24 hours' worth of data for use at the beginning of this chapter, get1pct.py, is a modified version of the program provided by this App.

The first panel in the dashboard presents the top hashtags in the last 15 minutes. Based on the search we built for the most popular words earlier in this chapter, we can formulate this search as:

```
index=twitter
| makemv text
| mvexpand text
| eval hashtags = mvfilter(match(text, "#+"))
| top hashtags
```

In this case we use the eval command to extract the hashtags. The match function compares the word with the regular expression "#+", which stands for "starts with a # and has one or more characters." The mvfilter function filters a multivalued field, such as text, based on a Boolean expression like the one we used with the match function.

When you put it all together, the eval command matches the words that start with a "#". This is pretty simple, but Twitter has an entities object associated with each tweet that makes it even simpler. This object contains metadata extracted from the tweet itself (the text field) such as:

- Hashtags, which have been parsed out of the tweet text.

- Media, which is a comprehensive object that represents media elements uploaded in the tweet, such a photos.

- URLs, which contain the original URLs in the tweet and the corresponding URL converted to the t.com form.

- Mentions, which are the screen names of other users mentioned in the tweet, that is, names starting with a "@" and are actual Twitter users.

Each of these elements includes the location within the tweet represented as the beginning and the end of the element. There are different ways of counting and presenting the element depending on which one it is. Details can be found in the Twitter documentation. The following is an example of an entities object with a couple of hashtags, a URL, and a mention:

```json
"entities":{
  "user_mentions":[
    {"name":"Twitter API",
     "indices":[4,15],
     "screen_name":"twitterapi",
     "id":6253282,
     "id_str":"6253282"}
  ],
  "hashtags":[
    {"indices":[62,66],
     "text":"lol"}
    {"indices":[79,85],
     "text":"rotfl"}
  ],
  "urls":[
    {"indices":[32,52],
     "url":"http:\/\/t.co\/IOwBrTZR",
     "display_url":"youtube.com\/watch?v=oHg5SJ\u2026",
     "expanded_url":"http:\/\/www.youtube.com\/watch?v=oHg5SJ\u2026"}
  ]
}
```

Using the Twitter entities object, the search to find the top hashtags changes from the previous one to:

```
index=twitter
| rename entities.hashtags{}.text as hashtags
| fields hashtags
| mvexpand hashtags
| top hashtags
```

As mentioned earlier, we suggest as a best practice that multilevel JSON fields be renamed to avoid problems with the eval and where commands, even when they are not used in the search. Then there is the convenience of typing a shorter field name, which can also be used as a label for the *y*-axis of the chart. Using entities.hashtags{}.text is a Splunk wildcard specification for a JSON format field to get the text of all the hashtags as opposed to specifying individually entities.hashtags{0}.text, entities.hashtags{1}.text, and so on. The fields command is used to optimize the search. As there are so many fields in a tweet, we are excluding all of them but the hashtags, which is all we care about in this search. The mvexpand command expands the values of a multivalue field into separate events so that the top command can count them. If you wanted to create a panel that monitors a hashtag of your interest for, say the last minute, a possible search could be:

```
index=twitter
| rename entities.hashtags{}.text as hashtags
| fields hashtags
| mvexpand hashtags
| where like(hashtags, "your_hashtag")
| stats count
```

You can change the visualization from a chart to a radial gauge, because you are only presenting a single number, and you specify the time range to be "from -1m@m to now." An example of the resulting panel can be seen in Figure 12-8.

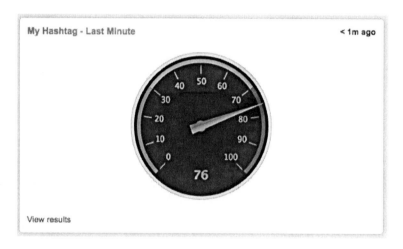

Figure 12-8. *Counting a specific hashtag*

The panel that presents the top mentions uses the same search as the top hashtags, except that it changes the field names to be entities.user_mentions{}.screen_name as mentions. The time zones panel uses the following search:

```
index=twitter
| rename user.utc_offset as z
| search z=* z!=null
| eval z=round(z/3600)
| stats count by z
| sort +z
```

As you saw earlier in this chapter, a high percentage of tweets have null time zones, so to avoid a problem in the calculation the search command is used to weed out all those tweets. Remember the implicit AND Boolean in the search command. The calculation breaks down the time zones by full hours. As discussed earlier, there are a few time zones that are offset by 30 and 45 minutes. Dividing by 3,600 will have the count of the odd time zones roll into the next full hour. We tried visualizing the results with the eval command removed, but there were so many time zones that the different types of charts we tried were illegible, so we can call this one the pragmatic time zone chart.

The next panel, which shows the top user agents, will allow us to introduce a couple of new commands. A user agent is an App or client that is used to create a tweet. This information is quite valuable for marketing-oriented companies, as it not only presents the actual App but it usually includes the platform it runs on. The field that contains this information is source. As you might remember, we were having some conflicts between this field and the Splunk default field source when we were verifying the loaded data after indexing. We will clarify these conflicts as we explain the search. The content of the Twitter source field is slightly different to the other fields, as in most cases it will contain the URL where you can download the App. The only exception is when the Twitter web site is used, in which case the value is simply "web". This is a typical example:

```
"<a href=\"http://twitter.com/download/android\" rel=\"nofollow\">Twitter for Android</a>"
```

The contents of the source field for other Apps have the same structure, the link destination for the download and the name of the App. All we are interested in is the name of the App, as in this example "Twitter for Android", so we have to come up with a regular expression that extracts it, but we have to consider the exception case: "source":"web". We can break the contents of field in three parts: the first one goes from the beginning up to the actual name of the App, the second one is the name of the App, and then the rest. Based on this, the regular expression can be:

```
(<[^>]*>)?([^<]*)
```

Some of the regular expression concepts were covered in "Modifying the Timestamp Processor" in Chapter 9. The first group of the regular expression, which is enclosed in between parenthesis, indicates that it starts with a "<" followed by any character except for a ">" zero or more times, and this group can happen zero times or once. The second group captures the name of the App by simply stating that it takes any character, except for a "<" zero or more times. We don't have to worry about the rest. Now we can have a look at the search:

```
index=twitter
| spath source
| fields source
| rex field=source "(<[^>]*>)?(?<source>[^<]*)"
| top source
```

The spath command allows you to extract fields from structured events based on JSON or XML. We have not needed to use this command in this chapter because the Twitter events are fully structured JSON; therefore, Splunk made all the fields automatically available. This command is mainly used when events include some parts that are structured as JSON or XML (there are some log files that combine plain text with XML structures in the same event). In this particular case, we are using spath to disambiguate the source field in the event from the Splunk default field source. By using the spath command, we force Splunk to use the source field we want. Once again, we use the fields command to improve performance by limiting the rest of the search to this field only.

The rex command is used to extract fields based on a regular expression. Note that we have modified the regular expression by adding ?<source> at the beginning of the second group. In the rex command this means that the match of that group will be placed in the field called source. In normal regular expression parsers the matches of each group would be available as $1, $2, and so on. In the Splunk rex command the matches of a group can be placed in a field that is defined at the beginning of the group using the ?<field_name> syntax. This is a pretty handy Splunkism.

The pie chart that displays the results of this search can be seen in Figure 12-7. It shows that the majority of the clients are the web, with over 26 percent, followed closely by the Apps for the iPhone and Android, each with about 23 percent, and then the BlackBerry App with about 16 percent. With the rise of Android and iOS devices, we decide that to better understand this phenomenon we will analyze the geographic location of the BlackBerry devices based on the time zone names. For this, we formulate a search to run it on the 24-hour file that we processed earlier in this chapter as we have more tweets than we have collected using the real-time App. Excluding the tweets without a time zone, which we saw earlier account for over 27 percent, we look at the time zone names with 15,000 or more tweets. These are the top six, as shown in Table 12-2. Our original expectations were that the majority of the BlackBerry phones would be concentrated in some foreign region, but as you can see three of the top six are in the United States, and over half of the top six are in the United States.

Table 12-2. *Top 6 BlackBerry time zones*

Time Zone	Count
Pacific Time (US & Canada)	74,607
Bangkok	47,761
Jakarta	34,064
Central Time (US & Canada)	20,636
Amsterdam	17,743
Hawaii	17,231

The fifth panel of the dashboard presents the text of the tweets for the last 30 seconds on a real-time basis. This is done by specifying the time range as "from rt-30s to rt." The search is extremely simple:

```
index=twitter text=*
| table text
| sort - _time
```

By specifying text=* we ensure that the tweet text field contains something, whereas the table command presents the selected fields as a table. This is finally sorted in decreasing order using the Splunk internal field _time (with one underscore), which contains the timestamp and has nothing to do with the __time field (with two underscores) that was created by the Python program that taps into the Twitter sample stream. We use _time because it is more efficient than using any other available field. Please note that fields that start with a single underscore symbol (_) are usually meant for Splunk internal use, but are available to the user. The search for the panel that presents the first tweets is also very simple:

```
index=twitter user.statuses_count=1
| rename user.screen_name as screenname
| table screenname text
| sort - _time
```

Suppose that as well as keeping track of the count of a specific hashtag, which we showed earlier in this chapter, you also want to display the tweets that contain it. The search would be a slight variation of the one used to count the hashtag:

```
index=twitter
| rename entities.hashtags{}.text as hashtags
| rename user.screen_name as account
| fields account,hashtags,text
| mvexpand hashtags
| where like(hashtags, "your_hashtag")
| table account, text
| sort - _time
```

We rename the user screen name as part of the best practices and restrict the fields to only those we use to improve the performance. Once we find a tweet that contains the hashtag we are looking for, we display the tweet next to the name of the user.

Summary

In this chapter we reviewed the data structure of the Twitter 1 percent sample stream, how to collect it, and how to process it either on a real-time basis or from a historic perspective. We went over the timestamp issues for indexing tweets and discussed some key fields and their contents. We did a number of interesting searches that allowed us to introduce a few new commands, including some that are specialized in handling multivalue fields. We also used an external tool to create visualizations that are more appropriate for dealing with words and tags.

CHAPTER 13

■ ■ ■

Analyzing Foursquare Check-Ins

Location-based services—especially on mobile phones—are growing dramatically. With an ever increasing number of GPS-enabled smartphones in use, location data is adding a whole new dimension of applications, including advertising, recommendations of nearby businesses, and weather. In this chapter, we will focus on some of the basic concepts related to geographical (geo) location. We also discuss the intricacies of handling multiple time zones. For this, we will use the Foursquare service as the base of our examples.

Foursquare can be described as a location-centric social networking game. It allows you not only to check in at locations you want to share with friends but also to earn rewards, most of them virtual. Every check-in will give you points. With points you earn badges, which can be described as formal bragging rights. The badges can go from the simple, a newbie badge for checking in for the first time at a venue, to an explorer badge for checking in to 25 different venues. You have some more complicated badges, such as the JetSetter, which is attained after checking into five different airports, or the Zoetrope, which is obtained at the tenth movie theater check-in. We should not forget those badges related to food such as the Pizzaiolo, with 20 check-ins at pizza restaurants, and the Foodie badge, which you get after five check-ins at Zagat rated spots. The badge list is very long and always changing, with many badges promoted by businesses, such as the Zagat-based badge.

Another bragging right considered important is becoming the mayor of a venue, which is attained by being the person that has checked in the most in the last 60 days into a specific location. This status can be hotly contested in very popular venues. Another feature of Foursquare is the ability of users to leave tips about or reviews of a venue, which can then be seen by other users.

We haven't completely described all of the features offered by Foursquare, but you can quickly realize the commercial potential it has. From businesses engaging with customers by giving the mayor of a venue a free drink or a special discount, all the way to complete marketing campaigns that promote a new product or location, the possibilities are endless. The Foursquare web site has some good case studies on how businesses are reaching out and engaging with people.

In a similar fashion, there are a growing number of mashups, that is, applications that combine data or functionality from two or more sources to create a new service. In this case, the mashups take Foursquare data and combine it with other services or applications, such as Twitter, LinkedIn, YouTube, Flickr, and others. There are also applications that are built on top of the Foursquare API, for example the 4sqwifi App, which finds nearby venues with Wi-Fi and provides their passwords. This particular App takes advantage of the user-generated venue tips to provide some of the information.

The Foursquare API comprises two basic functionalities, the core API and the real-time API. The core API allows an application or mashup to do all the things that can be done with the mobile applications provided by Foursquare or its web site. The real-time API provides two kinds of push notifications, a venue push, which notifies venue managers when users check into their venues, and a user push, which notifies developers when the users of their applications check in anywhere.

The Check-In Format

A check-in generated by the real-time API is based on JSON. As you will remember from Chapter 12, when we first introduced JSON, it provides for defining a multilevel structure by using curly brackets "{}" and multiple values for a field by using square brackets "[]". Also, remember that the fields do not necessarily come in the same order all the time and not all fields appear in all check-ins. At a high level, a typical check-in generated by the user push API is almost identical to one generated by the venue push API. Using the headquarters of Foursquare as the sample venue for a check-in, the format looks like this:

```
{
"checkin": {
    "id": "4e6fe1404b90c00032eeac34",
    "createdAt": 1315955008,
    "type": "checkin",
    "timeZone": "America/New_York",
    "user": {[+]},
    "venue": {
        "id": "4ab7e57cf964a5205f7b20e3",
        "name": "foursquare HQ",
        "contact": {
            "twitter": "foursquare"
        },
        "location": {[+]},
        "categories": [+],
        "verified": true,
        "stats": {[+]},
        "url": http://foursquare.com
    }
}
}
```

We will review the most important fields of this structure. To begin with, there is the time the check-in was created, which is a Unix timestamp, and then the name of the time zone. These are two important fields, which we will be focusing on when discussing how to load the data into Splunk. After that, using Splunk's notation to indicate that a field can be expanded into another level, is the user field. When expanding this field, these are the contents:

```
"user": {
    "id": "1",
    "firstName": "Jimmy",
    "lastName": "Foursquare",
    "photo": "https://foursquare.com/img/blank_boy.png",
    "gender": "male",
    "homeCity": "New York, NY",
    "relationship": "self"
},
```

The gender field can have three values: female, male, and none. Next is the venue field, which is the core of the check-in, specifically the location and categories fields. Let's review the location field:

```
"location": {
    "address": "East Village",
    "lat": 40.72809214560253,
```

```
    "lng": -73.99112284183502,
    "city": "New York",
    "state": "NY",
    "postalCode": "10003",
    "country": "USA"
},
```

It is really nice that Foursquare provides so much location information, especially when a mobile device only sends the latitude and longitude of the location from which you are checking in. The rest of the information is generated by a process called reverse geocoding, which we explain later in this chapter. The categories field expands as follows:

```
"categories": [
    {
        "id": "4bf58dd8d48988d125941735",
        "name": "Tech Startup",
        "pluralName": "Tech Startups",
        "shortName": "Tech Startup",
        "icon": "https://foursquare.com/img/categories/building/default.png",
        "parents": [
            "Professional & Other Places",
            "Offices"
        ],
        "primary": true
    }
],
```

Unlike Twitter, Foursquare does not provide a firehose stream with all the check-ins, or even a sample stream of check-ins. This situation puts us in a bind, as we want to illustrate how to work with this data. One option is to create an application that uses the Foursquare API focused on either offering some user-focused features or functionality for managing venues, but this is out of the scope of this book. Instead, we decided to create a set of fake, but very realistic, check-ins based on real venues, which will be used as the base of this project. To simplify things, the structure we used for the data we created is a subset of the official Foursquare structure previously described:

```
{
"checkin": {
    "id": "4e6fe1404b90c00032eeac34",
    "createdAt": 1315955008,
    "timeZone": "America/New_York",
    "user": {
        "gender":"male"
    },
    "venue": {
        "id": "4ab7e57cf964a5205f7b20e3",
        "name": "foursquare HQ",
        "location": {
            "lat": 40.72809214560253,
            "lng": -73.99112284183502
        },
        "categories": {
            "id": "4bf58dd8d48988d125941735",
```

```
        "name": "Tech Startups",
        "fullName": "Professional & Other Places:Offices:Tech Startups",
        "icon": "https://foursquare.com/img/categories/building/default.png",
      }
    }
  }
}
```

Note that we combined the `parents` field and the `pluralName` field of the original format into a new field we call `fullName`, separating each value with a colon (:). The categories defined in Foursquare can have as many as three levels, as can be seen in this example. A comprehensive list of Foursquare categories can be found at `http://aboutfoursquare/foursquare-categories`.

Using this simplified format, we created two sets of check-ins, one for the New York area, with 541,238 check-ins, and another one for the San Francisco Bay area, with 271,662 check-ins. Both data sets extend over a period of one week, from Saturday, April 6, at midnight UTC until Friday, April 12, at 23:59:59 UTC. These files, along with a smaller version of them, that only contain 10,000 check-ins each, are available as part of the download package of this book.

The fact that the check-ins use the Unix timestamp, which implicitly means they use the UTC time zone, will allow us to discuss some interesting issues related to handling time zones. We will do so later in this chapter, but first we will discuss reverse geocoding.

Reverse Geocoding

Reverse geocoding is the process of taking a point location defined by latitude and longitude and converting it into a human-readable address. As simple is this might look, it is a rather complex and compute-intensive process. Although we will not go into the details of how to create an application to do this, reverse geocoding is important enough that it deserves we explain the basics.

Latitude is defined as the angular distance of a place north or south of the equator, that is, the imaginary lines that run parallel to the equator. When north, the distance is indicated in positive numbers or simply with the word `north`, whereas negative numbers or the word `south` indicate a distance south of the equator. In a similar fashion, longitude is defined as the angular distance of a place east or west of Greenwich, England. This defines a set of imaginary lines that run from the North Pole to the South Pole called meridians, and the one at Greenwich is known as meridian zero. To the east of meridian zero the word `east` or a positive number is used, as is the word `west` or a negative number used to denote a position to the west.

Note that both definitions refer to angular distance. This is because planet Earth is a sphere, although not perfect as it's a bit flatter on the poles; thus, specialists describe it as an oblate spheroid. Latitude and longitude are expressed in degrees, minutes, and seconds, so the center of New York City is located at 40° 44' 27.6498", –74° 0' 14.0616," and the center of Cape Town in South Africa is at –33° 51' 40.6548," 18° 24' 56.9982." As you can imagine, doing any math calculation with these kind of numbers is very difficult, so there is a decimal representation for degrees, minutes, and seconds. Converting these geolocations to the digital format (and there are plenty converters in the Internet), the center of New York City is located at 40.741014, –74.003906 and Cape Town is at –33.861293, 18.415833. Whereas maritime and aeronautical navigation charts still use the degrees, minutes, seconds format, GPS systems mainly use the decimal representation.

Now that you have a latitude and longitude, which are always provided in that order, there are a number of things that you might be interested in:

- The nearest address, that is, a formal address based on the rules of the corresponding country

- The nearest intersection of streets, defined as the nearest street and the next crossing street

- The nearby postal code, within a given radius

- The nearest weather station, and the most recent weather observation

- The country or country administrative subdivision; for example, the county of Surrey in England, or the state of California in the United States

- The time zone, which usually has its boundaries defined politically, not necessarily based on meridians

- And probably the most important for a service such as Foursquare, a nearby place within a given radius

The simple words `nearest` and `nearby` completely hide the complexity involved in calculating these items, which are usually based on the Haversine formula. This formula is used to calculate the distance and bearing between two points with specified latitude and longitude. It is heavy with trigonometry, that is, plenty of `sin`, `cos,` and `atang` functions, and if that is not enough, the angles have to be given in radians. As you can see, it's a rather complex proposition, which additionally requires a database with places, streets, weather stations, cities, countries, time zones, or whatever you are looking for.

Something as apparently simple as the nearest address requires a calculation of a radius, which then is compared with a database to see which record is closest to the current location. For example, the address of Splunk's headquarters is 250 Brannan St, San Francisco, and shows up in the various databases as 37.783007, −122.391184, which is very likely the entrance to the building. Now consider somebody that checks in to an event such as a Splunk Developers Forum in a conference room in Splunk's headquarters with a geolocation of 37.782971, −122.390935; this is a different latitude, longitude pair, or location pair as it is also known. If the conference room is large enough, people seated at opposite sides of the room will produce different location pairs. Additionally, to complicate things, location pairs do not include elevation. So you could be in the living room of your apartment on the 20th floor and the nearest location is the bar on the ground floor of your building.

Trying to find if a location pair that is within the boundaries of a postal code or a city is even more complex, as these kinds of boundaries are irregular. They are represented as polygons, and calculating if a location pair is within a polygon is far more complex than calculating a radius.

Now you can understand why we are so grateful that Foursquare processes the location pairs and provides the address, including postal code. You can also see the amount of effort that goes into providing recommendations for nearby places, along with the tips left by other users.

If your project requires the use of reverse geocoding, there are a couple of options you can consider. The first is to use a commercial service that provides you with the information that you require based on a location pair you send them. The second option is to write your own reverse geocoding application. In this case, you can use commercial databases that are updated almost on a daily basis or free databases that tend to be not as accurate and typically out of date. There are plenty of free reverse geocoding services in the Internet, but all of them have restrictions, mainly on the number of location pairs that can be processed on a daily basis.

Time Zone Considerations

As we mentioned earlier, we have two files with check-ins, one for the New York metropolitan area and another one for the San Francisco Bay area. Both areas are in different time zones and the check-in uses a Unix timestamp for the creation time. Unix timestamps are based on the UTC time zone (Coordinated Universal Time), which is located on the Greenwich meridian, so sometimes it is referred to as Greenwich Mean Time (GMT). UTC is the successor of GMT, but both are essentially the same and you will see both of them used interchangeably. We bring this up because we need to handle time zones correctly to get proper results on the analysis we will do.

Given the type of data that we are going to analyze, time zones are a big consideration, as check-in patterns to venues need to reflect the appropriate time of the day for the corresponding location. Say we want to analyze Sunday brunch patterns in New York and San Francisco, and we define that Sunday brunch starts at 11:00am. That means that for New York it will be 15:00 GMT, whereas it's 18:00 GMT in San Francisco. This of course depends on whether

Daylight Saving Time is active or not. To make a better decision on how to handle this, we need to understand the precedence rules that Splunk applies when assigning time zones to events as they are being indexed:

1. Splunk uses any time zone specified in the event, for example, PST or -0800

2. Splunk uses the TZ attribute defined in the properties file. This is a great way tie a time zone with a specific host, source, or source type

3. If the previous rules fail, Splunk will assign the time zone of the server that indexes the event

Even though the check-in events do have a separate field that contains the time zone, Splunk will ignore it as it recognizes the `createdAt` field as a Unix timestamp; therefore, the time zone is automatically set to GMT. When using a Unix timestamp, the time zone cannot be changed with the TZ attribute. This means that we will have to calculate the local times of day based on the difference between the local time zone and GMT.

Because the check-in events include the `timeZone` field, we have a couple of options on how to handle time zones as we load the data into Splunk. In the first option we can load both data sets into a single index and then deal with time zones as part of the searches; that is, every time we want to narrow a search to either metropolitan area we will have to start the search specifying the time zone, for example:

```
checkin.timeZone="America/New_York" | ...
```

This is easy enough, but the problem with combining both data sets into the same index is that it becomes difficult to handle different times when formulating searches such as the Sunday brunch example we referred to earlier in this chapter. The second option is loading each data set into a different index, which can then be combined when doing searches that go over both metropolitan areas. This one seems to be easier to handle as times and dates can be selected using the `earliest` and `latest` attributes of the `search` command. The data from both indexes can be combined using the append command with a subsearch, but you must be aware that there is a limitation on the number of results a subsearch will return, currently at 50,000. Normally this is not an issue, as subsearches return aggregated information that usually is substantially smaller than the limitation.

Loading the Check-Ins

We start by creating two indexes, appropriately called ny and sf. As usual, we go to the data preview option of the new data input in the user interface. Here we will create a new source type since Splunk does not have a predefined Foursquare source type. Once we select the file to work with, we notice that Splunk has broken every event correctly and also has already found the Unix timestamp, so we don't even have to go through the process of defining where it is, as can be seen in Figure 13-1.

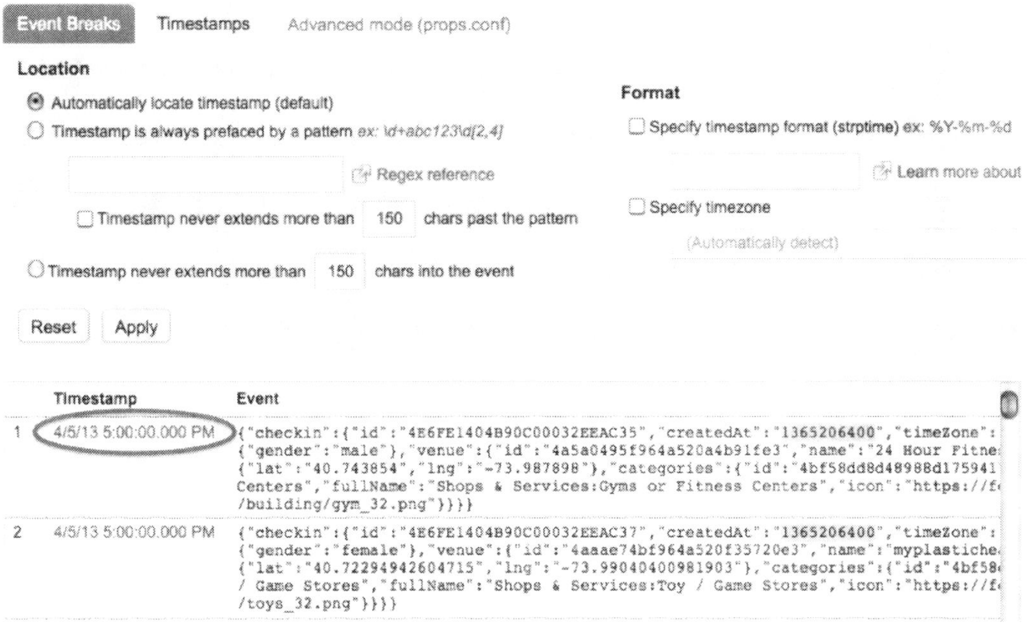

Figure 13-1. *Defining the Foursquare source type*

As you can see in Figure 13-1, the timestamp of the first event, 4/5/13 5:00:00.000 PM, is a date and time earlier than Saturday, April 6, at midnight, the beginning of our data. This raises a concern, because the Unix timestamp of that first event, 1365206400, converts to Saturday, April 6 2013, 00:00:00 GMT, which is different than what Splunk is presenting on the data preview.

The reason for this discrepancy is that Splunk presents the timestamp of the events in the time zone defined for the Splunk user, in our case US Pacific. The difference of seven hours is correct, as GMT/UTC does not change for Daylight Saving. The Splunk user time zone can be modified under the "Your account" heading in the Manager tab of the user interface.

◼ **Caution** When using a Unix timestamp, the default datetime fields (date_hour, date_mday, etc) contain information based on the GMT/UTC time zone, not the Splunk user time zone.

As all of this looks correct, we proceed to save the new source type with the name foursquare and load the New York data into the ny index. Loading the San Francisco data into the sf index is even easier, as we don't have to define a new source type; we just use foursquare, which is already defined, and because of this there is no need for a preview.

The next step is to verify that the data was indexed correctly. First we review the number of events that were indexed by looking at the Index activity overview screen, which can be found in Figure 13-2. In this figure we can see that the event count for each index matches the number of check-ins we had in the original data sets. This information does not appear in the Summary tab, as that tab only displays information about the main, or default index.

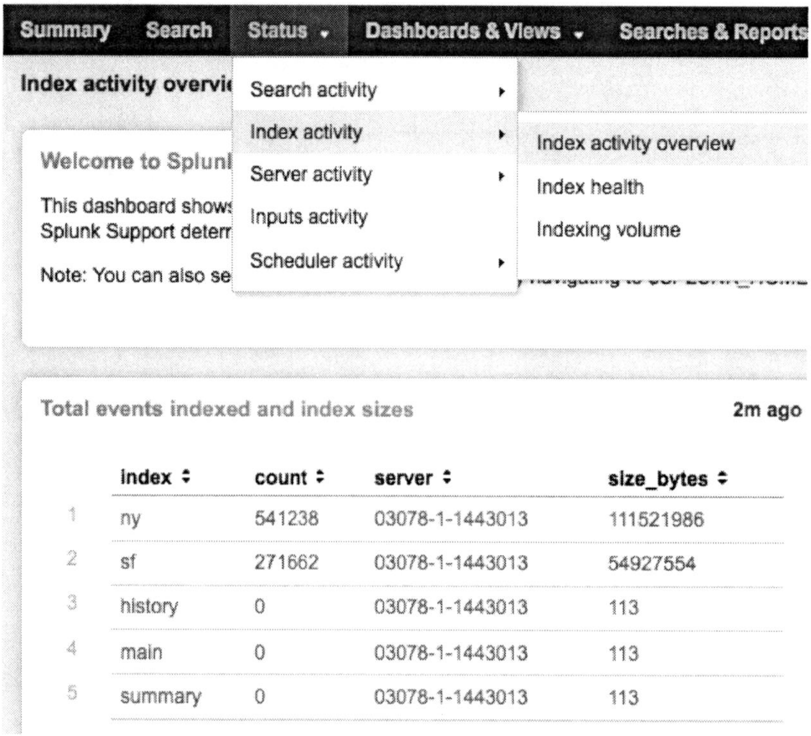

Figure 13-2. *Index activity overview screen*

Now we review the events. We are looking for a few things, such as verifying that the Unix timestamp in the createdAt field matches the timestamp Splunk presents. We are also looking for the proper fields to show up in the left side bar, as well as the host, source type, and source file to show up correctly at the end of every event. To display a few events we just type the following search index=ny | head, which generates the output that can be partially seen in Figure 13-3. The time stamp matches correctly, the 13 fields that make up the check-in event are present on the left side and they contain appropriate information. The fields at the bottom of every event also present the appropriate information. With this we feel comfortable that the data was properly loaded into Splunk. We repeat the process for the San Francisco data successfully. With this phase completed, we can start with the analysis.

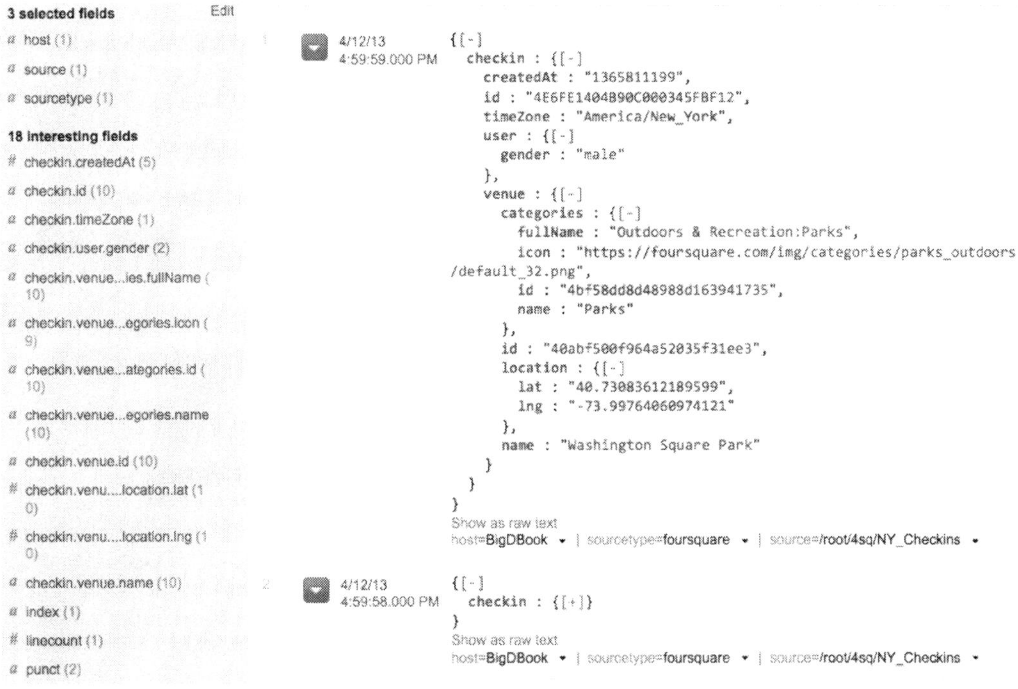

3 selected fields   Edit

a host (1)

a source (1)

a sourcetype (1)

18 interesting fields

checkin.createdAt (5)

a checkin.id (10)

a checkin.timeZone (1)

a checkin.user.gender (2)

a checkin.venue...ies.fullName (10)

a checkin.venue...egories.icon (9)

a checkin.venue...ategories.id (10)

a checkin.venue...egories.name (10)

a checkin.venue.id (10)

checkin.venu....location.lat (10)

checkin.venu....location.lng (10)

a checkin.venue.name (10)

index (1)

linecount (1)

a punct (2)

```
1    4/12/13           {[-]
     4:59:59.000 PM        checkin : {[-]
                              createdAt : "1365811199",
                              id : "4E6FE1404B90C000345FBF12",
                              timeZone : "America/New_York",
                              user : {[-]
                                gender : "male"
                              },
                              venue : {[-]
                                categories : {[-]
                                  fullName : "Outdoors & Recreation:Parks",
                                  icon : "https://foursquare.com/img/categories/parks_outdoors
         /default_32.png",
                                  id : "4bf58dd8d48988d163941735",
                                  name : "Parks"
                                },
                                id : "40abf500f964a52035f31ee3",
                                location : {[-]
                                  lat : "40.73083612189599",
                                  lng : "-73.99764060974121"
                                },
                                name : "Washington Square Park"
                              }
                            }
                          }
                          Show as raw text
                          host=BigDBook ▾ | sourcetype=foursquare ▾ | source=/root/4sq/NY_Checkins ▾

2    4/12/13           {[-]
     4:59:58.000 PM        checkin : {[+]}
                          }
                          Show as raw text
                          host=BigDBook ▾ | sourcetype=foursquare ▾ | source=/root/4sq/NY_Checkins ▾
```

Figure 13-3. *Verifying the Foursquare check-in events*

Analyzing the Check-Ins

In this section we will be formulating a number of searches that will deal with handling time zones, geolocation, and some complex searches that expose limitations of some search commands. We will also discuss some visualization items. We start with the Sunday brunch search that was mentioned earlier in this chapter.

The Sunday Brunch Search

With this search we are trying to compare the most popular spots for Sunday brunch in the New York and San Francisco metropolitan areas. There are a couple of challenges with this search. The first one is that we have to handle events from two different indexes, and the second one is that we have to handle two different time zones. If we define that brunch is a check-in into a food venue between 11:00am and 2:00pm local time, then we will have to do the same search with different local times. Let us start by formulating the search for New York with the correct date, time, and venue.

There are two ways to handle local time. On the first one you can use the default datetime fields, which we already know are based on the GMT time zone. Given the time difference between New York and GMT this time of the year is four hours, we want events that go from 15:00:00 to 17:59:59 GMT on Sunday, April 7, 2013. Using the datetime fields, the first clause of the search would be:

```
index=ny date_wday=sunday date_hour>=15 date_hour<18
```

Remember that the ANDs are implicit in the search command. This clause works for our data set because we only have one Sunday in the data and the times do not straddle over two days. The other option, which is more generic and covers the limitations we just explained with the datetime fields, is using the earliest and latest attributes of the search command.

■ **Caution**   The earliest and latest attributes of the search command are based on the time zone of the Splunk user, not the time zone of the data.

Remember that the latest attribute is exclusive, so using these attributes and knowing that our Splunk user time zone is US Pacific, then 11:00am in New York is 8:00am in California, thus the first clause of the search is:

```
index=ny earliest=4/7/2013:08:00:00 latest=4/7/2013:11:00:00
```

The next item is which categories we should search for. After analyzing the list of categories in the aboutfoursquare.com web site mentioned earlier, we could look for the word Restaurant in the checkin.venue.categories.name field, but that would exclude places such as bagel shops and bakeries. The alternative is to look for the word Food in the checkin.venue.categories.fullName field, this way we catch any venue that is food-related. A typical entry for this field looks like "Food:American Restaurants", so we will have to make sure that we specify the match is with the word Food at the beginning of the field, followed by anything else. The initial search is the following, and the output can be seen in Figure 13-4.

```
index=ny earliest=4/7/2013:08:00:00 latest=4/7/2013:11:00:00
| rename checkin.venue.categories.fullName as Type
| where like(Type, "Food%")
| top Type
```

	Type ⬍	count ⬍	percent ▾
1	Food:Coffee Shops	787	12.626344
2	Food:American Restaurants	768	12.321515
3	Food:Italian Restaurants	303	4.861223
4	Food:Cafés	302	4.845179
5	Food:Diners	269	4.315739
6	Food:Bakeries	248	3.978822
7	Food:Bagel Shops	230	3.690037
8	Food:French Restaurants	227	3.641906
9	Food:Breakfast Spots	204	3.272902
10	Food:Burger Joints	203	3.256859

Figure 13-4. *Top 10 Sunday brunch venue categories in New York*

Note that the like function uses the SQL notation %, which means any character any number of times. Also, remember that we have to rename multilevel JSON fields before using them with an eval or a where command, otherwise these commands will interpret the dots in the fields as a concatenation directive. To verify the results of this search we first select randomly an event and make sure it is within the allocated time frame; then we verify that the count of one of the restaurant types is correct. To examine a random event, we change the last clause of the search from top Type to head 1, which should produce the most recent event in that time frame. In the output, which can be found in Figure 13-5, you can see that the Splunk timestamp is 4/7/13 10:59:58.000 AM, which is 1:59:58 PM in New York, so that is correct.

```
4/7/13              {[-]
10:59:58.000 AM     checkin : {[-]
                        createdAt : "1365357598",
                        id : "4E6FE1404890C000334EECEB",
                        timeZone : "America/New_York",
                        user : {[+]},
                        venue : {[-]
                          categories : {[-]
                            fullName : "Food:American Restaurants",
                            icon : "https://foursquare.com/img/categories/food/default_32.png",
                            id : "4bf58dd8d48988d14e941735",
                            name : "American Restaurants"
                          },
                          id : "4c0c5587340720a166978993",
                          location : {[+]},
                          name : "Westville Chelsea"
                        }
                      }
                    }
                Show as raw text
                host=BigDBook  ▾ | sourcetype=foursquare  ▾ | source=/root/4sq/NY_Checkins  ▾
```

Figure 13-5. *Sample check-in event for the selected time period*

To verify the counts we formulate a search that only looks for the most popular venue check-ins, Coffee Shops, for the time period of our interest. This is a very simple search based on the checkin.venue.categories.name field instead of the fullName field we used in the search we are now verifying:

```
index=ny earliest=4/7/2013:08:00:00 latest=4/7/2013:11:00:00
        checkin.venue.categories.name="Coffee Shops"
| stats count
```

The result is 787, which matches the result of the search we are verifying, so now we know that it is correct. Now that we have the events related to food, we want to find the most popular spots, which can be done very easily by modifying the last clause of the search as follows, and the output can be seen in Figure 13-6.

```
index=ny earliest=4/7/2013:08:00:00 latest=4/7/2013:11:00:00
| rename checkin.venue.categories.fullName as Type
| where like(Type, "Food%")
| top 5 checkin.venue.name
```

	checkin.venue.name ⇕	count ⇕	percent ⇕
1	Starbucks	310	4.971933
2	Dunkin' Donuts	94	1.507618
3	Shake Shack	50	0.801925
4	The Smith Restaurant	43	0.689655
5	McDonald's	38	0.609463

Figure 13-6. *Top five brunch venues in New York*

Remember that the percentages calculated in Figure 13-6 are based on all the food category events for the defined time period, not only the top five. Another thing to note is that this particular search matches venues by name, so all the Starbucks locations are grouped together. We can do a search based on the individual location of a venue, which very likely will provide a different perspective, as can be seen in Figure 13-7. The search counts based on the venue id number, which is unique per location. After that, the results are based on the name of the venue.

```
index=ny earliest=4/7/2013:08:00:00 latest=4/7/2013:11:00:00
| rename checkin.venue.categories.fullName as Type
| rename checkin.venue.name as Venue
| where like(Type, "Food%")
| stats count by checkin.venue.id, Venue
| table Venue, count
| sort -count
```

	Venue ‡	count ‡
1	The Smith Restaurant	27
2	Food Truck Festival	24
3	Jing Fong Restaurant 金豐大酒樓	23
4	Cafe Orlin	21
5	Peels	20

Figure 13-7. *Top five brunch locations in New York*

As can be seen in Figure 13-7, Starbucks is no longer in the top five list. As a matter of fact, for the specified period of time, the Starbucks location with the most check-ins has only six.

This was an interesting diversion, but let's get back to the main brunch search. We are at a point where we can now add the San Francisco Bay area with the following search that produces the results seen in Figure 13-8. All we do is add a subsearch with the appropriate time window and those events are appended to the events that match the search in the first clause.

```
index=ny earliest=4/7/2013:08:00:00 latest=4/7/2013:11:00:00
| append [
          search index=sf earliest=4/7/2013:11:00:00 latest=4/7/2013:14:00:00
         ]
| rename checkin.venue.categories.fullName as Type
| where like(Type, "Food%")
| top 5 checkin.venue.name
```

	checkin.venue.name ‡	count ‡	percent ‡
1	Starbucks	474	5.340243
2	Dunkin' Donuts	94	1.059036
3	Shake Shack	50	0.563317
4	McDonald's	49	0.552050
5	The Smith Restaurant	43	0.484452

Figure 13-8. *Top five brunch venues in New York and San Francisco*

This search has a problem: it does not distinguish between the metropolitan areas, so the counts are a total for both areas. As can be seen in Figure 13-8, the count for Starbucks increased from 310 to 474 because it just added the check-ins from San Francisco.

There are two ways we can think of to distinguish between the metropolitan areas. The first one is by looking at the longitude of the event, the second one is by looking at the time zone of the event. With the first option we need to select a meridian somewhere between New York and San Francisco. New York is approximately at –75 degrees, whereas San Francisco is at about –122 degrees. Because we like round numbers, we will choose the –100 meridian. This way, any longitude that is greater than –100 is in New York, whereas any longitude that is less than –100 is in San Francisco. Note that we are working with negative numbers, so a longitude that is greater than –100 includes numbers like –75. As we add the clauses that calculate at what side of the –100 meridian is and clean up the search, the new version of the search is as follows:

```
index=ny earliest=4/7/2013:08:00:00 latest=4/7/2013:11:00:00
| append [
        search index=sf earliest=4/7/2013:11:00:00 latest=4/7/2013:14:00:00
        ]
| rename checkin.venue.categories.fullName as Type
| rename checkin.venue.name as Name
| rename checkin.venue.location.lng as Lng
| where like(Type, "Food%")
| eval MetroArea = case(Lng> -100, "New York", Lng < -100, "San Francisco")
| stats count by Name, MetroArea
| sort 10 -count
```

The only new item in this search is the case function of the eval command, which is very similar in functionality to the case statement in many programming languages. The function takes pairs of arguments, if the first one, which is a Boolean, evaluates true, then the second argument of the pair is returned.

The second way to distinguish metropolitan areas is to look at the timeZone field of the check-ins. For this, we modify the search as follows:

```
index=ny earliest=4/7/2013:08:00:00 latest=4/7/2013:11:00:00
| append [
        search index=sf earliest=4/7/2013:11:00:00 latest=4/7/2013:14:00:00
        ]
| rename checkin.venue.categories.fullName as Type
| rename checkin.venue.name as Name
| rename checkin.timeZone as TZ
| where like(Type, "Food%")
| eval MetroArea = if(like(TZ, "%New_York"), "New York", "San Francisco")
| stats count by Name, MetroArea
| sort 10 -count
```

In addition to renaming the field that contains the time zone, the change is in the eval command, where now we use the if function. This function evaluates the first argument; if it is true it returns the second argument, else it returns the third (if present), otherwise it returns null. The like function just looks for anything followed by the string New_York, as denoted by the SQL notation %. The results for both searches are exactly the same and can be seen in Figure 13-9.

Name ⇕	MetroArea ⇕	count ⇕
1 Starbucks	New York	310
2 Starbucks	San Francisco	164
3 Dunkin' Donuts	New York	94
4 Shake Shack	New York	50
5 The Smith Restaurant	New York	43
6 McDonald's	New York	38
7 Peet's Coffee & Tea	San Francisco	34
8 Philz Coffee	San Francisco	30
9 In-N-Out Burger	San Francisco	29
10 Food Truck Festival	New York	24

Figure 13-9. *The top 10 brunch venues by metropolitan area*

As we have already verified the underlying searches, and notice that adding the counts of Starbucks for New York and San Francisco produces the same number as the one we got on the search that produced the results in Figure 13-8, we feel comfortable that this search is producing dependable results. We can now move on to the next interesting search.

Google Maps and the Top Venue

Because we are working with data that contains geographical location information, we might as well play a little bit with the Google Maps App, which we introduced in Chapter 4. Let us first run a very simple search to find out the top five venues in New York for the week's worth of data we have. Based on this, we will pick the number one venue, and explore the locations using the Google Maps app for Splunk. The following produces the table that can be seen in Figure 13-10.

```
index=ny
| top 5 checkin.venue.name
```

checkin.venue.name ⇕	count ⇕	percent ⇕
1 Starbucks	9960	1.840226
2 John F. Kennedy International Airport (JFK)	3802	0.702464
3 LaGuardia Airport (LGA)	3656	0.675488
4 New York Penn Station	3525	0.651469
5 New York Sports Club	2699	0.498672

Figure 13-10. *Top five venues in New York*

Based on the results of the search presented in Figure 13-10, the top venue in the New York metropolitan area is by far Starbucks. The second closest, JFK airport, has slightly more than one-third of the check-ins of Starbucks. Given that Starbucks has many locations, it is a great example for using the Google Maps App to display them.

In Chapter 4, the Maps App was used by providing an IP address to the geoip command. The geoip command takes the IP address and looks it up in a table, then it returns a location pair, which is then presented in the map. Because in this chapter we are not dealing with IP addresses, but straight location pairs, the way to call the Maps App is slightly different. The Maps App expects a latitude, longitude pair in a field called _geo, which contains them in the form of latitude, longitude, which we have to build ourselves. Nothing complicated. The search within the

Google Maps App to show all the Starbucks locations in the New York metropolitan area for the week's worth of data we have available follows. The map, presenting a selected section of the New York metropolitan area, is shown in Figure 13-11.

```
index=ny checkin.venue.name=Starbucks
| rename checkin.venue.location.lat as Lat
| rename checkin.venue.location.lng as Lng
| eval _geo=Lat+","+Lng
```

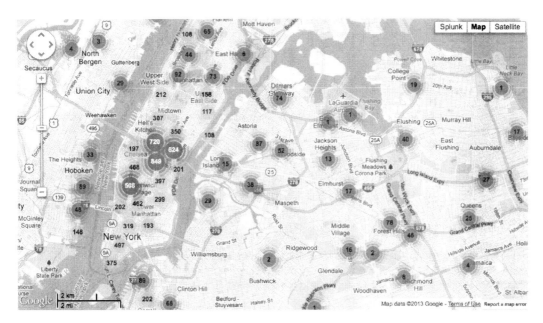

Figure 13-11. *Map of Starbucks locations in the New York metropolitan area*

As with Google Maps, you can zoom in, which will provide a more granular view of the counts of Starbucks locations. It seems that the greatest concentration of Starbucks check-ins occur in midtown Manhattan. When we zoom into that area, we can see that the most popular Starbucks location for the week of our analysis is the one located in the corner of 5th Avenue and West 33rd Street, as can be seen in Figure 13-12.

Figure 13-12. Map of the most popular Starbucks in Manhattan

Check-Ins Patterns of a Venue

The searches for finding out the check-ins patterns for a venue are extremely simple, but we still wanted to share a couple of options with you and also comment on a limitation of the `stats` command. This time we will focus on the top location in the San Francisco Bay area. The search to find this is:

```
index=sf
| rename checkin.venue.name as Venue
| stats count by checkin.venue.id, Venue
| table Venue, count
| sort -count
```

In this search we count the individual locations using the `id` field of the venues but present the results by their names. You can see that the `stats` command includes the Venue field, which you would think is superfluous. The issue is that the `stats` command is a transforming command; as such, the only fields left after invoking it are the split-by fields and the aggregates computed per split. Basically, if you want a field that you created before the `stats` command to pass on to the next clause, you have to use it in the `stats` command, even if it's doing nothing. In this example, if we don't include the Venue field in the stats command, it would not be available for the `table` command. The output of the previous search is presented in Figure 13-13.

	Venue ⇕	count ⇕
1	San Francisco International Airport (SFO)	5880
2	AT&T Park	883
3	Norman Y. Mineta San José International Airport (SJC)	856
4	Oakland International Airport (OAK)	848
5	Splunk Developers Forum	775

Figure 13-13. Top five locations in the San Francisco Bay area

Now that we know the top location, the easiest way to find the check-in pattern for the San Francisco airport is by just searching for all of the events that have it in their name. The Splunk timeline will present the pattern as can be seen in Figure 13-14, along with the search.

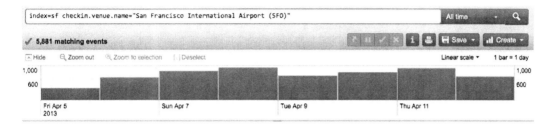

Figure 13-14. *Daily check-in pattern for the San Francisco airport*

However, the problem with the timeline presented in Figure 13-14 is that the granularity is not good enough. It shows the results by day and we would like them by hour. To achieve this, we formulate another search that uses the `timechart` command and specify the time span to be hourly. The search follows along with the results in Figure 13-15. One of the nice features of the charts in Splunk is that when you hover over the chart, it will display the details of that particular data point.

```
index=sf checkin.venue.name="San Francisco International Airport (SFO)"
| timechart span=1h count
```

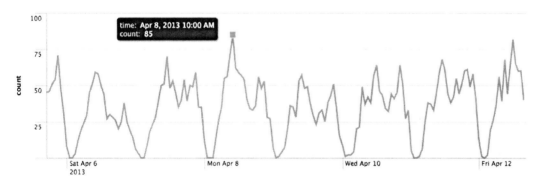

Figure 13-15. *Hourly check-in pattern for the San Francisco airport*

Venues by Number of Check-Ins

A question we often hear is related to the number of check-ins that venues have over a certain period of time: specifically, how many venues have one check-in per week, how many have two, and so on. This can provide another interesting perspective of the Foursquare world. The basis of this search is counting the check-ins for every location, and then counting the locations by number of check-ins. This implies two stats commands, one after another:

```
index=sf
| stats count as NumberOfCheckins by checkin.venue.id
| stats count as Locations by NumberOfCheckins
| table NumberOfCheckins, Locations
```

This is simple enough, and the limitation we explained earlier of the stats command is not an issue, as the field we generate in the second clause is immediately used in the third clause. Verifying the results of this search is a more complicated proposition. First we run the same search, but for a substantially shorter period of time, say, for one hour. The output can be seen in Figure 13-16.

	NumberOfCheckins ⬍	Locations ⬍
1	1	1445
2	2	109
3	3	28
4	4	14
5	5	1
6	7	1
7	8	1
8	14	1
9	35	1
10	85	1

Figure 13-16. Venues by number of check-ins, one hour

As it turns out, the search on a shorter period of time produces exactly 10 results. Now we have a more manageable number of results. The challenge is finding a different search that allows us to verify the results presented in Figure 13-16. In those results, we can see that one venue has 85 check-ins, one has 35, and one has 14. If we formulate a search that counts by venue name, these venues must show up in the results as single location venues. The verification search is the following, with partial results presented in Figure 13-17.

```
index=sf earliest=4/8/2013:10:00:00 latest=4/8/2013:11:00:00
| stats count by checkin.venue.name
| table count, checkin.venue
| sort -count
```

	count ⬍	checkin.venue.name ⬍
1	85	San Francisco International Airport (SFO)
2	70	Starbucks
3	35	Splunk Developers Forum
4	22	Peet's Coffee & Tea
5	15	Philz Coffee
6	14	Oakland International Airport (OAK)

Figure 13-17. Results of the verification search

The venue with 85 check-ins is the San Francisco airport, the one with 35 is the Splunk Developers Forum, and the one with 14 is the Oakland airport. The other venues that show up in the results shown in Figure 13-17 are multilocation venues, so it makes sense that they did not show up in the results presented in Figure 13-16. We feel comfortable that the results of the initial search are correct for the whole period of time.

It is interesting to note that Foursquare has a Swarm badge, which is granted when 50 people or more check in to the same location in a three-hour period. Based on the results displayed in Figure 13-17, the people checking in the San Francisco airport should obtain this badge, and it seems quite possible that they could obtain the Super Swarm badge, which is the same as the Swarm badge, but for 250 people or more.

The results of the initial search, which has now been verified, are quite lengthy. When graphed as a line chart, the labels of the x-axis are so many, they are not displayed. To make the results more readable, we limited the output to the first 30, which are shown in Figure 13-18, where you can see that the majority of the locations have a small number of check-ins over the week of our analysis.

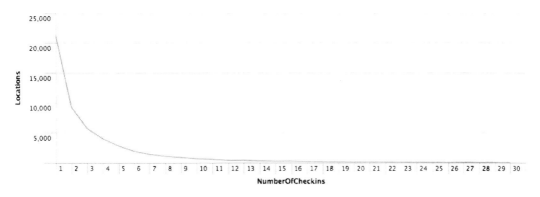

Figure 13-18. *Venues by number of check-ins in San Francisco*

Next, we complicate this search by stating that we want the results as a percentage of check-ins instead of a count, which is what we have right now. The main issue with this challenge is that we need the total number of check-ins, which is easy to calculate, but given the limitation of the stats command that we mentioned earlier, it will be a difficult proposition. The trick we have used to get around it, which is using the field you want to keep in the calculation of the next stats command, does not always work. For example:

```
index=sf
| stats count as Total
| stats sum(Total) as Total, count as NumberOfCheckins by checkin.venue.id
| stats count as Locations by NumberOfCheckins
```

In this example, we obtain the total by just counting the events and place it in the Total field. In the next clause, we do nothing with Total by calculating a sum that should not affect the value of Total. However, because the stats command uses a by argument, it does affect the value of Total. There is no way that we can pass the total through all of the stats commands that we need. There is however, another way to solve this problem, by just multiplying the values of the Locations field with that of the NumberOfCheckins field for each location and then adding them up. This gives us the total number of events. Based on this, the search is:

```
index=sf
| stats count as NumberOfCheckins by checkin.venue.id
| stats count as Locations by NumberOfCheckins
| eventstats sum(eval(NumberOfCheckins * Locations)) as Total
```

```
| eval PCT=round(Locations / Total * 100, 2)
| table NumberOfCheckins, Locations, PCT, Total
```

The eventstats command is used to calculate the total as described earlier and then it adds the Total field to all the search results that come from the previous clauses of the search. With this, we can now calculate the percentage and present the results in table form, which makes it easy to debug any problems in the search. Verification of this search is very easy. As we already have the count of venues by number of check-ins and the total number of events, we can easily calculate the percentage. If the percentage of locations with only one check-in is 7.87 percent, the search worked correctly. As you can see in Figure 13-19, the search is successful.

	NumberOfCheckins ⇕	Locations ⇕	PCT ⇕	Total ⇕
1	1	21391	7.87	271662
2	2	9301	3.42	271662
3	3	5740	2.11	271662
4	4	4020	1.48	271662
5	5	2837	1.04	271662

Figure 13-19. *Percentage of venues by number of check-ins*

Now we clean up the search by just leaving the number of check-ins and the percentages and limit the output to the top 25. This allows us to produce a nice column chart that answers the question of what percentage of locations have how many check-ins during the period of our analysis in the San Francisco metropolitan area. The chart can be seen in Figure 13-20.

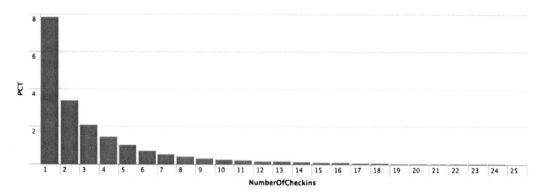

Figure 13-20. *Percentage of venues by number of check-ins—column chart*

Analyzing Gender Activities

It would be rather interesting to analyze check-ins by gender. For this, we will have to keep a count by gender for each category name. As we have seen earlier, this is not easy with the stats command, so we will shift our focus to the eventstats command, which allows us to add new fields to events. Furthermore, we can add these counts for each gender in separate sections of a search by using a subsearch. The initial search looks at San Francisco in the middle of the week, Wednesday, April 10, from 7:00am to 10:00pm:

```
index=sf earliest=4/10/2013:07:00:00 latest=4/10/2013:22:00:00
        checkin.user.gender=male
| eventstats count(checkin.id) as Male by checkin.venue.categories.name
| append [
          search index=sf earliest=4/10/2013:07:00:00 latest=4/10/2013:22:00:00
                checkin.user.gender=female
          | eventstats count(checkin.id) as Female by checkin.venue.categories.name
          ]
| rename checkin.venue.categories.name as Category
| stats count(Male) as Men, count(Female) as Women by Category
| table Category, Women, Men
| sort 25 -Women, -Men
```

	Category ⬍	Women ⬍	Men ⬍
1	Coffee Shops	679	920
2	Offices	652	1152
3	Homes (private)	596	567
4	Grocery Stores	414	420
5	Airports	328	550

Figure 13-21. *Activities by gender in San Francisco*

With the eventstats commands, we calculate the count of check-ins by category name and place it in a field, which is then tacked to the results of the searches, in this case all the male and female events. Then the stats command counts those new fields by category to obtain the final count. Partial results of this search can be seen in Figure 13-21.

The verification of this search is quite simple, as we can zoom directly into any of these categories and obtain the count. For example:

```
index=sf earliest=4/10/2013:07:00:00 latest=4/10/2013:22:00:00
        checkin.user.gender=male
        checkin.venue.categories.name=Offices
| stats count
```

The result for this search is 1,152, which matches the corresponding result shown in Figure 13-21. We do another verification search, just to make sure that things are working well. This time we use airports for females, and the result matches; thus, we can state that this search works correctly. Ideally, we should present the results as a percentage, as that will provide more clarity, but we don't want to complicate the search. Interestingly enough, there is a way to present these results as a percentage without having to calculate them by using either a bar or column chart and selecting the 100% stacked option. We show the results of the search in Figure 13-22. As there would not be any space in a column chart to display the names of the categories, we use a bar chart.

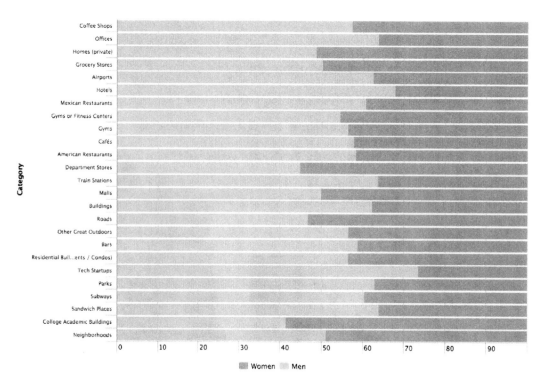

Figure 13-22. Activities by gender in San Francisco—bar chart

It is really interesting to see in Figure 13-22 that the San Francisco Bay Area is focused on technology, with about 75 percent of the male check-ins coming from tech startups, and the second category hotels, probably visitors to the tech startups. For women, the first category is colleges and the second is department stores. This type of chart works very nicely to convey a comparison of results, such as the ones we just produced. For a more informal comparison, we can do a search for the top activities by category name for males and females in New York over the whole week of data. The results, using a word cloud generated with IBM's Many Eyes web site, are displayed in Figure 13-23 (females) and Figure 13-24 (males).

Figure 13-23. *Activities for women in New York*

Figure 13-24. *Activities for men in New York*

Summary

In this chapter, we used check-ins to the Foursquare service to learn about geolocation pairs, reverse geolocation, and used the Google Maps App of Splunk. We also discussed how to handle time zones and all of the issues involved with this. We introduced the eventstats and the append commands and learned about the limitation of the stats command. We also explored visual aids to better convey the results of the searches, some of them rather complicated.

## CHAPTER 14

Sentiment Analysis

Sentiment analysis is usually associated with social media and, by extension, with big data. In this chapter you will learn what sentiment analysis is about and go through some practical examples where we analyze the sentiments of tweets. The project for this chapter is the World Sentiment Indicator, which strives to provide an indicator of the general world sentiment based on analyzing the news headlines from a large number of web-based news outlets worldwide.

Opinions, Views, Beliefs, Convictions

Sentiment analysis can be defined as the process of examining a text or speech to find out the opinions, views, or feelings the author or speaker has about a determined topic. (This definition applies to computer systems, as when a human does it, it's simply called reading.) All of the words in this section's title are closely related to sentiment and describe highly subjective and ambiguous concepts. This makes it quite a challenge for a computer program to be extracted from some text, be it a large document or a tweet. The challenge increases in complexity when you consider that certain words or expressions have different meanings depending on the knowledge domain, or domain of expertise. For example, "go around" in a normal conversation has a totally different meaning than it does for an airplane pilot, for whom it implies to abort the landing procedure and try again. Compounding the issues related to domains of expertise is the slang that is usually associated with them.

The documents or speeches of some knowledge domains tend to be much more difficult to analyze than others and politics is usually recognized as the most complex domain. Trying to determine if a political speech is in support or not of a certain issue is difficult, as those discussions are full of sarcasm, quotations and references, potentially complex, to other organizations or people. We can easily conclude that sentiment is contextual and domain dependent.

Given the ambiguity of the concepts involved and human nature, it can also be expected that different people have different understandings and draw different conclusions from the text being examined or the speech being viewed, the latter with the advantage of seeing body language. The challenges described so far exist independently of the language used and are presented in the context of human beings doing the work of extracting the desired information.

Trying to automate the extraction of information with a computer system increases the complexity, as software programs require well-defined boundaries that eliminate ambiguity. Inevitably, this tends to specialize the analysis by domain and, additionally, the media being used. For example, the treatment of a news article will be different from that of a tweet. An article tends to follow the grammatical rules of the language it's written in, typically uses proper words, and does not have orthographical mistakes. By contrast, a tweet lacks sentence structure, likely uses slang, includes emoticons(☺, ☹) and sometimes words are lengthened by repeating letters to increase the emotional charge ("I looooooove chocolate"). Also, because of the nature of the devices used to create the tweets, the frequency of misspellings is higher.

As you can see, sentiment analysis is difficult. Later in this chapter we will go into more detail on the various techniques to automate this analysis, but first we will see why you might want to use it.

Commercial Uses

Starting with a historical perspective and following the evolution of technology and social media, we can break down the uses of sentiment analysis, or opinion mining as it is also known, into two broad categories. The first one is when it is used as input to the decision-making process. What other people think has always been a very important component in this process, especially when deciding to purchase an item. In the past, being able to get another opinion was limited to family, friends, and specialists who published their reviews in trade magazines or other media. Consumer Reports, a magazine dedicated to rate all sorts of products that range from blenders to cars, is the ultimate example of specialists that offer the "what other people think."

As our digital footprint keeps on growing and our commercial transactions continue to increase on the Internet, we find more and more digital forums that give people the ability to share their experiences and opinions, this way satisfying the need to know "what other people think." The forums range from the traditional specialists that use the Internet as another medium to publish their opinions, to very well organized web sites dedicated to capture and share opinions from anybody. Sites such as Yelp, Angie's List, and Rotten Tomatoes do an excellent job at aggregating and ranking commercial establishments, merchants, tradesmen, and movies based on public feedback. Additionally, there are the reviews that people can leave on the web sites where they purchase the products, such as amazon.com. Finally, you have the individual opinions, which range from blog postings to tweets and can be considered the equivalent of "word of mouth."

Digital forums that deal with specialized topics, such as stackoverflow.com (a must for developers), and those created by corporate support departments offer simple yet powerful ways that allow participants to rate the quality of the answers and implicitly the credibility of the people who provide the answers.

This category of decision support has exploded in the last few years, and it is very easy and quick for anybody to find information on pretty much anything they might be interested in. Of course, they still have to weed out what is a respected opinion from an incomplete or useless review, and also identify a plain old advertisement disguised as a specialized review.

The second broad category has to do with the feedback that companies can obtain about their products or services. Historically, companies had to conduct focus groups and polls or review service center notes or warranty claims to gain insight about the perception of their brands, products, and services. Now they can monitor the social media outlets and gain an understanding of what is happening on an almost real-time basis, not only with their products but also those of competitors. Using sentiment analysis, companies can monitor consumer trends and market buzz. They can get faster to the root causes of customer behaviors in their web sites and also find new opportunities and threads a lot faster. Ultimately, all this leads to the management of the most important thing that a company has, the reputation of its brand.

As we have seen, sentiment analysis is not easy, but when it's done correctly it can be very powerful. In a 2010 paper entitled "From Tweets to Polls: Linking Text Sentiment to Public Opinion Time Series" by O'Connor et al., the authors make a very compelling statement: "We analyze several surveys on consumer confidence and political opinion over the 2008 to 2009 period, and find they correlate to sentiment word frequencies to contemporaneous Twitter messages. While our results vary across datasets, in several cases the correlations are as high as 80%, and capture important large-scale trends. The results highlight the potential of text streams as a substitute and supplement for traditional polling."

Of course, when sentiment analysis is not well done, it can lead to some rather interesting situations. An example of this is described by Dan Mirvish in a post entitled "The Hathaway Effect: How Anne Gives Warren Buffet a Rise" (http://www.huffingtonpost.com/dan-mirvish/the-hathaway-effect-how-a_b_830041.html). The author suspects that the robotic trading programs on Wall Street include sentiment analysis, and every time that Anne Hathaway makes the headlines, the stock of Warren Buffett's Berkshire-Hathaway (symbol BRK.A) goes up. Dan provides a list of landmark dates in Anne Hathaway's career and the corresponding increases in the share price of BRK.A. Of course, the article is based on a very simplistic analysis, but it's a nice example.

It is clear that there are many commercial uses for sentiment analysis and there are plenty of commercial and open source products with diverse claims. But before you decide between building or buying, let's have a look at the technical aspects of sentiment analysis.

The Technical Side of Sentiment Analysis

When you search the Internet looking for information on sentiment analysis, you will typically find either a large number of scientific papers plagued with complex probabilistic formulas or marketing documents explaining how their secret sauce is better than others. At the risk of oversimplifying and possibly misrepresenting certain concepts, we present an explanation of the technical side of sentiment analysis so that you can get a decent idea without needing to have a degree in statistics. This part of the chapter is loosely based on the seminal paper "Opinion Mining and Sentiment Analysis" by Bo Pang and Lillian Lee (2008).

Sentiment analysis is in essence text categorization; thus, the results from such analysis usually fall into two classes:

- Polarity, which is a straight positive, neutral, or negative result

- Range of polarity, which are ratings or rankings, such as star ratings for movies, and can sometimes extend into strength of opinion, for example, a scale from 1 to 10

Finding the polarity or range of polarity can be done at various levels: word, phrase, sentence, or document. It is common to use the results of one level as inputs for the analysis of higher levels. For example, you can apply sentiment analysis to words, and then use the results to evaluate phrases, then sentences, and so on.

Classifying text based on its sentiment has many challenges, which we will review as we present various algorithms to extract and categorize polarity based on what are called features:

- Frequency. This is probably the most obvious and intuitive of all the features. The idea is that the words that appear most often in a document reflect the polarity of the document. As you can imagine, this premise is not always right.

- Term presence. The inverse of frequency. Instead of paying attention to the most frequent words, you look for the most unique ones and base the polarity on them. Again, this premise is not always accurate.

- N-grams. The position of the word determines, and sometimes reverses, the polarity of a phrase, from positive to negative or vice versa. The analysis is done trying to understand the context of a word within unigrams (only one word), bigrams (two consecutive words), trigrams (three consecutive words), and so on.

- Parts of speech. Traditional English grammar classifies words into eight parts of speech: the verb, the noun, the pronoun, the adjective, the adverb, the preposition, the conjunction, and the interjection. In this particular case, adjectives are considered very good indicators of sentiment in text; thus, they are used to define the polarity. This feature implies breaking down a sentence into all the different parts of speech.

- Syntax. There have been attempts to incorporate syntactic relations within feature sets, but there are still discussions about the merits of using this feature.

- Negation is one of the most important features in sentiment analysis. Although the bag-of-words representations of "I like this book" and "I don't like this book" are considered to be very similar by most commonly used similarity measures, the only differing word, the negation term, forces the two sentences into opposite polarities. However, not all of the appearances of explicit negation terms reverse the polarity of the enclosing sentence.

As you can see, these algorithms have their strengths and weaknesses, and there is no single one that can address all of the issues. In general, text classifiers use combinations of feature extractors to assign polarity. However, for these feature extractors to assign contextual polarity implies that they must have a base polarity for the words. There are a couple of ways to obtain the base polarity of words:

- Using a lexicon. The most basic lexicons just provide a polarity for each individual word, but they can get really fancy with additional labels that can denote, for example, degrees of specific emotions.

- Using training documents. These each contain a number of sentences and each document is classified with a specific polarity or range of polarity. The polarity of an individual word is calculated based on combination of feature extractors and its appearance in the different classifications. This calculation is referred to as training and the results are placed in a model. Obviously the more sentences in each of the classified documents the more accurate the resulting categorization will be.

Text classifiers that use training documents fall under the category of machine learning tools. Arguably, the most popular sentiment analysis tools are based on machine learning classifiers, which typically use one of the following approaches:

- Naïve Bayes Classifier. A simple probabilistic classifier based on Bayes' Theorem. The feature extractors include words or unigrams, bigrams and trigrams, as well as parts of speech in each sentence. The Naïve Bayes Classifier can include counts of positive and negative words, as well as counts of polarities of sequences of words. It can also include counts of parts of speech combined with polarity information.

- Maximum Entropy. Whereas Bayes assumes that each feature is independent, maximum entropy does not. This allows the use of bigrams and phrases without worrying about overlapping words.

- Support Vector Machines. Based on vectors that contain and are as long as the number of feature extractors used for the analysis. Multiple functions, such as linear, polynomial, sigmoid, and others can be applied to the vectors. The simplest analysis is based on a single word combined with a term of presence.

A lot of research and many comparisons have been done with regard to these approaches. The conclusions can be generalized as each approach can slightly outperform another depending on the domain of the text being analyzed. The common theme is that with high-quality training documents, all approaches have presented accuracies of over 80 percent on specific domains, such as movie reviews. Because it's simpler to program, the Naïve Bayes approach tends to be the most popular.

As you can see, the key to obtaining good results in sentiment analysis when using machine learning tools is to have high-quality training documents. This implies that you should train your system with the same data that you will be analyzing; otherwise, there is a high chance when training with documents from another domain you will reduce the accuracy of the results. For example, "go read the book" reflects a positive sentiment for the domain of book reviews, but it's negative in the domain of movie reviews.

The Internet has a large number of resources that can be used as training materials. For example, the movie reviews in imdb.com are already labeled with a range of polarity—the number of stars assigned. Each review contains much more than just a few sentences and there are a large number of them. Other web sites such as amazon.com and cnet.com also have a large number of reviews that are already categorized. Obviously, using these reviews only makes sense if you are going to analyze review data.

There are different ways of obtaining training data. You will hear about supervised and unsupervised training, which is based on how much you participate in the categorization of the training data. There is also the concept of bootstrapping the training data, where you take some existing data and transform it to your needs. There are many imaginative ways to create training data. A good example of this is explained in an excellent paper from Go et al. (2009), entitled "Twitter Sentiment Classification using Distant Supervision." The authors describe how they used emoticons to categorize the polarity of a tweet and then used those classified tweets as training data. The result of this work is available at http://www.sentiment140.com. On this web site, you can input a keyword and it will analyze tweets that contain it, and then produce a couple of charts with sentiment information as well as the actual tweets color-coded based on their polarity. Figure 14-1 shows a screen capture with the results of searching for the keyword splunk.

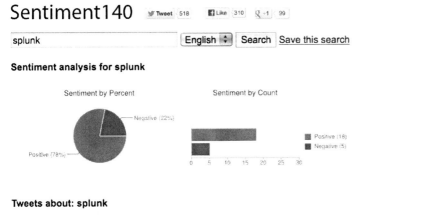

Sentiment analysis for splunk

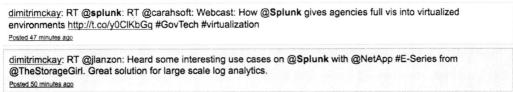

Tweets about: splunk

dimitrimckay: RT **@splunk**: RT @carahsoft: Webcast: How **@Splunk** gives agencies full vis into virtualized environments http://t.co/y0ClKbGq #GovTech #virtualization
Posted 47 minutes ago

dimitrimckay: RT @jlanzon: Heard some interesting use cases on **@Splunk** with @NetApp #E-Series from @TheStorageGirl. Great solution for large scale log analytics.
Posted 50 minutes ago

Figure 14-1. *Twitter-based sentiment analysis of the word splunk using* `sentiment140.com`

The Sentiment Analysis App

You can find a sentiment analysis App in Splunkbase. This App is based on the Naïve Bayes approach and offers a set of interesting commands and a couple of dashboards. Additionally, the App includes the necessary tools to train and test data, and two trained models based on Twitter and imdb movie reviews. It can be found at:

`http://splunk-base.splunk.com/apps/57214/sentiment-analysis`

The App, which was written by David Carasso, an engineer at Splunk, is available for free under the Creative Commons BY 3.0 license. The steps to install it from within Splunk are the following:

- From any screen in the user interface, click on the Apps pull down menu located on the upper right corner and then select "Find more apps online"

- Search for the sentiment analysis App; the search box is on the upper right part of the screen, just type sentiment

- Locate "Sentiment Analysis" and click on "Install free"

The App is now installed and ready to use. In addition to a couple of dashboards, which we will leave for you to explore, the App includes three commands:

- `sentiment`, which calculates the sentiment based on a specific model created with the training tool provided in the package

- `language`, a specific command to predict the language of the contents of a field

- `token`, which makes it easier to tokenize terms and phrases for analyzing text

Globally Enabling Commands

For the examples in this chapter, we will only be using the sentiment and language commands. This provides us with an opportunity to explain how to use commands from one App in other Apps. By default, these commands are restricted for use in the sentiment analysis App only. To enable them to be used globally within Splunk, follow these steps:

- From any screen in the user interface click on the manager button on the upper right side

- Click on "Advanced search," which is located on the bottom left side of the options

- Now click on the "Search commands" option

- From the pull down menu entitled "App context" select "All," as presented in Figure 14-2

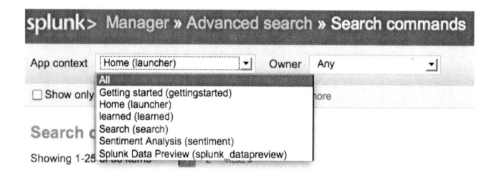

Figure 14-2. *Selecting the context of the App commands*

- Locate sentiment on the first column, entitled "Command name." In our case it is in the second page. Click on the permissions link, as shown in Figure 14-3

Command name ⇕	Command script filename ⇕	Owner ⇕	App ⇕	Sharing ⇕	Status ⇕
return	return.py	No owner	search	Global | Permissions	Enabled | Disable
runshellscript	runshellscript.py	No owner	search	Global | Permissions	Enabled | Disable
scrub	scrub.py	No owner	search	Global | Permissions	Enabled | Disable
searchtxn	searchtxn.py	No owner	search	Global | Permissions	Enabled | Disable
sendemail	sendemail.py	No owner	search	Global | Permissions	Enabled | Disable
sentiment	sentiment.py	No owner	sentiment	App | Permissions	Enabled | Disable

Figure 14-3. *Selecting sharing permissions of a command*

- Under "Object should appear in," select "All apps," as shown in Figure 14-4. Then click on "Save," located on the bottom right corner. This is also the place where you can grant role-based permissions for the commands

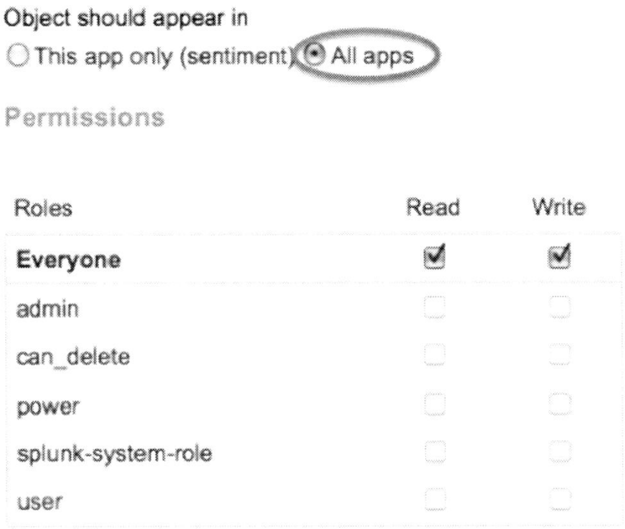

Object should appear in

○ This app only (sentiment) ◉ All apps

Permissions

Roles	Read	Write
Everyone	☑	☑
admin	☐	☐
can_delete	☐	☐
power	☐	☐
splunk-system-role	☐	☐
user	☐	☐

Figure 14-4. *Enabling a command globally*

Follow the same steps to globally enable the language command. From this point on we will be working in the context of the search App.

Finding Sentiments

Using the Twitter data that we worked with in Chapter 12, which is sample day with almost five million tweets, we will explore the use of the sentiment and language commands. Our natural curiosity leads us to first review the sentiment of tweets that include the word "love." We would expect that the resulting sentiment to be on the positive side.

Before we formulate the appropriate search, we have to explain that sentiments are expressed as numbers. When they are a straight polarity, positive or negative, they are typically represented as 1 for positive and –1 for negative, with 0 considered neutral. Ranges of polarity are represented as sequences of numbers defined by the person that trains the data. For example, 1 through 5 can be used to represent the popular five-star rating system used by many movie review web sites.

Note All of the examples and training data in this chapter are based on the English language. Machine learning tools are flexible enough to handle any language. The appropriate training data for the desired language has to be used during the training.

The search is quite simple, as all we want is to select the tweets that are in English and contain the word "love," then pass them on to the sentiment command, and, finally, calculate the average of the results:

```
index=twitter lang=en
| where like(text, "%love%")
| sentiment twitter text
| stats avg(sentiment)
```

The `sentiment` command has two arguments. The first one is the model to be used to calculate the sentiment. In this case we used the model called `twitter` that comes with the sentiment App. The second argument is the name of the field that contains the text to be analyzed. The result of the sentiment analysis is placed in a field named `sentiment`, which is used by the `stats` command to calculate the average sentiment of all of the analyzed events. The result of this search is 0.4666, which as we expected leans on the positive side. This result also illustrates one of the challenges the feature extractors have: that not all text with a powerful positive word such as love is positive. With the scale going from –1 to 1, the result can also be interpreted as "about 75 percent of all the tweets that contain the word love are positive." To understand this statement, you can look at it as follows: a score of zero (neutral) means that 50 percent of the tweets are positive and the other half negative. A score of 1 implies that 100 percent of the tweets are positive; therefore, a score close to 0.50 can be explained as 75 percent of the tweets are positive.

From the visualization standpoint, it is quite a downer to see this result as a plain number. We feel that a more appropriate visualization is a radial gauge, but when we select it, Splunk presents a gauge, shown in Figure 14-5 with a scale that goes from 0 to 100 and we need a scale from –1 to 1. Additionally, the color coding is not suitable, as green is assigned from 0 to 30, yellow from 31 to 70, and red from 71 to 100. The ideal color coding for our case is red for negative, green for positive and yellow, for neutral.

Figure 14-5. *Default radial gauge*

Unfortunately, the formatting options offered do not include changing the scale. The way around this is by creating a dashboard. When editing a panel, you can choose to edit the visualization. By selecting "Color ranges" and clicking on "Manual," you can modify the scale and colors. By default, the dialog box presents 3 ranges, which can be increased or decreased as desired, and each one has a default color. The first range has green, the second one yellow, and the last one red. Because the values of the color ranges have to be entered from lowest to biggest and we are dealing with negative numbers, we had to change the order of the colors. As can be seen in Figure 14-6, we chose to set the first range, for negative sentiment, from –1 to –0.25, with red. We set the second range, neutral, which goes from –0.25 to 0.25, with yellow. The final range, on the positive side, goes from 0.25 to 1, with green. The resulting radial gauge can be seen in Figure 14-7.

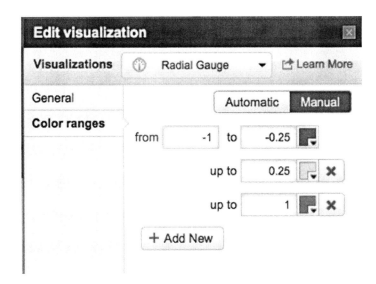

Figure 14-6. *Color range selection dialog box*

Figure 14-7. *Customized radial gauge*

It is only fair that if we analyze the sentiment of tweets that include the word love, we should also explore those that include the word hate. In Figure 14-8, you will see partial results, as the searches are still running, presented in an enhanced dashboard. It is interesting to see that the results are for any practical purpose complementary.

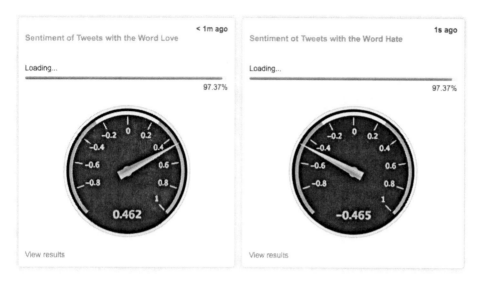

Figure 14-8. *Love and hate sentiment dashboard*

Just to complete our exploration of sentiments, we decided to analyze a set of tweets that we are suspicious are highly positive. Because Justin Bieber is one of the top three on Twitter based on the number of followers, we can only think that his fans would tweet positively about him, even more so when they use the #Beliebers hashtag. The results, which can be seen in Figure 14-9, in an even more enhanced dashboard, confirm our suspicions. The search we used is based on a similar one reviewed in Chapter 12.

```
index=twitter lang=en
| rename entities.hashtags{}.text as hashtags
| fields text, hashtags
| mvexpand hashtags
| where like(hashtags, "Beliebers")
| sentiment twitter text
| stats avg(sentiment)
```

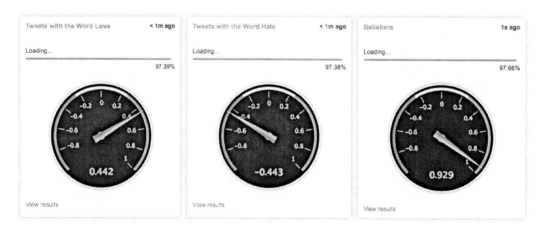

Figure 14-9. *Love, hate, and Justin Bieber*

Dealing with Languages

Trying to guess a language from a small text is quite a challenge. When you consider the nature of a tweet, which has a higher frequency of misspellings and a lot of slang, in addition to user names prefaced with the at sign (@), plenty of links in the form of http://t.co/... and word lengthening, the challenge increases dramatically.

In Chapter 12 we saw that a tweet object contains two fields for language. One is the declared language of the user and the other is what we think is a prediction from Twitter about the language of the actual tweet. As you might remember, we found that Twitter's guess at the language of the tweet was not very accurate. This is not surprising when there are tweets like:

```
@cutiemz1 Lol...I knw jor
#crawleycrowd 5850
@arturo_p hahahahahahahahahahahahahahahahahahahahsjjahahajasjhsjshjshash
RT @carlitos898: #teamfollowback
```

Knowing how challenging this is, let us give the language command a try by reviewing the first 35 tweets with English as the tweet language. At this point, we must mention that the sentiment analysis App is still in an experimental stage, and as of this writing there are some minor bugs for which we offer workarounds. Let's review the search, which is pretty simple:

```
index=twitter lang=en
| head 35
| language text
| rename language as Lng
| stats count by Lng
```

The language command has only one argument, which is the field that has to be examined to predict the language. The result is placed in a field called language. The minor bug at this writing is that the resulting field gets lost (maybe in translation) and the way to get around this is by renaming the field. The results of this search states that 34 tweets are in English and one is in Polish. The tweet categorized as being in Polish is the first one in the examples we listed earlier. For us, that is a pretty good approximation.

When we change the language to be analyzed to Spanish, things get a little more interesting. Just looking at the first 10 tweets with Spanish as the tweet language, the results state that five are in Spanish, four are in English, and one is in Catalan. After reviewing the tweets, our count is six tweets in Spanish, one in Portuguese, and three undetermined. Given the similarity between Spanish and Portuguese, this miscategorization could be understood.

By being a little bit more restrictive, we can get better accuracy. We tried again with the tweets in Spanish, but this time we not only specified that we wanted the tweets to be in Spanish but also that we wanted the declared language of the user to be Spanish. The results show seven tweets in Spanish, two in English, and one in Catalan. Our inspection counted eight tweets in Spanish and two undetermined.

After these simple tests we feel that the language command is pretty much up to par with the guesses that Twitter offers for the tweets. We are sure that with a larger text than the 140-character limit of a tweet, the language command included with the sentiment analysis App will do a much better prediction job. If you are planning to use this command you should test it more carefully and specifically to the use cases you are considering. This way, you can set the right expectations for the results.

Training and Testing Data

We have mentioned before in this chapter that the quality of the training data is key to a good sentiment prediction. Based on the issues surrounding feature extraction we reviewed earlier in this chapter, we know that sentiment analysis is not an exact science. Even with scientists and academics involved, the best accuracy is in the 80 percent to 85 percent range. In this section we will go in detail how to build training data sets and test their quality.

One of the first things you have to define is how you will handle polarity. Will your results be based on simple polarity or a range of polarity? The sentiment App comes with two models already trained. The Twitter model, which we already used in previous examples, is based on a simple polarity, positive or negative. The other model that comes with the App is based on imdb movie reviews, which uses a range of polarity from 1 to 10.

In the case of simple polarity, you have to provide a couple of sets of documents, one with examples of positive sentiment and another with negative sentiment. If your domain has a lot of neutral sentences, you should also provide a set with those samples. In the case of a range of polarity, you have to provide a set of documents for each rating in the range. In general, these documents are called the training corpus. When you are working with multiple models, it's said that you have a corpora.

The machine learning software will take the corpus, process it, and create a model, which will then be used to predict the sentiment. The bigger your corpus, that is, the number of samples in your training data, the better quality of the model. In our previous examples, the `sentiment` command was using a model called `twitter`. The corpus of this model is based on 190,852 positive tweets and 37,469 negative tweets, for a total of 228,321 tweets. The corpus of the `imdb` model is based on 11 movies reviews for each of the 10 ratings, and each review averages about 200 words. That model is available to the `sentiment` command under the name of `imdb`. In a similar fashion as the examples we used earlier, where results were either a 1 or a −1, using the imdb model will return a number between 1 and 10 for each analyzed event.

To know how well the model you created works, you test it. For this you require test data, which has to be different than the data you used for training. You classify the test data in the same way as you did the training data; this way, the testing program knows what the result should be. The testing program will then run the model and compare the predicted result with the expected result. The bigger your test data corpus is, the more reliable the accuracy of the model will be. It's customary to separate about 10 percent of the training data for testing purposes exclusively.

The twitter model included with this App also has test data, classified as 2,000 positive tweets and 2,000 negative tweets. When we run the test program for the Twitter model using this test data, it states an accuracy of 69.09 percent, which can be considered pretty decent. Additionally, the test program will also state the margin of error of their calculations. In this case, it is 1.11 percent.

To better understand how this training and testing works, we will procure a corpus and go through the process. After searching the Internet, we found an interesting corpus candidate for Twitter. Created by Niek Sanders, it consists of 5,513 tweets classified by the author into positive, negative, neutral, and irrelevant. Additionally, the corpus is labeled by topic, in this case Apple, Google, Microsoft, and Twitter, meaning that the tweet contains one of those words. Because of the restrictions in the Terms of Service of Twitter, the actual tweets cannot be distributed with the sentiment corpus, but the author provides a Python program to download the tweets according to Twitter's rules, in this case, one tweet every 28 seconds, for a total duration of about 43 hours. In the documentation, the author has a good explanation for the classification criteria. The corpus can be found at:

```
http://www.sananalytics.com/lab/twitter-sentiment/
```

We downloaded the corpus and the tweets. Then we manually extracted the positive, negative, and neutral tweets, and deleted the labels (topic, sentiment, tweet id, and timestamp). The next step is to create three files for testing. As the corpus is relatively small, we decided to create the test files ourselves by taking tweets from the data that we have been using, and classified them into the three categories with ten tweets each.

Now that we have the necessary files with the training and test data, we will explain the steps to create and test the Sanders model. As the training and testing program is written in Python, it can execute in pretty much any operating system. We illustrate the process on a Linux system. Given that the App is at an experimental stage at this writing, the program is not very user friendly. Assuming that $SPLUNK_HOME contains the location you installed Splunk, change directory and create the new subdirectories as follows:

```
cd $SPLUNK_HOME/etc/apps/sentiment/training_data
mkdir sanders
mkdir sanders/train
mkdir sanders/test
```

Copy the positive training file into the sanders/train subdirectory with the name 1.txt, and the neutral training file with the name 0.txt. Also copy the negative training file into the sanders/train subdirectory with the name -1.txt. Watch out when doing this, as many Unix commands don't like a minus sign (-) at the beginning of a file name. If you are in the same directory as the file, use the following notation every time you refer to this file: ./-1.txt.

Repeat the copy process with the test files, but copy them into the sanders/test subdirectory. Remember to name them accordingly, 1.txt, 0.txt, and -1.txt. Change directory and run the program to train the data as follows:

```
cd $SPLUNK_HOME/etc/apps/sentiment/bin
python train.py train sanders m_sanders
```

The train.py program can be used for both training and testing, so we specify that we are going to train with data in the sanders subdirectory and it will create a model called m_sanders. The output messages of this program can be found in Figure 14-10.

```
Training Directory: ../training_data/sanders/train

+0k
0k
++
loaded 3307 reviews from ../training_data/sanders/train
Original Token Count: 11079
After removing rare: 5806
After removing weakly correlated: 240
Wrote lookup approximation to "../local/m_sanders_lookup.csv"
Wrote learned model to" ../local/m_sanders.json"
```

Figure 14-10. Output of the training program

As can be seen in Figure 14-10, the number of tokens is reduced by removing rare words and words that are weakly correlated. These actions can be selected as options in the training program, and will be reviewed later in this chapter. You should note that the training program creates not only a model but also a lookup table that contains the tokens on the training data, each with the expected sentiment rating. Consider this lookup table as a bonus that can be used for basic sentiment analysis or any other related purpose.

We are very interested in learning how this model compares to the one included with the App, so we proceed to test it. For this we type python train.py test sanders m_sanders, and the output can be seen in Figure 14-11.

```
test dir: ../training_data/sanders/test

+0k
0k

loaded 33 reviews from ../training_data/sanders/test
Correct: 21.2121212121% Error: 0.937436866561 Guesses: 0.0%
```

Figure 14-11. Output of the testing program

With an accuracy of 21.21 percent, we are not very happy about this corpus. The low accuracy can be attributed to the corpus being rather small when compared to the Twitter corpus included with the sentiment analysis App, which is almost 70 times bigger. The other issue to be considered is that our test data is minimal when compared with the 4,000 tweets of the twitter model. Additionally, the setting of internal options in the training program can have an effect, but this will be discussed later in this chapter.

We can test the Sanders model with the test data of the Twitter model and see if the accuracy changes. This will be interesting, as the test data is bigger than the training data. To test, we type the following command at the Unix prompt:

```
python train.py test twitter m_sanders
```

The results are even more discouraging, as the accuracy drops to 11.67 percent. Please note that all we did was change the test data, not the model. This shows you how sensitive the models are to the amount and quality of the training and testing data. Out of curiosity we ran again the Love, Hate, and Justin Bieber dashboard, this time using the Sanders model. The results, which are radically different, can be seen in Figure 14-12. As compared to the results presented in Figure 14-9, love and Justin Bieber are neutral and hate is slightly negative.

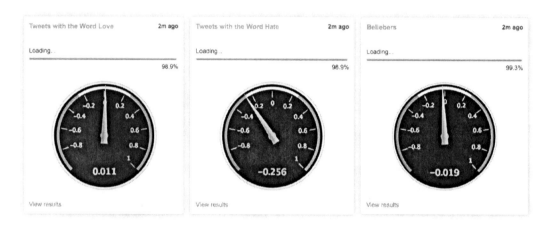

Figure 14-12. *Love, hate, and Justin Bieber using the Sanders model*

To understand the impact of a model in the predicted sentiment, we ran a quick comparison between the Twitter model and the Sanders model using the `sentiment` command. As the latter model leans heavily toward technology topics, we chose to analyze the first 1,000 tweets with the word twitter. Using the included Twitter model, the sentiment was 0.35, whereas using the Sanders model it was -0.08. This is a big difference, which goes from the low end of positive into the negative side of neutral. This is another example of how sensitive the models are.

In order to see if a bigger corpus will increase the accuracy of the model, we searched the Internet again. This time we stumbled on the blog of Ibrahim Naji, who combined the Sanders corpus with a very large corpus created in a Kaggle competition at the University of Michigan. The URLs for both sources are the following:

```
http://thinknook.com/twitter-sentiment-analysis-training-corpus-dataset-2012-09-22/
```

```
http://inclass.kaggle.com/c/si650winter11
```

After downloading the corpus, deleting the labels, and creating the positive and negative files, we extracted 50,000 records from each file to be used as test data. The final tally is that the corpus has 740,185 positive tweets and 738,442 negatives tweets, for a total of 1,478,627 training tweets. After creating the model, we tested it with the three sets of Twitter test data we have available. The results can be seen in Table 14-1.

Table 14-1. *Accuracy of the University of Michigan Twitter corpus*

Test Data	Accuracy	Margin of Error
University of Michigan	72.49%	1.05%
Twitter (included with App)	68.79%	1.12%
Sanders	60.61%	0.76%

Based on these results, it seems that the bigger the corpus the better the accuracy of the model, assuming the polarity classification is of high quality. But before we declare a winner, we should review the accuracy of the Twitter model with the University of Michigan test data. For comparison purposes, we present the results of testing the Twitter corpus shipped with the sentiment analysis app with the three test data sets we have in Table 14-2.

Table 14-2. *Accuracy of the Twitter corpus included with the sentiment analysis App*

Test Data	Accuracy	Margin of Error
University of Michigan	64.41%	1.19%
Twitter (included with app)	69.09%	1.11%
Sanders	57.57%	0.84%

Again, it seems that bigger is better, and it's interesting to note how each model performs better with their own test data. The issue is that a bigger model is inherently slower and takes more memory. These are points to ponder when making a decision. In closing this section, we remind you that the key to good predictions is the quality of the training data.

The World Sentiment Indicator Project

After reviewing how to measure sentiment based on Twitter and seeing how popular it is, we thought that we should come up with a project that does not involve tweets to measure sentiment. Arguably, newspapers can be considered the original social media, and as such it becomes an interesting alternative to measure sentiment. Pretty much all newspapers, periodical publications, and television channels have a web site, which offer the news for free or based on a paywall that provides various degrees of restricted access. Additionally, there are plenty of news outlets that exist only in the Internet.

Most news web sites offer a service where news headlines are constantly updated and made publicly available. This service is based on a format know as RSS, which stands for Rich Site Summary, also known as Really Simple Syndication. An RSS document, typically referred to as an RSS feed, or simply a feed, uses a standardized XML format, which we will review in a moment. The use of RSS feeds is not limited to news web sites alone; any web site that wants to distribute or syndicate content, such as blogs and other online publishers, use RSS as well to distribute not only news but also other media such as podcasts and videos. The web sites that offer this service display an RSS link, usually with a small icon and the acronyms RSS, XML, or RDF, to let you know that a feed is available. An example of this typically orange icon can be seen in Figure 14-13.

Figure 14-13. *The RSS icon*

RSS feeds can be read using programs called RSS readers, feed readers, or aggregators. A user can instruct a reader to subscribe to a number of feeds. What this means is that the reader program will read the RSS document at the desired web site on a regular basis, for example, every hour. These reader programs, which range from the very simple to fancy ones, allow the users to organize the headlines as they want, pretty much creating their own publications. The RSS feeds only carry the headlines and some additional information. To read the actual news article, the user clicks on the included link, which takes him to the original web site.

Every news web site is organized differently, and so are RSS feeds. Some sites have only one feed for the latest news, whereas others provide multiple feeds that categorize news by topic, such as national, business, world, sports, and so on.

What is interesting and makes this project feasible is that a large number of news web sites around the world whose main language is not English do offer one or more feeds in English. Granted, the way a news item is written is in the hands of the person that writes the article and the headline, but the variations on the same news item can make the perception of sentiment change from one polarity to another one. We will discuss this in more detail when we get to the training corpus of this project.

An RSS document contains a number of fields describing the feed, technically called a channel, and another set of fields for each news item. Although there are two standards, RSS and Atom, not all web sites fully comply with them, especially when it comes to what is placed as content in each field. The typical news item fields found in an RSS field are listed in Table 14-3.

Table 14-3. *Typical RSS item fields*

Field	Description
title	The news headline
description	A summary of the news item. The content varies a lot, ranging from nothing, to repeating the title to the whole news article
content	Not always present. As with the description field the content varies a lot
link	The link to the article in the web site
pubDate	Timestamp of the news item publication by the web site

There are other fields, such as author, copyright, category, and image, but they are not widely used. For the purposes of this project, the title field should be enough, as that is all we need to predict sentiment. By selecting RSS news feeds in English from around the world we should have a pretty good sample to predict a general world sentiment. The steps we need to do for this project are:

- Collect the RSS feeds
- Index the headlines into Splunk
- Define the sentiment corpus
- Create a visualization of the results

Collecting the RSS Feeds

The ideal solution would be to have a Splunk App that would do this, but it would require that we write the App, probably using the Python language, which is outside the scope of this book. Thus, we need to find alternate ways of collecting the RSS feeds. We looked at various RSS readers and aggregators, but none of them offered a way to store the headlines into a file, which could then be used to index in Splunk. Expanding our search, we found an open source application called Feedory, which collects RSS feeds and stores them in a database. The application is written in PHP and can be found at:

```
http://sourceforge.net/projects/feedory/
```

The installation is very simple, and connecting it to our MySQL database was a breeze. The only negative thing we found is that it generates a warning for every news item it reads, because it uses an old version of one of the libraries. It's annoying, but it does not seem to affect the application. You can see a screenshot of the main page in Figure 14-14.

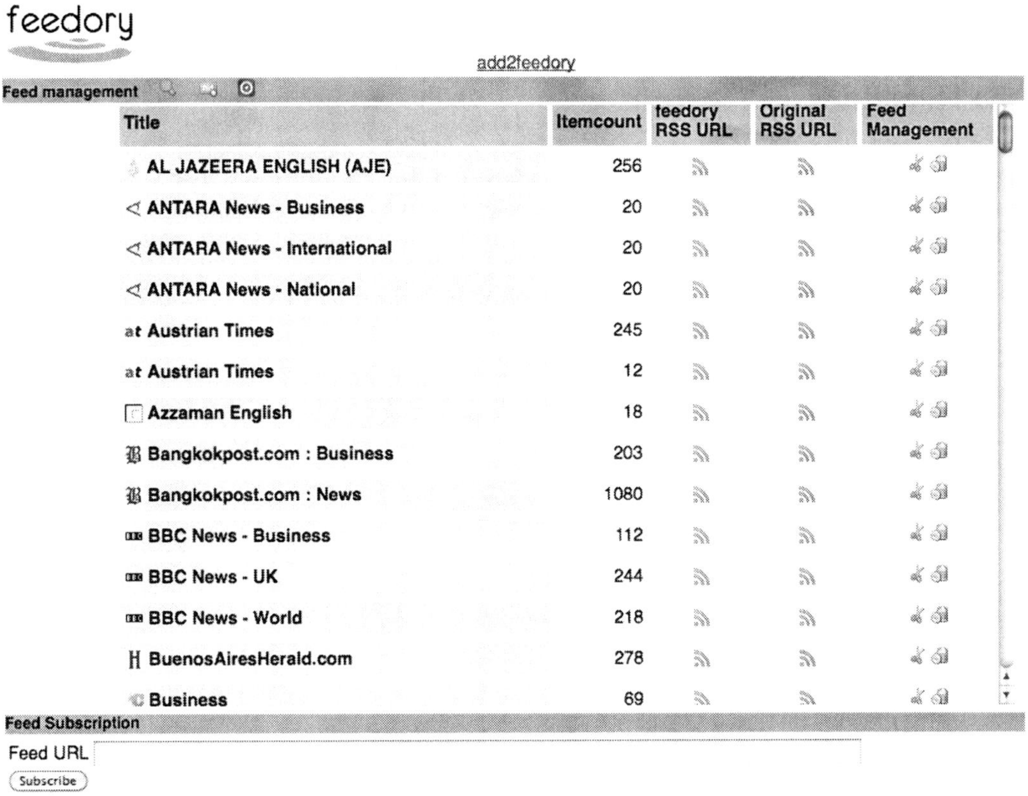
feedory
add2feedory
Feed management
Title
Itemcount
feedory RSS URL
Original RSS URL
Feed Management
AL JAZEERA ENGLISH (AJE) 256
ANTARA News - Business 20
ANTARA News - International 20
ANTARA News - National 20
Austrian Times 245
Austrian Times 12
Azzaman English 18
Bangkokpost.com : Business 203
Bangkokpost.com : News 1080
BBC News - Business 112
BBC News - UK 244
BBC News - World 218
BuenosAiresHerald.com 278
Business 69
Feed Subscription
Feed URL
Subscribe

Figure 14-14. *Feedory's main page*

As you can see in Figure 14-14, the naming of the feeds or channels is quite inconsistent. The Antara News agency of Indonesia and the Bangkok Post from Thailand clearly specify the topic of their RSS feeds, whereas the Austrian Times does not. As mentioned earlier, some web sites only have one general feed in English; in this example, Al Jazeera from Qatar, Azzaman from Iraq, and the Buenos Aires Herald from Argentina. You can also see the last entry, which has a logo and the word Business; this is not very informative. As it turns out, this feed is from the Houston Chronicle, but we only figured it out because we hovered the mouse over the RSS logo on the fourth column, which corresponds to the original URL, and it presented the one for that feed. These inconsistencies are visible in the other fields that make up the RSS feed.

We went through the process of subscribing Feedory to over 150 news feeds from about 100 web sites from all over the world. Our choices were national, world, and business news; when these were not available, we just subscribed to the general feed. The list of feeds is available in the download package of the book in a file called `world_sentiment.opml`. The OPML format is specially designed for exchanging the URLs of RSS feeds. Feedory or any RSS reader can read this file and automatically subscribe to those feeds.

To obtain the RSS feeds on a regular basis, we used the Unix crontab command to execute the Feedory update program every 15 minutes:

```
0,15,30,45 * * * * /usr/bin/php /var/www/html/feedory/feedstore.php
```

Indexing the Headlines into Splunk

We already explained how to index from a database in Chapter 9, when we loaded the airline data from our MySQL database. The difference is that in that chapter we did a one-time load and this time we will do a "tail" load, that is, we will go to the database on a regular basis and pick up the most recent news headlines. Before we do that, we need to review the tables and fields created by Feedory in the database. Then we can decide which fields we want to load into Splunk. Feedory created two tables, `fy_feeds`, which contains a list of the RSS feeds we subscribed to, and `fy_items`, which holds the actual news items. As we are only interested in the news items, we will focus on the fields of the `fy_items` table, which are presented in Figure 14-15.

```
mysql> describe fy_items;
+-------------+----------------+------+-----+---------+----------------+
| Field       | Type           | Null | Key | Default | Extra          |
+-------------+----------------+------+-----+---------+----------------+
| id          | int(10) unsigned | NO | PRI | NULL    | auto_increment |
| title       | tinytext       | NO   |     | NULL    |                |
| description | text           | NO   |     | NULL    |                |
| content     | text           | NO   |     | NULL    |                |
| link        | tinytext       | NO   |     | NULL    |                |
| author      | tinytext       | NO   |     | NULL    |                |
| postid      | tinytext       | NO   |     | NULL    |                |
| date        | datetime       | NO   |     | NULL    |                |
| feedname    | varchar(45)    | NO   |     | NULL    |                |
+-------------+----------------+------+-----+---------+----------------+
```

Figure 14-15. *Description of the fy_items database table*

The id field will be key for our project, as it will be used by the DB Connect App to keep track of which was the last item that was loaded into Splunk. The `title` field contains the news headline we want. We will ignore the description and content fields but bring over the link field as we can extract the web site the news item comes from. As mentioned earlier, the `author` field is not widely used, so we will not load it. The `postid` field contains a variant of the link to access the article, so we will ignore it, too.

As usual, the `date` field will require a more detailed examination. The first thing that we have to review is how the timestamps are set in the RSS feeds. Then we need to understand if the database does any conversions on the timestamp. After reviewing a dozen randomly selected RSS feeds, we concluded that they present a well-formed timestamp in the pubDate field; the date has the following format: Wed, 13 Feb 2013 01:22:59 +800. The only differences were in the time zone format, but they always complied with the standard. Next, we reviewed specific news items from the previously selected RSS feeds and compared the original timestamps with the ones in the database; we noticed that they had been normalized to the GMT time zone by the database. Based on this, we decided to use the date field in the database as the timestamp for the headline in Splunk. Because we are working with information that comes from all over the world, it will easier to use the GMT time zone. As we know which fields we want, we can proceed with the necessary definitions in Splunk. We start by defining the database connection under name RSS, which can be seen in Figure 14-16. You can reach this screen from any page in the user interface by clicking on the "Manager" tab on the upper right corner, then select the "External Databases" option and click on "Add new."

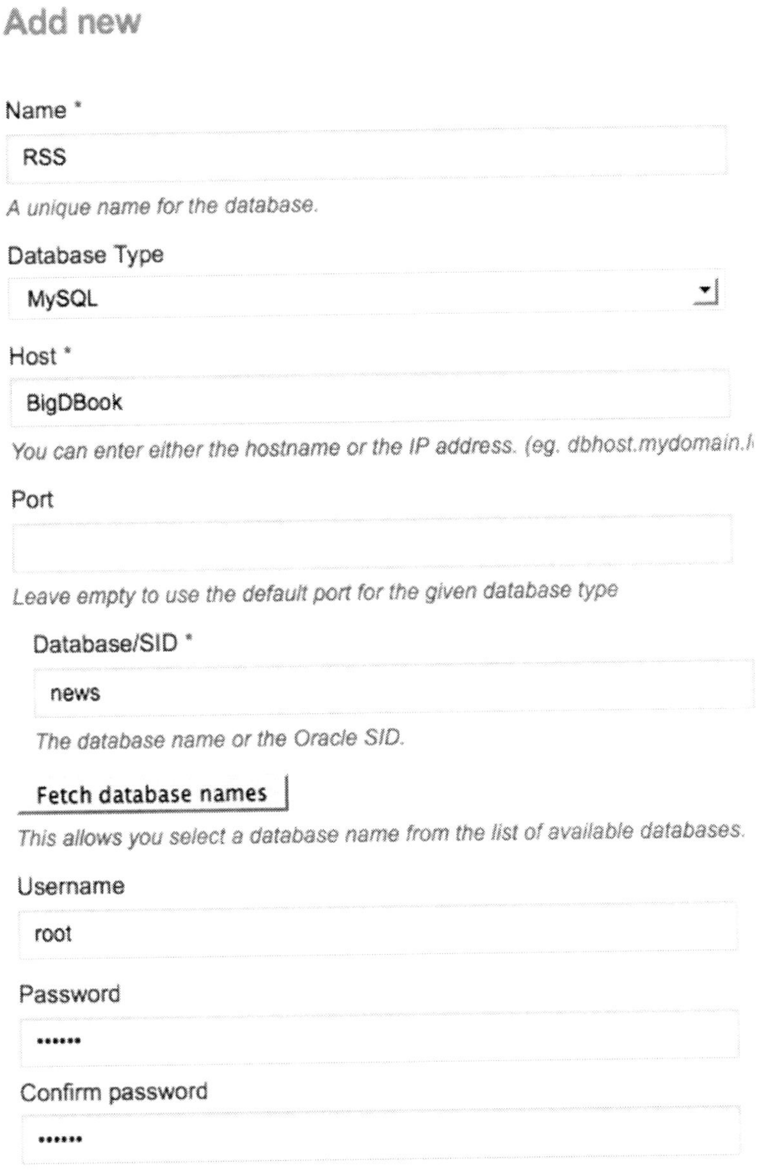

Add new

Name *

    RSS

A unique name for the database.

Database Type

    MySQL   ▼

Host *

    BigDBook

You can enter either the hostname or the IP address. (eg. dbhost.mydomain.l

Port

Leave empty to use the default port for the given database type

Database/SID *

    news

The database name or the Oracle SID.

Fetch database names

This allows you select a database name from the list of available databases.

Username

    root

Password

    ••••••

Confirm password

    ••••••

Figure 14-16. *Defining the database connection*

273

Now that we have a connection that links us directly with the Feedory RSS database, we can define the Splunk data input that will pick up the news on a regular basis. During our tests with Feedory we noticed that it picks up about 100 news headlines every hour from all the RSS feeds we are subscribed to, so there should be enough data to tell Splunk to update every 15 minutes.

To reach the data input page, click on the "Manager" tab located on the upper right corner of the user interface, then select "Data inputs;" on that screen click on the "Add new" link located to the right of the "Database Inputs" option. This will take you to the screen shown in Figure 14-17.

Add new

A Database input will fetch data from a SQL database.

Name *

```
News_Feed
```

Input Type

```
Tail (Follow based on increasing value)          ▾
```

Database

```
RSS                                              ▾
```

☑ Specify SQL query

SQL Query *

```
select id,date,title,link from fy_items {{where id > ?}} order by id
```

You can specify the SQL query that is executed against the database yourself. For information on how to s|
SELECT * FROM my_table {{WHERE $rising_column$ > ?}}

Tail input settings

Rising Column *

```
id
```

Choose a column with an increasing value. Such as a creation or modification timestamp or a sequential

Figure 14-17. *The SQL query to extract the RSS news iems from the database*

Figure 14-17 presents the first half of the definition screen. As you can see, we named the input News_Feed and defined it as a tail type. The query selects the fields that we want to load into Splunk in addition to the id field, which will be used as the rising column to keep track of the last record that was loaded. The DB Connect App will replace the question mark (?) in the query with the corresponding number. The double curly brackets ({}) enclose the part of the query that will not be used the first time the query is run. The rest of the information required to define this input is presented in Figure 14-18.

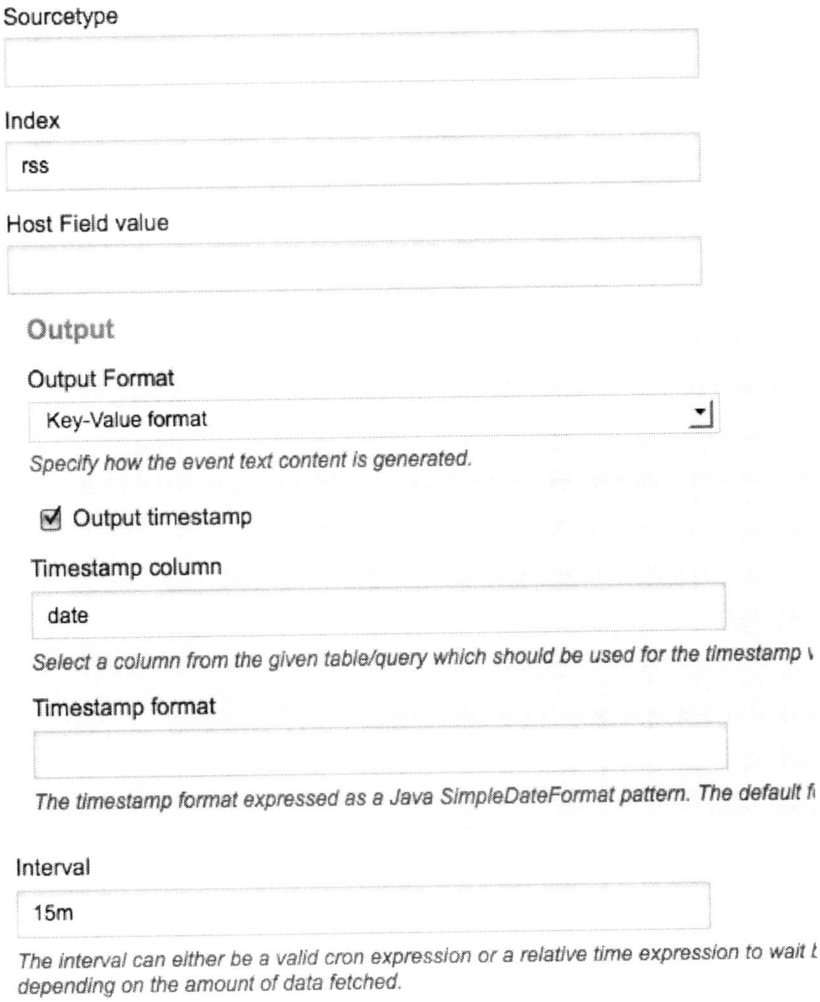

Sourcetype

Index

rss

Host Field value

Output

Output Format

Key-Value format

Specify how the event text content is generated.

☑ Output timestamp

Timestamp column

date

Select a column from the given table/query which should be used for the timestamp ∖

Timestamp format

The timestamp format expressed as a Java SimpleDateFormat pattern. The default fᵢ

Interval

15m

The interval can either be a valid cron expression or a relative time expression to wait ⌊
depending on the amount of data fetched.

Figure 14-18. Additional definitions

Figure 14-18 presents the second part of the data input screen, where we can define the source type, index, and host field value. In this case we don't have to define the source type because we will be using the key-value output option. This will provide a field name for every value we extract from the database. As mentioned in Chapter 9, this is the most efficient way to provide data for Splunk to index. In this screen we also define the field that contains the timestamp and the interval at which we want Splunk to go pick up the data, which is every 15 minutes.

Once we saved these choices, Splunk connected to the Feedory database, extracted all of the records, and loaded them into the rss index. From that point on, it only extracts and loads the records that are new since the last time. As usual, we review the events in Splunk to make sure that we have the fields we are expecting. We type index=rss | head in the search bar of the user interface, and the output is shown in Figure 14-19.

```
1    [▾]  2/14/13        2013-02-13T23:43:27.000 id=46699 title="S. Korean Bond Yields on Feb. 14, 2013"
          11:43:27.000 PM  link=http://yonhapnews.feedsportal.com/c/35025/f/647118/s/288c9fdd
                         /l/0Lenglish0Byonhapnews0Bco0Bkr0Cnews0C20A130C0A20C140C0A20A0A0A0A0A0A0AAEN20A130A
                         /story01.htm
                         host=FeedoryDatabase ▾ | sourcetype=dbmon:kv ▾ | source=dbmon-tail://RSS/News_Feed ▾

2    [▾]  2/14/13        2013-02-14T17:17:00.000 id=47211 title="Public Warehouse challenged to conduct
          5:17:00.000 PM  random inspections of rice" link=http://www.nationmultimedia.com/national/Public-
                         Warehouse-challenged-to-conduct-random-insp-30200023.html
                         host=FeedoryDatabase ▾ | sourcetype=dbmon:kv ▾ | source=dbmon-tail://RSS/News_Feed ▾

3    [▾]  2/14/13        2013-02-14T17:13:00.000 id=48117 title="IAEA committed to continuing negotiations:
          5:13:00.000 PM  Herman Nackaerts" link=http://www.mehrnews.com/en/NewsDetail.aspx?NewsID=1816990
                         host=FeedoryDatabase ▾ | sourcetype=dbmon:kv ▾ | source=dbmon-tail://RSS/News_Feed ▾
```

Figure 14-19. *The RSS news headlines in Splunk*

The timestamps are correct and every field we want is present, with the name before the value. The values are those expected for each field. The hostname is correct and the source type reflects that we chose to use key-value output. With this quick review, we feel comfortable that the RSS news headlines are being indexed into Splunk correctly.

Defining the Sentiment Corpus

Based on what we learned earlier in this chapter, this is the critical section of the project. The issue with news headlines is that the writers and editors try to create them so that they are compelling to potential readers. In doing this they can change the perceived sentiment of the news item. For example, Table 14-4 contains a few headlines of the same news item and our personal sentiment classification.

Table 14-4. *Sentiment examples of a news item*

Headline	Sentiment
Pope Benedict XVI announces resignation	Neutral
Pope 'too frail' to carry on	Negative
Pope steps down as head of Catholic Church	Neutral
Pope quits for health reasons	Negative

As you can see, sentiment classification is an extremely subjective proposition. The news of Pope Benedict XVI's resignation in February 2013 can be generally considered neutral for multiple reasons, all of them debatable. From our perspective, this news is of interest to a portion of the world population and irrelevant to the rest; therefore, we classify it as neutral. The two items we categorized as negative are because bad health can be generally considered leaning toward a negative sentiment.

As part of this process it will help us to understand the characteristics of news headlines we are working with. Unlike tweets, headlines are written by professional writers, so there should not be spelling mistakes or slang. Also, there should not be any strange characters, word lengthenings, or emoticons. In general, a headline is as clean as we can expect from a text. To find out the average size of a headline in number of characters, we do the following search:

```
index=rss
| eval chars=len(title)
| stats avg(chars)
```

The result of this simple search is 47.79 characters. To find the average size of a headline in number of words, we type the following search:

```
index=rss
| makemv title
| eval words=mvcount(title)
| stats avg(words)
```

This search converts the headline into a multivalue field, in which every value is a word, as the default separator of the makemv command is a single space. The eval command uses the mvcount function to count all the values, words in this case. The result presents an average number of 7.6 words per headline. When we compare these numbers with the findings of Go et al. in the paper mentioned at the beginning of this chapter (14 words or 78 characters per tweet), our headlines are a lot smaller, almost half the size of a tweet. Now that we have this information, we can consider a few choices for the training corpus:

- Create our own corpus based on the news headlines

- Use an existing Twitter corpus

- Use a movie review corpus

For the first option, we spent a few hours creating the training and testing files. It was a tedious exercise, in which we only produced 100 headlines for each training file and 10 for each test file. We can also confirm the conventional wisdom that during the period we worked on this, positive news was scarce and mainly found in the business news section. Another characteristic we found in the news headlines is that a large percentage of them can be considered neutral, so ideally our corpora should include a neutral file in addition to the positive and negative files. Based on the tests we did earlier in this chapter, we can predict that the quality of this corpus is not going to be very good. In order to make the best of this, we will play with some options the training program has available and are described in Table 14-5.

Table 14-5. *Training options*

Name	Description
phrases	Defines the use of phrases ("United States") or just single tokens ("United", "States")
remove_stopwords	Exclude words that are in the stop-word list
dedup	Defines if duplicate words should be eliminated from the text, therefore giving them less weight
stem	Should a word be stemmed, that is, reduced to its root form. For example, the words stemmer, stemming, and stemmed would be reduced to stem.
punct	Defines whether to pay attention to the "!" or "?" characters
emo	Include emoticons
min_term_percent	Defines the how rare of a term we are willing to accept. The default is to accept a term that appears in at least 1 of 200 headlines, tweets, or reviews (5 percent).

At the time of this writing, these options have to be changed in the training program. As the combinations are endless, we will focus on a few of them and start by defining dedup and stem as false, because we want to have as many words as possible. The reason for this is because our corpus is rather small. We will also define punct and emo as false because the headlines don't have emoticons and we personally think that "!" and "?" don't really affect the sentiment. Finally, we accept all words, no matter how rare they are, by setting the minimum term percent variable to none. At the risk of basing this on a wrong premise, we proceed to train and test our minimal news corpus by changing the phrases and remove_stopwords options one at a time. After training and testing the four combinations of the options, we found that they have exactly the same accuracy of only 38.89 percent. Although expected and frustrating, these results make sense, because the headlines are so short that the chances of any word being in the stop-word list are minimal. For the same reason, we would expect that analyzing words as a phrase or individual token won't make much of a difference.

Our second option is using an existing Twitter corpus. We have the two that we tried earlier, the one that is included with the sentiment analysis App and the University of Michigan corpus. Although not the same domain or format, but slightly similar because they are based on short text messages, we need to test them just to satisfy our curiosity. We trained both corpora with their corresponding data and tested them with the training data we used for our news model. The result using the Twitter model revealed a 40.80 percent accuracy, whereas the corpus of the University of Michigan presented an accuracy of 43.81 percent. It is obvious that a bigger corpus will yield better accuracy, even if it's a slight one. However, the low accuracy also demonstrates that the corpus has to be based on the same domain and format of the data the sentiment analysis will be done.

Just to complete the research cycle, we will take advantage of the movie review corpus available from the experiments done by Pang and Lee for the paper mentioned earlier in this chapter. In addition to the range of polarity corpus, there is a version based on simple polarity. You can find the corpora at:

```
http://www.cs.cornell.edu/people/pabo/movie-review-data/
```

We downloaded, prepared, trained, and tested the polarity dataset version 2.0, which has 1,000 each of positive and negative reviews. Our expectation was that we would obtain a higher accuracy than with the Twitter corpora, even though it's not the same domain or format. This is because the reviews are based on a large collection of words, which potentially might be a closer approximation of a news headline. Surprisingly, the accuracy dropped to 36.79 percent, the worst of all the corpora.

Given that our oversimplistic corpus with 100 each of positive, negative, and neutral are producing a respectable accuracy when compared to the others, we decided to spend another few hours increasing the size of the corpus. The newer version has 200 news headlines for each of the categories. When we tested this new model, we were surprised to see that the accuracy grew almost nine percentage points to 47.22 percent. This proves that having a corpus trained with the same data that will be analyzed is the best option of all. We present a final tally of the results of the corpora tests in Table 14-6.

Table 14-6. *Accuracy of the news models*

Model	Accuracy	Margin of Error
News (300 headlines)	38.89%	1.02%
News (600 headlines)	47.22%	1.05%
Twitter	40.80%	1.16%
University of Michigan	43.81%	1.11%
Movie reviews	36.79%	1.23%

Obviously, we will use the model based on our expanded corpus. If the difference between the models would have been small, say within a 3 percent margin, we probably would have chosen our news model because it's smaller and faster. Just to give our model a try, we run it on a sample of 10,000 headlines and the result of –0.35 matches our expectations. Based on our extensive review of headlines while we were creating the training corpus, it is mainly neutral with a slight negative tendency.

Visualizing the Results

There are many pieces of information that can be presented for the World Sentiment Indicator. In addition to displaying the sentiment for a recent period of time, we can also present a historic chart, which can provide a view of the past trends. We can also present some of the most recent headlines, along with our predicted sentiment, and the source of the news. Finally, we could also have a fine print section, which states the accuracy of the predictions and the number of headlines used to calculating the most recent sentiment indicator. All of these visualizations can be placed in a dashboard.

To display the sentiment, we can use the gauge we built earlier in this chapter. For this, we will calculate the sentiment on the headlines for the last hour. For the historic sentiment trends we can use a line chart based on the last 24 hours. So far, these are some very simple searches, but it gets a bit more complicated for the table that shows the most recent headlines. The complication is in defining the regular expression that will extract the domain of the web site from the link of the news article. The search is the following:

```
index=rss
| rename title as Headline
| sentiment news Headline
| eval Sentiment=case(sentiment==1, "Positive",
                      sentiment==0, "Neutral",
                      sentiment==-1, "Negative")
| rex field=link "http:\/\/(?<Source>[^\/]*)"
| table Headline, Sentiment, Source
```

We start by renaming the `title` field so that it shows up nicely as a header of the table. We use the `case` function of the `eval` command to convert the sentiment numeric values into something more readable, and finally we extract the domain with the `rex` command by using a regular expression that captures into the `Source` field anything after `http://` and before the next slash (/). Please note that the slash character has special meaning in a regular expression; therefore, we inhibit this by placing a backslash character (\) before it.

For the fine print we will use a visualization that is only available on dashboards, which is known as a single value display. When you define a panel in a dashboard, within the list of visualizations you can select "Single Value." The dialog box is presented in Figure 14-20. Here you can see that in addition to the title of the panel, you can define text that will be placed before and after the single value.

Figure 14-20. *Single value visualization dialog box*

The search we use is very simple, and can be seen in Figure 14-21, which displays the dialog box for the search. In this image you will also see the selected time range, which goes from –1h to now, that is, the last hour starting from the moment the search is executed.

Figure 14-21. Single value search dialog box

When we put together all of these panels in a dashboard called WorldSentimentIndicator, we get the final visualization of the project, as shown in Figure 14-22, where you can also see the single value display with the not so fine print.

Figure 14-22. *The World Sentiment Indicator dashboard*

The only thing left is that we want to automatically refresh this dashboard every 15 minutes. For this, we have to edit the XML that defines the dashboard. This is not as difficult as it might sound. Just follow these steps. From the search screen in the user interface, select the "Manage views" option from the "Dashboards & Views" pull-down menu, as can be seen in Figure 14-23.

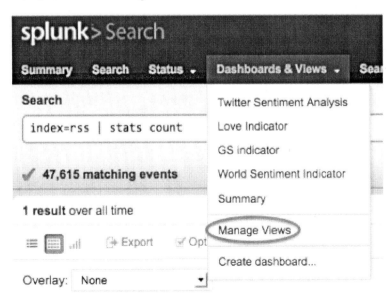

Figure 14-23. *Selecting to manage the views*

Once you are in the "Manage Views" page, you will see a list of all the views. Locate WorldSentimentIndicator and click on it. This will present you with a page that contains the XML definition of the configuration of the dashboard. The second line contains the string <dashboard>, which you have to modify by adding the following before the end: <dashboard refresh=900>. The resulting change can be seen in Figure 14-24. This directive states that we want the dashboard to be automatically refreshed every 900 seconds, which is 15 minutes.

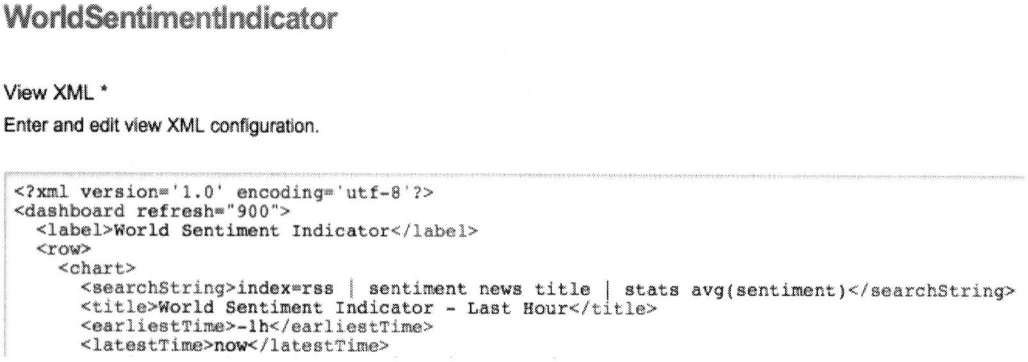

WorldSentimentIndicator

View XML *

Enter and edit view XML configuration.

```
<?xml version='1.0' encoding='utf-8'?>
<dashboard refresh="900">
  <label>World Sentiment Indicator</label>
  <row>
    <chart>
      <searchString>index=rss | sentiment news title | stats avg(sentiment)</searchString>
      <title>World Sentiment Indicator - Last Hour</title>
      <earliestTime>-1h</earliestTime>
      <latestTime>now</latestTime>
```

Figure 14-24. Adding the automatic refresh to the dashboard

Once you click on the save button at the bottom right corner of the page, the dashboard will start refreshing automatically every 15 minutes. You can find the most recent version of this project at http://www.opallios.com, where we continue increasing the size of the training corpus, trying to achieve a higher accuracy of the World Sentiment Indicator.

Summary

In this chapter you learned about the reasons to analyze sentiment. Then we downloaded and used the sentiment analysis App. You also learned about all the issues surrounding the creation of sentiment models and training corpora, and the importance of using the same data as that which will be analyzed to obtain a higher accuracy in the prediction rates. Next we examined the World Sentiment Indicator project, which took us through the collection of RSS news feeds from around the world, indexing them into Splunk by connecting on a regular basis to the database that contains the news headlines. Then we created our own training corpus and compared it to other corpora, where once again we found that the best training data is the one based on the data that will be analyzed. Finally, we created a dashboard that automatically refreshes every 15 minutes to visualize the results of the different searches we created.

■ ■ ■

## CHAPTER 15

Remote Data Collection

So far we have been indexing data into Splunk that comes from files that are located on the same server where Splunk is installed, or from databases using the DB Connect app. With the latter, it's pretty much transparent where the database is located, as the DB connect app—or any database client, for that matter—can easily communicate with a database over the network. In this chapter, we review how to collect data from remote servers using the Splunk forwarders. We will also go over the Splunk deployment server, which allows updating remote Splunk instances, including forwarders, and the deployment monitor app, which presents the state of all Splunk instances.

Forwarders

A simple definition of a Splunk forwarder is a program that collects data locally and sends it over the network to a Splunk indexer or another forwarder, which we will call a receiver. A forwarder can collect data using the methods (reviewed in Chapter 2) that are available to a Splunk indexer:

- From files and directories, either by monitoring them or uploading the contents only one time

- From network ports, using the UDP or TCP protocols

- From Windows systems

As you will see, there are several ways that a forwarder can be used, but it is generally associated with the collection of machine data, such as log files and metrics related to the performance and monitoring of computer servers.

Traditional methods of data collection include methods that rely on a central data repository, by using NFS or SMB to mount common disks across multiple servers where log files are placed. In this case, a Splunk indexer can monitor the directories that contain those files and index them. Considering that data centers can have hundreds or thousands of servers, this method does not really scale very well, in addition to increasing latency when there are many files involved. Similar topologies can be considered when using syslog on Unix-based systems, but it gets more complicated when dealing with Windows systems. Splunk forwarders have a set of features that make them a very attractive alternative for remote data collection:

- Caching. A Splunk forwarder has the capacity to store data locally in a queue, either in memory or in disk. This becomes extremely useful when the receiver is not available due to maintenance or a crash, as the data being collected will not be lost. The size of the queue can be configured, but it defaults to 500 KB.

- Receiver acknowledgment. A forwarder can be configured to wait for an acknowledgment to make absolutely sure that a block of data was received. If there is a failure, the forwarder sends the data block again. This feature is disabled by default.

- Compression. Data can be compressed by the forwarder and later decompressed by the receiver. This operation slightly increases the CPU usage of both servers, but the data takes less time to travel from one end to the other.

- Security. The whole data stream between the forwarder and the receiver can be encrypted using SSL.

- Load balancing. When the forwarders are sending information to a set of indexers configured for distributed search, the forwarders will distribute the data between all the indexers based on a defined period of time. This will be discussed in more detail in Chapter 16.

- You can also use a forwarder to send information to a third-party system.

Forwarders run not only on Unix-based systems, including MacOS, but also on Windows systems. There are other features that apply depending on the type of forwarder that is being used. Because there are so many different uses of forwarders, Splunk has three types available:

- Universal forwarder. This is the preferred forwarder, given its very small memory footprint and minimal CPU consumption.

- Light forwarder. This one is found mainly on legacy situations. Under normal circumstances, there is no reason to use it.

- Heavy forwarder. This is an actual instance of Splunk with certain features disabled, such as the searching and alerting capabilities, and Splunk Web, the user interface.

The general rule of thumb is always to use the universal forwarder, unless you need to parse events, either to route or filter them. The latter makes sense only if you are filtering more than 50 percent of the events, which is a typical situation when sending data over a Wide Area Network (WAN). The heavy forwarder is also useful for "store-and-forward" situations.

Popular Topologies

Now that you have a good idea of the features the forwarders offer, we can share some of the most typical topologies that involve the use of forwarders. In the first example in this chapter, we mentioned an NFS-shared disk as a repository of log files. If this topology (presented in Figure 15-1) is something that already exists in your network, the easiest way to index these files into Splunk is by installing a universal forwarder on the server that hosts the disk, or disks that are shared with the other servers.

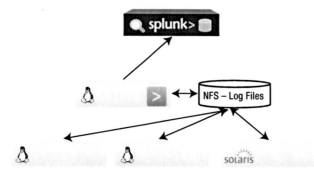

Figure 15-1. *Shared disk topology*

As discussed earlier, the legacy topology presented in Figure 15-1 does not scale very well, and it limits the inputs of the forwarder to files and scripted inputs that exist in the specific server that hosts the forwarder. The ideal situation is to have a forwarder deployed on each server. Not only is this topology more scalable and offers a better performance; it also can handle all the traditional Splunk inputs on each and every one of the servers where the Splunk forwarder is installed, as can be seen in Figure 15-2. Additionally, if the Splunk deployment is a cluster of indexers, then the forwarders will also load balance the data to be indexed.

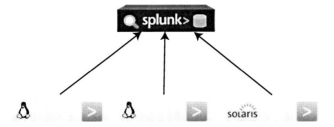

Figure 15-2. *Universal forwarders on each server*

A word about Windows systems: you cannot directly index from a Windows server to a Splunk instance running on a Unix-based system. You can load data from a Windows server to an instance of Splunk running on Windows, or, alternatively, you can install a forwarder on a Windows system and send the data to an instance of Splunk running on a Unix-based system. The following Windows specific inputs can be handled by Splunk:

- Event logs. These can be collected by a forwarder, either locally or remotely, that is on another Windows server, but it requires valid domain credentials.

- Performance monitoring. As with event logs, you can collect PerfMon data or WMI metrics in general, either locally or remotely.

- Registry monitoring. Where you can monitor the Windows registry for any change. This can only be done locally.

- Active Directory (AD) monitoring. As with the registry, this can only be done locally.

Whereas having a Windows server running a forwarder, which monitors remotely a number of other Windows servers, is a feasible topology, it is not recommended due to performance, deployment, and management considerations. Moreover, with remote monitoring you will not be able to monitor registries and active directory. A more popular topology is simply to install universal forwarders on all of the Windows servers. A typical topology, with a mix of Windows- and Unix-based servers, can be seen in Figure 15-3.

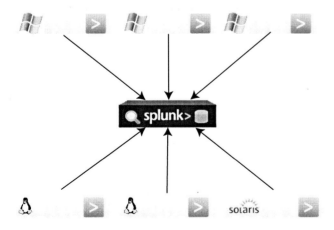

Figure 15-3. *A typical mixed operating system topology*

Whereas the data collected by forwarders contains the standard Splunk fields (host, source and source type), which makes for easy categorization, these fields might not be enough to catalog data. For example, when collecting Windows WMI events, you might want to separate hardware from software events, so that they can be analyzed by the appropriate technical people. This is a good example of where you should use a heavy forwarder, as the categorization goes beyond the standard fields and you actually have to examine an event in detail to determine the WMI code and decide where it should be handled. Assuming that each technical group has its own Splunk instance, which is monitoring and alerting issues, Figure 15-4 shows a topology to handle this situation.

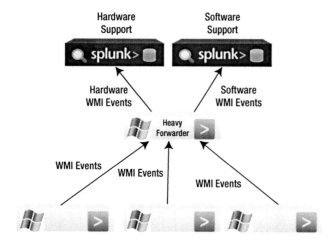

Figure 15-4. *Routing topology based on a heavy forwarder*

The installation of the forwarders depends on the type and operating system. Whereas the heavy forwarder is essentially a normal Splunk instance, a universal forwarder has the following options:

- On Windows, it can be installed using the installer GUI or from the command line on a local system

- For Unix-based systems, the forwarder can only be installed using the CLI on a local server

- For either system, you can remotely deploy the forwarder

- You can make the forwarder part of the system image

Installing a Forwarder

We will go through the exercise of installing a forwarder on a Unix system using the CLI on a local server, and then installing one on a Windows system using the GUI. For either case, you will have to start by enabling the receiver to accept data streams from the forwarders. This can be done by either using the Splunk CLI or Splunk Web. In the first case, it is very simple. For example, using Linux, just type the following from the bin directory of Splunk:

```
./splunk enable listen <port> -auth <username>:<password>
```

Here port is the network port on which Splunk will listen for the data stream coming from the forwarder(s). Port 9997 is typically used, but you can use any other unused port, usually above the number 1024. If you don't specify the -auth argument, Splunk will prompt you to input the user name and password. To disable receiving, just change the word enable for disable.

To enable the receiver using Splunk Web, just go to the "Forwarding and receiving" section in the "Data" area of the Manager. To the right of the line that states "Configure receiving", click on the "Add new" link. This will take you to a screen like the one shown on Figure 15-5. All you have to do here is specify the port to which Splunk should listen for the data streams coming from the forwarder(s) and save the configuration.

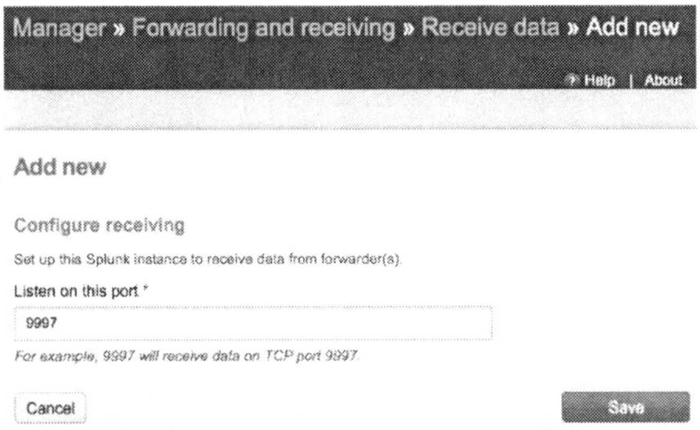

Manager » Forwarding and receiving » Receive data » **Add new**

⊕ Help | About

Add new

Configure receiving

Set up this Splunk instance to receive data from forwarder(s).

Listen on this port *

9997

For example, 9997 will receive data on TCP port 9997.

Cancel   Save

Figure 15-5. *Configuring receiving*

Remember that you have to open the selected port in the firewall, otherwise you will have obscure problems on the forwarder side that will not point to a closed port on the receiver. Now that the main Splunk instance in our example is ready to receive data from forwarders, we can go ahead and install a forwarder. First, we will provide the instructions for installing a universal forwarder locally on a Unix-based system:

- Download and install the universal forwarder on the desired server

- We strongly recommend that you have the forwarder start every time the system boots up. You can specify this by typing the following command from the `bin` directory of Splunk: `./splunk enable boot-start`

- Start the forwarder by typing the following command from the `bin` directory of Splunk: `./splunk start`

- Configure the forwarder to connect to the receiver, in our example `bigdbook`, by typing the following from the `bin` directory of Splunk: `./splunk add forward-server bigdbook:9997`

- Verify that the forwarder has connected with the receiver. Type the command shown in Figure 15-6 and expect an output similar to the one shown in that figure.

```
[root@03078-1-1443013 bin]# ./splunk list forward-server
Active forwards:
        bigdbook:9997
Configured but inactive forwards:
        None
[root@03078-1-1443013 bin]# █
```

Figure 15-6. *Verifying the connection with the receiver*

- Now you can define the data that you want to collect with the forwarder. For our example, we will have the forwarder collect all of the logs in the `/var/log` directory of the machine where it is installed. This is done with the following command: `./splunk add monitor /var/log`. If you want to send the data to a specific index on the receiver, the command will look like this: `./splunk add monitor /var/log -index your_index`

To install a universal forwarder on a Windows system, just follow these steps:

- Download the appropriate .msi file (32 or 64 bits)

- Double click on the .msi file; this will bring up a screen with the Install Wizard, a common tool to install software on Windows systems. Click on "Next", which will take you to a screen with the software license agreement. Select the radio button to accept the agreement and click on "Next".

- The wizard now asks for the location of the installation. Unless you prefer to install the forwarder on a different directory, just click on "Next".

- The next screen gives you the option to specify a deployment server. For the purposes of this example, you will skip this step, so just click on "Next".

- Now you can input the information of the receiver. For this example, you use the same information as you used for the previous example for a Linux server, that is, bigdbook:9997, and click on "Next".

- You can optionally provide information regarding the SLL certificate in the next screen. For this example we skip this by clicking on "Next".

- Now the wizard asks if you want to obtain the data locally or remotely. We suggest that you install a forwarder on every Windows server you have and collect data locally. For this example, you select "Local Data Only" and click on "Next".

- Because you have chosen to monitor data locally, the next screen presents the data that is available to monitor locally, as can be seen in Figure 15-7. For this example, chose to monitor the security log, PerfMon and Active Directory, and click on "Next" after that.

Figure 15-7. *Windows monitoring options for the Splunk universal forwarder*

- At this point, Splunk is ready to install the forwarder. Just click on "Install" and it will be done in a couple of minutes.

When working in a network with hundreds or thousands of servers, the deployment of a forwarder on every single server might seem a difficult task. However, most system administrators running large networks make use of configuration management tools such as Chef and Puppet for deployment and day-to-day management of large networks. Once the forwarders are installed and configured as desired, you can continue to use those tools, or you have the choice to use the Splunk deployment server tool for ongoing maintenance of the forwarders and Splunk instances in general.

Deployment Server

Splunk's deployment server is a tool to manage configurations of Splunk instances, be it remotely or locally. The instances can be indexers, search heads, or forwarders. The way the deployment server works is very simple. It holds a set of configuration files, which are considered the masters, and are copied over to the remote servers every time there is a change. These configuration files are known with the confusing name of deployment app; note that this is not a Splunk app. Deployment clients, which are Splunk instances enabled to receive content from the deployment server, periodically check to see if there are any updated versions of deployment apps. A deployment client can receive one or more deployment apps. Deployment clients can be grouped into what is called server classes. This allows for logical groupings based on similar configuration files. Clients can be part of more than one group.

▪ **Caution**   The deployment server cannot be used to install Splunk instances on remote servers. Its functionality is limited to update content, including licenses on existing remote Splunk instances.

To illustrate the concept of server classes, we will use an example. Suppose that you have a typical multitiered web facing application as depicted in Figure 15-8. The first layer is composed of Linux servers running the Apache web server. The second layer hosts the application that has all the business logic, which runs on the JBoss Java application server, which runs on top of the Linux operating system. The third layer has the Linux servers running the MySQL database. There is a Splunk instance, which is used to do searches and index the data coming from all the forwarders, one on each server. This Splunk instance is also the deployment server.

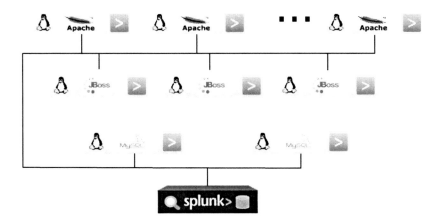

Figure 15-8. *Typical multitiered application*

Using the types of inputs as an example of what each forwarder monitors and collects, we can create a few server classes:

- Linux servers. This class applies to all the servers on our example. We want the forwarders to monitor all the Linux log files.

- Web servers. This class is only for the servers on the first tier, which are running the Apache web server. We want the forwarders to monitor the Apache log files, such as the access log.

- JBoss servers. This class is for the second tier servers running the JBoss Java application server. The forwarders should monitor the JBoss log files.

- Database servers. This class comprises the servers that run the MySQL database. The forwarders monitor the log files of the database.

With these four input-based server classes, we can update the Splunk instances as needed. For example, if the database has been upgraded and now has an additional log file we want to collect, all we do is update the master configuration files on the deployment server for the database server class. The next time a deployment client, which is a member of the database server class checks for updates, it will download the latest version of the corresponding configuration files. After doing this, the forwarder will restart in order to activate the new configuration files. If you update a configuration file in the deployment app of the Linux servers class, all of the servers in the example will be updated.

By organizing the servers of the example as we did, you can efficiently update configuration files, or deployment apps, depending on the type of inputs that are being collected. You can organize the server classes as best suits you: for example, you can organize by geographical location, data center, time zone, operating system, and so on.

To illustrate how to use the deployment server, we will use it to handle the forwarders on three Unix-based systems. To simplify the example, the main Splunk instance is also the deployment server, but be aware that when you have more than 30 forwarders you are better off having a deployment server on a separate instance of Splunk. You will have the forwarders collect all the log files in /var/log and send them over to the main Splunk instance. For this, you will create two deployment apps, one that specifies the files to be collected, the inputs, and another deployment app that specifies where the data will be sent to, the output. This way you can easily change the inputs or outputs independently. The topology of this example can be seen in Figure 15-9.

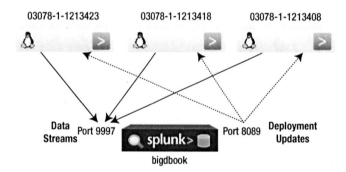

Figure 15-9. *Topology of deployment server example*

There are several ways that this setup can be done, and we don't claim that any one is better than the others. The approach to take for this example is first to work with the Splunk instance designated as the deployment server, and once you are ready then to go from server to server installing the forwarders. This last step can definitely be optimized by the use of tools such as Chef and Puppet, or creating server images, but for didactic purposes we will leave it as explained.

Configuring the Deployment Server

The configuration of the deployment server has to be done by editing the configuration files directly. Our example is quite simple, but will show all the necessary steps to accomplish this job. More complicated configurations can be done, and you can find explanations and examples in the Splunk documentation.

As it is the first time the bigdbook server will be set as a deployment server, we need to create a serverclass.conf file in the etc/system/local directory of the Splunk installation. This file defines the server classes, which in our example there is only one that we call CentOS6.2, the name and version of the operating system used on the Linux servers we have. This file also contains the names or IP addresses of the servers that are members of this server class. Additionally, the file includes the deployment apps available for this server class. As mentioned earlier, this example has two deployment apps, one that defines the inputs to the forwarders called collect, and another that specifies the receiver, which in our case is the bigdbook server. This last deployment app is called *send*. The file looks like this:

```
# Example serverclass.conf

# Global server class - Must exist!
[global]
whitelist.0=*

# The server class and its members
[serverClass:CentOS6.2]
whitelist.0=03078-1-1613423
whitelist.1=03078-1-1613418
whitelist.2=03078-1-1613408

# The members could also have been defined using wildcards, for example:
# whitelist.0=03078-1-16134*

# The Collect deployment app definition
[serverClass:CentOS6.2:app:collect]
stateOnClient=enabled
restartSplunkd=true

# The Send deployment app definition
[serverClass:CentOS6.2:app:send]
stateOnClient=enabled
restartSplunkd=true
```

The server class configuration file uses the concept of white lists and black lists to include or exclude servers, allowing for a high level of detail when managing servers. The stateOnClient property states if the deployment app is enabled or disabled on the client regardless of the state on the deployment server. The restartSplunkd property specifies if a client has to be restarted after being updated, which in our case is necessary for the changes to take effect. The next step is to create the directories where the configuration files for each deployment app will be placed:

```
mkdir -p $SPLUNK_HOME/etc/deployment-apps/collect/default
mkdir -p $SPLUNK_HOME/etc/deployment-apps/send/default
```

In the send/default directory we create the outputs.conf file, which contains the specifications of where the forwarders will send the data stream. In this example, we are sending it to port 9997 of the bigdbook server, but if in the future we have a cluster of indexers, all we have to do is modify this file on the deployment server and it will automatically be updated on the deployment clients of this server class, and each forwarder will automatically restart so the changes can take effect. The file looks as follows:

```
# Example outputs.conf file
[tcpout]
defaultGroup=receiver

# Here we specify the Splunk instance that receives the data
[tcpout:receiver]
server=bigdbook:9997
```

The next step is to create the `inputs.conf` file for the collect deployment app. This is done in the `collect/default` directory that was created earlier. The file contents are:

```
# Example inputs.conf file
[monitor:///var/log]
disabled=false
```

Admittedly, this is a very simple example of an `inputs.conf` file as it only monitors the `/var/log` directory. The inputs configuration file has a large number of attributes, which are detailed in the Splunk documentation. The reason for having three slashes (/) is that the first two are part of the format required by Splunk. The third slash is part of the path name of the directory (or file) to be monitored.

Because it is the first time that the deployment server has been set up, we need to restart the Splunk instance. This has to be done every time the `serverclass.conf` file is changed, otherwise it's enough to issue the reload command: `./splunk reload deploy-server`. Finally, we need to make sure that Splunk is ready and listening on port 9997 for the data streams that the forwarders will send, which can be done as presented in Figure 15-5.

Configuring the Forwarders

As mentioned earlier, there are better ways to install and configure the forwarders using other system administration tools, but to illustrate the steps to configure the forwarders, we will use this rather rudimentary procedure. Once you have downloaded the corresponding forwarder to the appropriate server, you expand it by using the `tar` command, for example: `tar xzf splunkforwarder-5.0.2-149561-Linux-x86_64.tgz`.

Now you can change directory to `splunkforwarder/bin` and configure the forwarder before starting it by typing the following command: `./splunk set deploy-poll bigdbook:8089 --accept-license`. This command specifies the name of the deployment server, along with the management port, which usually is 8089. As we are installing the forwarder for the first time, it will present and ask us to accept the license; we avoid having to go through this by accepting the license on the command itself. Now we can start the Splunk forwarder by typing `./splunk start`. Please be aware that if you are asked for a login and password, you must use `admin` and the default password `changeme`, as the password has not been changed. After it starts it will take a couple of minutes to connect with the deployment server, download the corresponding deployment apps and restart itself. One way to see if the process has completed is by typing the command: `./splunk list forward-server`, which should produce an output similar to that presented in Figure 15-6, when it is done. These steps are repeated for each Unix-based server where we want a forwarder, which in our example is just three servers.

Deployment Monitor

The deployment monitor is a free Splunk app that keeps track of all the Splunk instances, including forwarders, and provides early warnings of abnormal or unexpected behaviors. This app offers a wealth of information that ranges from weekly comparisons of the amount of data indexed and the number of connections, to license usage by source type. We show some of the pages the app produces to give you an idea of its functionality. In Figure 15-10, you will see some of the warning the app presents.

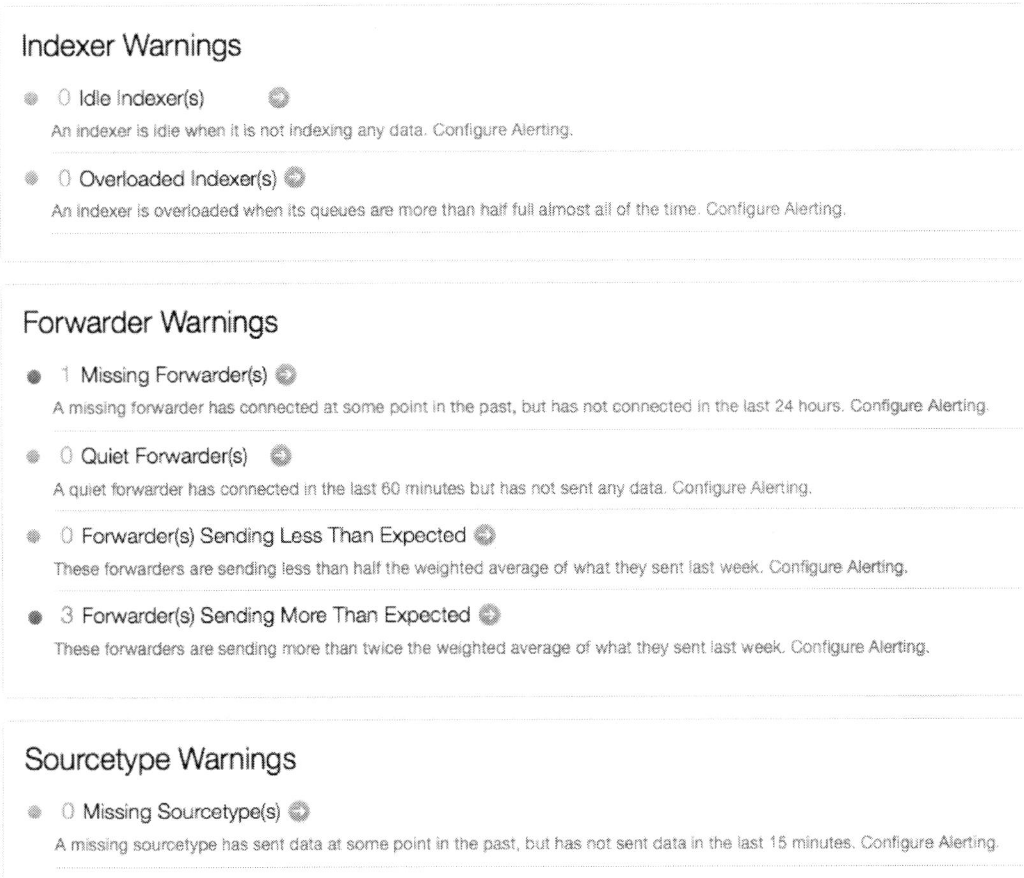

Indexer Warnings

● 0 Idle Indexer(s) ◎

An indexer is idle when it is not indexing any data. Configure Alerting.

● 0 Overloaded Indexer(s) ◎

An indexer is overloaded when its queues are more than half full almost all of the time. Configure Alerting.

Forwarder Warnings

● 1 Missing Forwarder(s) ◎

A missing forwarder has connected at some point in the past, but has not connected in the last 24 hours. Configure Alerting.

● 0 Quiet Forwarder(s) ◎

A quiet forwarder has connected in the last 60 minutes but has not sent any data. Configure Alerting.

● 0 Forwarder(s) Sending Less Than Expected ◎

These forwarders are sending less than half the weighted average of what they sent last week. Configure Alerting.

● 3 Forwarder(s) Sending More Than Expected ◎

These forwarders are sending more than twice the weighted average of what they sent last week. Configure Alerting.

Sourcetype Warnings

● 0 Missing Sourcetype(s) ◎

A missing sourcetype has sent data at some point in the past, but has not sent data in the last 15 minutes. Configure Alerting.

Figure 15-10. *Warnings from the deployment monitor app*

The warnings presented in Figure 15-10 are self-explanatory. In this particular case, there is one missing forwarder because we turned off the one we used as an example on how to install a forwarder on Windows using the GUI. Also, the warning about the three forwarders that are sending more data than expected is there because the previous week there were no forwarders. Another interesting page is the one that details the forwarders and their activity, which can be seen in Figure 15-11.

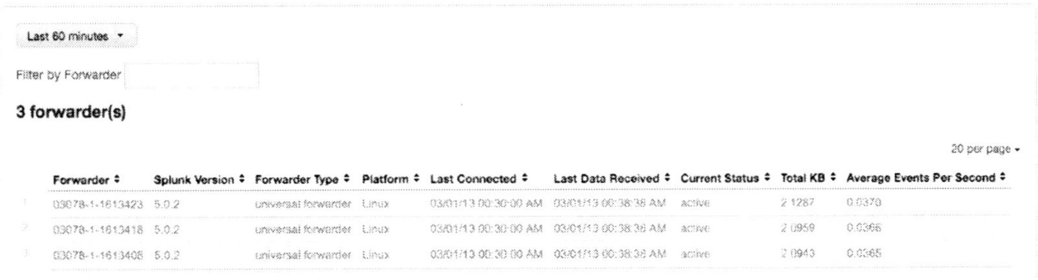

Last 60 minutes ▾

Filter by Forwarder

3 forwarder(s)

20 per page ▾

Forwarder ⇕	Splunk Version ⇕	Forwarder Type ⇕	Platform ⇕	Last Connected ⇕	Last Data Received ⇕	Current Status ⇕	Total KB ⇕	Average Events Per Second ⇕
03078-1-1613423	5.0.2	universal forwarder	Linux	03/01/13 00:30:00 AM	03/01/13 00:38:36 AM	active	2.1287	0.0370
03078-1-1613418	5.0.2	universal forwarder	Linux	03/01/13 00:30:00 AM	03/01/13 00:38:36 AM	active	2.0959	0.0356
03078-1-1613408	5.0.2	universal forwarder	Linux	03/01/13 00:30:00 AM	03/01/13 00:38:36 AM	active	2.0943	0.0365

Figure 15-11. *Forwarder details*

As can be seen in Figure 15-11, the information about the forwarders is comprehensive. Because we have only three forwarders, we don't need to use the filter option, but it is very handy when you have hundreds or thousands of forwarders. The deployment monitor app also presents detailed information on the indexers, but as we have only one, there is not much to be seen. In Figure 15-12, you can see the page that presents the information related to source types.

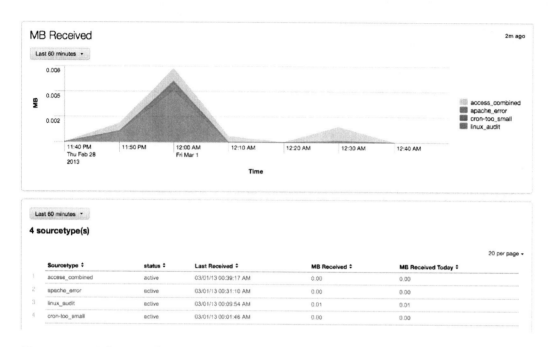

Figure 15-12. *Information by source type*

As you can see, the deployment monitor app can be extremely useful, especially when you have large numbers of forwarders, as it provides a very comprehensive set of information related to all the data that is being indexed, the indexers, and the remote sources of the data.

Summary

In this chapter, you have learned about Splunk forwarders and their use, configuration, and topologies. You have also learned about the deployment server and reviewed an example of its use, as well as the deployment monitor, used to capture information on the use of all of the Splunk instances and warnings related to abnormal situations.

CHAPTER 16

■ ■ ■

Scaling and High Availability

All of the examples and projects in the book so far have been based on using a single instance of Splunk, which acts as an indexer and searcher. In this chapter, we will review how to scale Splunk by adding more servers to handle the indexing and search components. You will also learn the basic principles of clustering, which increases the resiliency of Splunk by handling automatic failover of indexers. We will set up a sample cluster to illustrate the basic steps of this process.

Scaling Splunk

The functionality of Splunk can be roughly broken down into three basic areas:

- Collecting data

- Indexing the collected data

- Searching for information in the indexed data

One of the powerful features of Splunk is that you can separate each of these functional areas and scale them individually. In Chapter 15, you learned how you can expand the data collection to remote servers, be they Unix-based systems or Windows servers, with the use of forwarders; you also reviewed some of the typical topologies that involve the use of forwarders. In this chapter we will review how to scale the indexing and search functionalities, individually or in combination.

There are multiple reasons to scale either functionality. If you have hundreds of forwarders sending information to a single Splunk instance, depending on the volume of data they are sending, the single instance can quickly become a bottleneck, so you can add another Splunk instance to assist with the load. One advantage of doing this is that the forwarders can distribute the data across both instances, helping with scaling. An additional advantage is that this distribution implicitly creates a fail-over capability for the forwarders. When one Splunk instance goes down, the forwarders automatically go to the other instance. This topology can be seen in Figure 16-1.

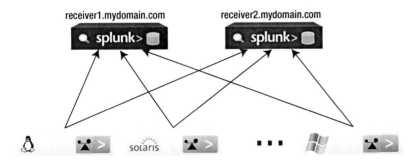

Figure 16-1. Many forwarders load balancing data on two indexers

Whereas the topology presented in Figure 16-1 is a valid one, it does raise the problem that each instance of Splunk will only have part of the indexed data. A search on one instance will only go over the part of the data that was indexed in that particular instance, not all of the data. In order to be able to search over all the indexed data on both instances, you have to use distributed search. This is not as difficult as it sounds, as all it takes is a two-step process, which involves activating this feature in the Manager in Splunk Web (see Figure 16-2), and then defining the search peers. To get to the activation screen, go to the Manager and select the Distributed search option from the Distributed Environment section. Once there, click on the Distributed search setup link.

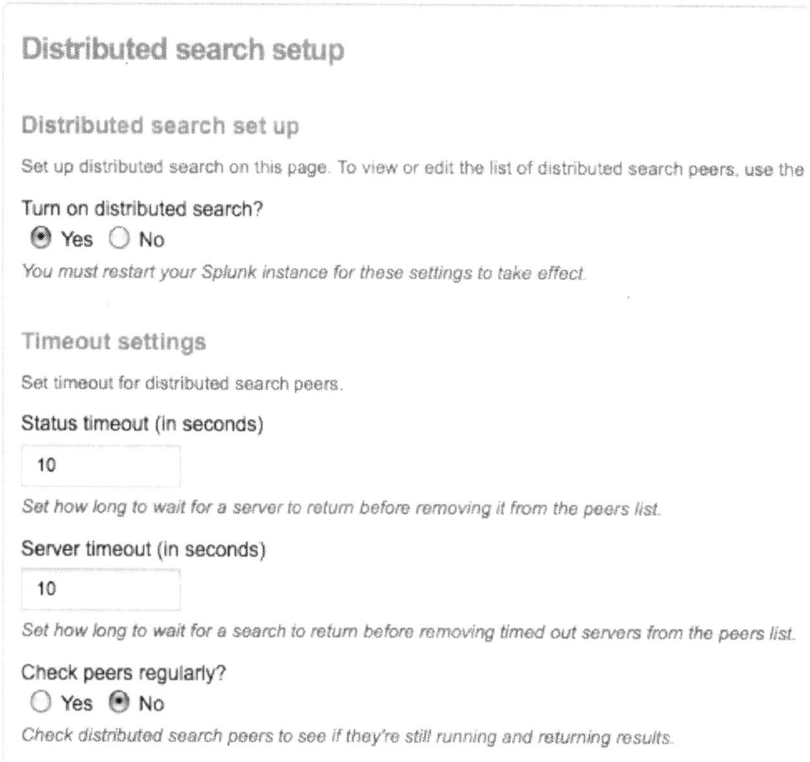

Distributed search setup

Distributed search set up

Set up distributed search on this page. To view or edit the list of distributed search peers, use the

Turn on distributed search?
◉ Yes ○ No
You must restart your Splunk instance for these settings to take effect.

Timeout settings

Set timeout for distributed search peers.

Status timeout (in seconds)

| 10 |

Set how long to wait for a server to return before removing it from the peers list.

Server timeout (in seconds)

| 10 |

Set how long to wait for a search to return before removing timed out servers from the peers list.

Check peers regularly?
○ Yes ◉ No
Check distributed search peers to see if they're still running and returning results.

Figure 16-2. Enabling distributed search

Once you have enabled the distributed search (we usually leave the other options at their default values), you will have to restart the Splunk instance. The second step in this process, which is the definition of the other Splunk instances that participate in the search, is also very simple, as can be seen in Figure 16-3, in which we define the second indexer as a peer of the first one.

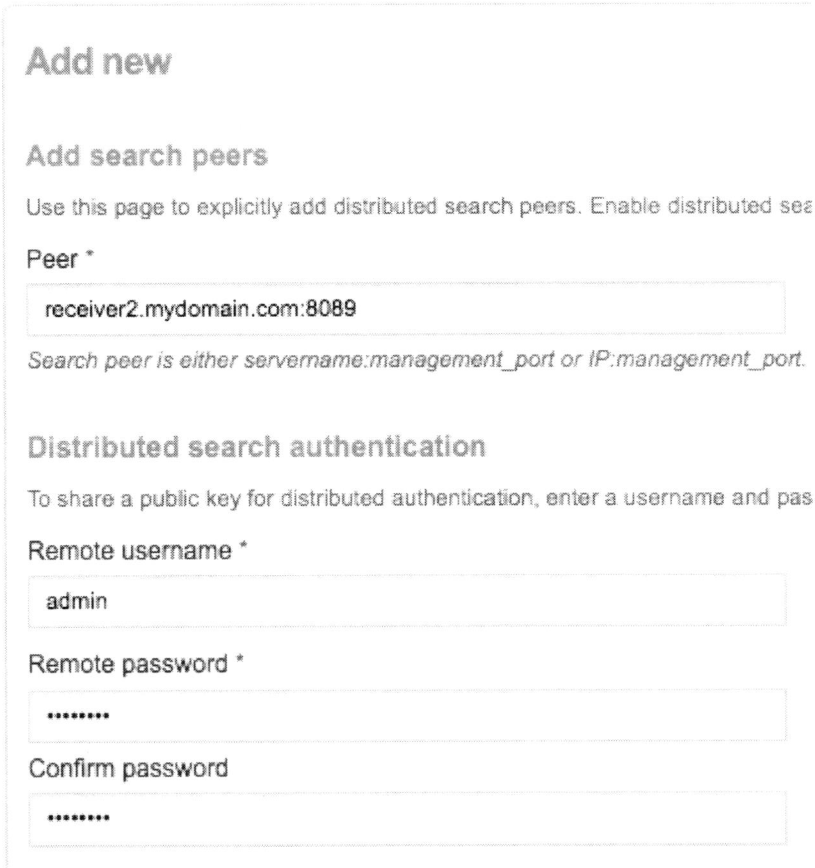

Add new

Add search peers

Use this page to explicitly add distributed search peers. Enable distributed sea

Peer *

```
receiver2.mydomain.com:8089
```

Search peer is either servername:management_port or IP:management_port.

Distributed search authentication

To share a public key for distributed authentication, enter a username and pas

Remote username *

```
admin
```

Remote password *

```
••••••••
```

Confirm password

```
••••••••
```

Figure 16-3. *Defining search peers*

■ **Caution**   You must change the initial password of Splunk when using distributed search. It will not work with the default password.

The search peer defined in Figure 16-3 is done from the first indexer. Then you must define the first indexer as a peer of the second one. Basically, you have to define every search peer for every indexer in the mix, and remember that you have to open the management port, usually 8089, on the firewall, or else the connections will not work.

On the forwarder side, there are a couple of things that have to be done. The first one is defining the set of receivers. This can be done using the CLI or by modifying directly the outputs.conf file. Either way, you can define a new A record on the DNS list for each receiver's IP address so that it looks like this:

```
receiver.mydomain.com        A      192.186.1.1
receiver.mydomain.com        A      192.186.1.2
```

Obviously, you would use your domain and appropriate IP addresses. When using this method, the `outputs.conf` file will look similar to this:

```
[tcpout]
defaultGroup=my_indexers

[tcpout:my_indexers]
disabled=false
server=receiver.mydomain.com:9997
```

An important point to make is that the forwarders distribute data based on time: that is, the load balancing happens by changing from one indexer to the next on a fixed period of time, which by default is 30 seconds. As Splunk is a time-series indexer, this is a critical point regarding the distribution of data. What this means is that when you do a search, each indexer will have slices of 30 seconds of the total data according to a round robin mechanism. Distributing data this way is highly efficient, and indexers will participate in a distributed search only if they have data for the time period over which the search is being done. The result usually is that pretty much all of the indexers will contribute in the search in a quite balanced fashion. There is rarely a reason to change the 30-second default, but if you would like to change it to, say, 20 seconds, just add the following directive in the `outputs.cong` file: `autoLBFrequency=20`. The other way to specify load balancing is using the CLI; in this case, you would issue the following commands at the Unix prompt:

```
./splunk add forward-server receiver.mydomain.com:9997 -method autobalance
```

This example assumes that you defined a DNS A record as described earlier. If you did not do this, you can specify each individual indexer on the forwarders as follows:

```
./splunk add forward-server receiver1.mydomain.com:9997 -method autobalance
./splunk add forward-server receiver2.mydomain.com:9997 -method autobalance
```

If you are using the deployment server, as you did in the example in Chapter 15, all you have to do is modify the `outputs.conf` file in the `send/default` directory, and the change will be automatically sent to the forwarders, which will restart after they get the update and will distribute the data between both Splunk instances on 30-second intervals.

By enabling distributed search, a user can go to any instance of Splunk to do a search and it will use the data indexed on all the Splunk instances, or both instances as is the case with this example. The resulting topology, where the searches are coordinated between indexers, is shown in Figure 16-4. In this example, the users access either Splunk instance as determined by a load balancer.

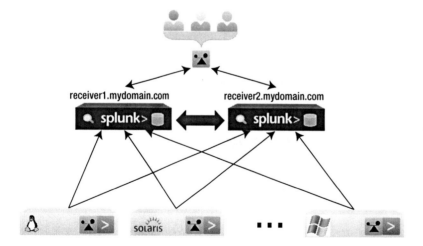

Figure 16-4. *Topology with distributed search enabled*

As a reminder, there are two network ports used in this setup: the management port, which is usually 8089, used by the Splunk instances to communicate between each other; and the receiver port, usually 9997, used by the receiver to get the data streams of the forwarders. Needless to say, both network ports have to be open in the firewall for this to work. We point this out because it is a common pitfall to forget opening the network ports in the firewall, and the error messages produced by Splunk are not related to the firewall.

Any saved search, Splunk app, or any knowledge object in general, that is, defined in one Splunk instance, is known as a knowledge bundle. In a distributed search setting, the knowledge bundle has to be copied to the other Splunk instances or search peers, so that the users can transparently use any instance to do their searches. If the knowledge bundle is rather large, copying it over to all the search peers can be a slow replication process. To increase efficiency, you can use what is known as a mounted bundle, where the actual knowledge bundle is placed on a shared disk between all the search peers, using NFS or SMB, this way you eliminate the need for replication. The Splunk documentation details the steps that have to be taken to enable a mounted bundle. The resulting topology that includes search peers using a mounted bundle can be seen in Figure 16-5. Be aware that this topology will work fine for what can be considered a mid-sized deployment. At a certain point the number of searches will quickly overwhelm both the mounted bundles and the search polling.

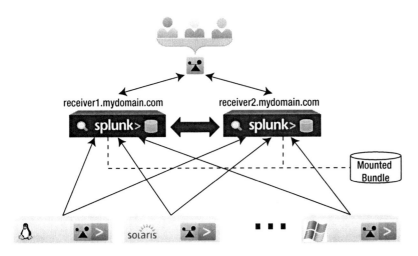

Figure 16-5. Distributed search topology with mounted bundle

In the examples we have used so far in this chapter, the Splunk instances are self-contained, that is, they have both the indexer and search components. We mentioned earlier that these components can be separated. A typical example is where the search component is installed on its own server, as shown in Figure 16-6, where it is called a search head. A search head does not receive and/or collect data, all it does is coordinates searches with all the indexers. Additionally, the receivers or indexers are no longer accessed directly by the users, as they have to go through the search head to do searches.

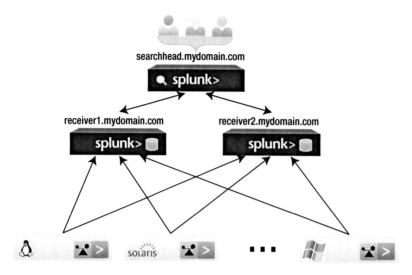

Figure 16-6. *Distributed search topology using a search head*

As your use of Splunk increases, you can count on the flexibility it offers, especially with the ability to separate the basic functions, as can be seen in the topology shown in Figure 16-7, where not only are there various indexers but also multiple search heads. This example is a rather complicated one, but it is also very realistic. Adding more search heads is driven by the number of concurrent searches, usually scheduled searches that are part of dashboards. As dashboards become more popular, it is normal to find search heads dedicated exclusively to service the searches used by dashboards. Additional search heads are also needed when there is a large population of users that interact with Splunk, be it with ad-hoc or saved searches. A discussion about when to add an indexer or a search head can be found in Appendix A.

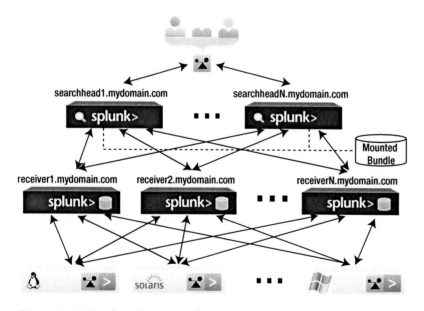

Figure 16-7. *Distributed search topology with multiple search heads*

The bottom line is that because of the flexibility that Splunk offers by allowing to separate searching from indexing, and the ability to define search peers and knowledge bundles, you can create a topology that best suits your unique needs. This flexibility is not limited to scaling but can also be used in other situations in which you need to manage data that is spread out organizationally or geographically. For example, you might have a corporation that handles most of its data at a local level, except in the state of California, where the offices in San Francisco and Los Angeles share the data. In addition to this, there is a search head at headquarters that can access the data of all the offices. This situation is depicted in Figure 16-8.

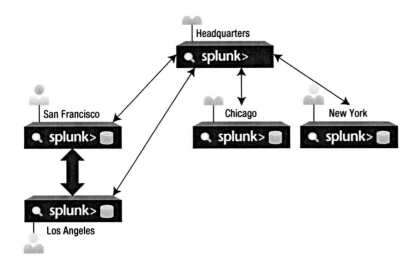

Figure 16-8. *Geographically dispersed topology*

As you can see from the example presented in Figure 16-8, the flexibility that can be achieved by separating the indexer and search functionality can meet most needs, be it for scaling to handle more data and searches, or to handle special organizational or geographical needs. More details about scaling, such as when to add an indexer or a search head, can be found in Appendix A.

Clustering

Clustering Splunk indexers is the way to achieve high availability. One of the issues with distributed search is that, when a search peer or indexer becomes unavailable, the data in that indexer is no longer available. Although this is not an issue for forwarders that auto load balance, as they will send their data streams to the next indexer, it becomes an issue for the person or process that does a search. The slice of data that is indexed in that Splunk instance is no longer available and the results of the search will be incomplete.

Clustering replicates data over multiple indexers so that when an indexer goes down, others that contain the replicated data can step in to provide completeness of data for the search. The coordination of the replication, failure, and other clustering specific items is done by a cluster master, which is a Splunk instance that only does this coordination function, nothing else. In Figure 16-9 you can see a cluster topology that includes a bunch of forwarders, four indexers, one search head, and the necessary cluster master, illustrating how the data flows.

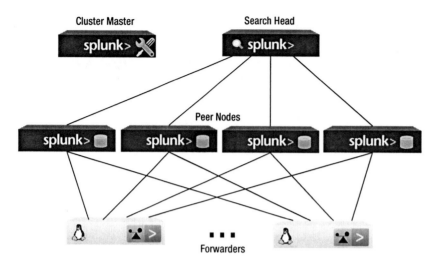

Figure 16-9. *A typical cluster topology depicting data links*

Because the cluster master only coordinates replication and failure activities between the peer nodes, Figure 16-9 does not show any connection between the master node and other instances. Before we explain how the master node handles its responsibilities and how the replication mechanism works, we must explain how the data is stored in Splunk. The data in Splunk is broken down in two parts:

- Raw data, which is not raw in the strict sense of the word, but more of a processed version that is already broken down into events. It is stored in compressed format.

- Index, which contains metadata that points to the raw data. This makes the searches very fast, as they start in the index and look into specific events in the raw data to retrieve the desired information, if it exists.

Both parts are stored as flat files, which together are generically known as a Splunk index and are organized into buckets. These buckets, which are nothing more than directories that contain those flat files, are arranged by age based on the timestamp of the event. Buckets have their own life cycle that takes them through various stages from hot to warm to cold and, finally, frozen. Data rolls from the hot to the warm stage based on size; from that point on, data rolls into the next stage based on size and age. Data rolled to the frozen stage is deleted, unless you archive it, in which case it can later be thawed. There can be, and usually are, many buckets in all the stages.

In a clustered environment, Splunk replicates buckets, and it does this in a quite efficient way. There are two items to be specified, the first one being the replication factor. This defines the total number of indexers or peer nodes that have the raw data of a bucket. The second item is the search factor, which defines how many of those replicas also contain the index or metadata. The default values are three for the replication factor and two for the search factor. What this means is that one indexer contains the primary bucket, raw and index; another indexer contains a replica of the primary bucket, with raw and index, and a third indexer contains only a replica of the raw data of the bucket. The primary bucket is the one used for the searches, whereas the others will be used only in the case of a failure.

The efficiency is based on the fact that the index part can be generated from the raw data at any time; therefore, there is a full secondary bucket available and, if for any reason that bucket is no longer available, the third partial replica can regenerate the index part and become a fully functional bucket.

Of course, you can define the replication factor and search factor to values that are most convenient to your particular situation, but the replication factor cannot be bigger than the number of peer nodes. In a similar fashion, the search factor cannot be bigger than the replication factor. Be aware that once you define these values and start a cluster, you cannot change them. In Figure 16-10, you can see an example of how indexing works in a cluster that uses the default values. The left-most indexer receives data and copies a full replica to another peer node and a partial bucket to the third node. The master node tells the indexer which peers to send over the replicated data.

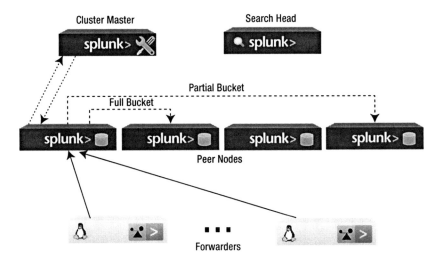

Figure 16-10. *A cluster topology showing the replication process of one indexer*

When using a cluster, it is highly recommended that you use forwarders to collect the data and enable the data acknowledgment option to make sure that the indexers receive the streams of data the forwarders send.

The master node receives a heartbeat signal from all the peer nodes every second. When it does not receive a heartbeat signal for 60 seconds (the default; although not recommended it can be changed), it assumes that a peer node is down and starts remedial actions. In addition to this, the master node has other important functions, which can be summarized as follows:

- It tells each peer node where to replicate the data. Whereas the peer assignment algorithm is not publicized, this obviously depends on the replication and search factors.

- It provides the search head(s) with a list of the available indexers

- Coordinates remedial actions when a peer is unavailable:

  - For every primary bucket in a failed peer, select a new peer node with a full bucket to become the new primary bucket. This usually takes just a few seconds.

  - For every full bucket in a failed peer, select a new peer node that will host a full bucket in order to maintain the search factor. In some cases, this might imply calculating the metadata from the raw data of a partial bucket. This can take a few minutes.

  - For every partial bucket in a failed peer, select a new peer node that will host a partial bucket in order to maintain the replication factor.

This high level review of the basic principles of a Splunk cluster should provide you with the essential knowledge needed to setup a cluster. We strongly suggest that you review the Splunk documentation on the subject before you set up a production cluster.

Setting up a Cluster

To illustrate the cluster concepts we just reviewed, we will go through the exercise of setting up a cluster similar to that depicted in Figure 16-9, which consists of four indexers, or peers nodes, one search head, and one master node. We will use the default values of three for the replication factor and two for the search factor.

The first step is to enable the cluster master node. Please note that a master node cannot simultaneously be a peer node or a search head, it can only be a master node and nothing more. Once you have installed and started the Splunk instance, go to the Manager and select the clustering option. Once there, you click on Configure, which will bring up the screen that is presented in Figure 16-11. There you can see that we are enabling clustering and making this instance the cluster master, using all the defaults. You can additionally provide a secret key, which will be used to authenticate the communications between the master and all the peers. It must be the same key on all the cluster instances. We choose to leave it blank.

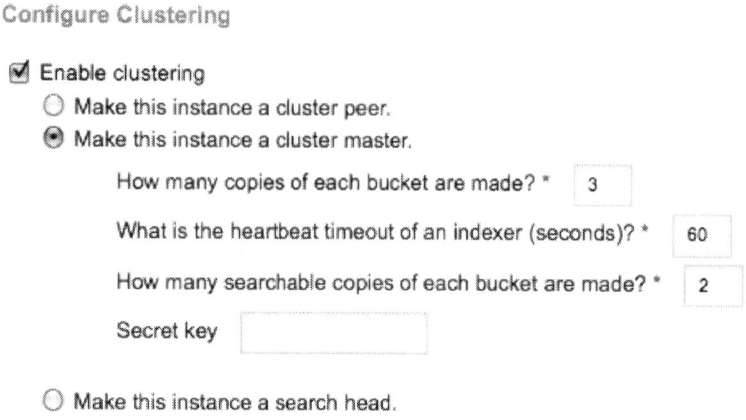

Configure Clustering

☑ Enable clustering

○ Make this instance a cluster peer.

◉ Make this instance a cluster master.

How many copies of each bucket are made? * `3`

What is the heartbeat timeout of an indexer (seconds)? * `60`

How many searchable copies of each bucket are made? * `2`

Secret key [   ]

○ Make this instance a search head.

Figure 16-11. *Enabling clustering for a master node*

After you save this configuration, you must restart Splunk for this to take effect. Once you restart the instance, it is ready to accept peer nodes. Note that it will block indexing on the peers until there are enough peer nodes to satisfy the replication factor. The next step is to enable the peer nodes. This is done by going to the same screen as we did for enabling the master node, but you select to make the instance a cluster peer, as shown in Figure 16-12. Here you have to provide the name of the master node and the management port, which is usually 8089. Remember to add https://, otherwise it will fail. Additionally, you have to provide the network port that the peer will use to receive the replicated data from the other peers. Any unused port will do, so we chose 8100. Because we did not define a secret key on the master node, we leave it blank here as well.

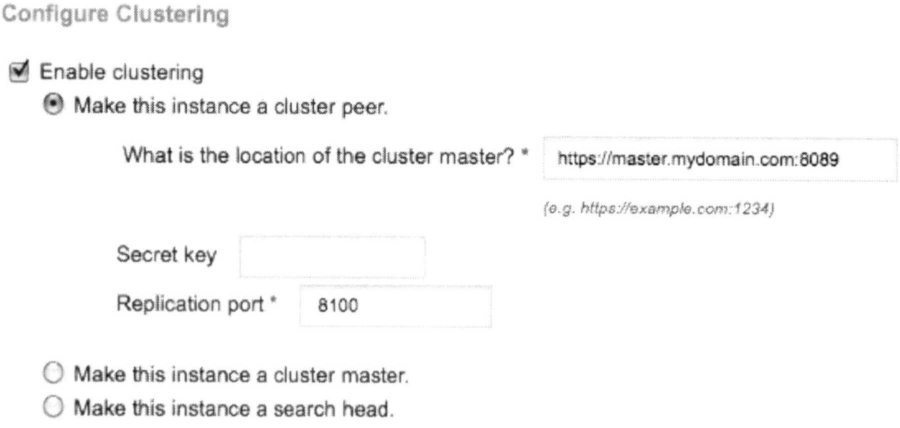

Configure Clustering

☑ Enable clustering

◉ Make this instance a cluster peer.

What is the location of the cluster master? * `https://master.mydomain.com:8089`

(e.g. https://example.com:1234)

Secret key [   ]

Replication port * `8100`

○ Make this instance a cluster master.

○ Make this instance a search head.

Figure 16-12. *Enabling a cluster peer*

After restarting the Splunk instance on the first peer node, we get the following warning message: "Indexing not ready; fewer than replication_factor peers are up", which is to be expected, as we only have one peer node in the cluster. In the master node you review the status of the cluster by going to the Manager, then selecting Clustering and then clicking on the Dashboard link. This dashboard can be configured, but out of the box has three sections of information. In Figure 16-13 we present the cluster overview, which shows the status we already commented on.

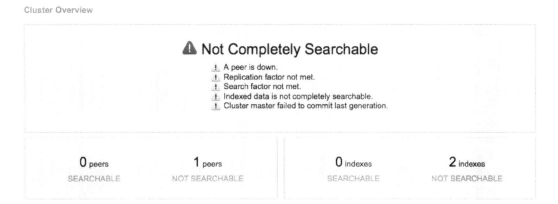

Figure 16-13. Cluster overview report

Once again, we must remind you about having all the necessary network ports open in the firewall; otherwise, the cluster will not work as all the peer to peer communications cannot happen. Once you have enabled all of the search peers, you can proceed to enable the search head. The process is the same as that of enabling a master or peer nodes, but you just select a search head. This will ask you for the URL of the master node and the secret key. After saving the configuration, you have to restart the Splunk instance. To verify that the search head is working correctly, you can view the search head dashboard, under the Clustering group of the Manager. The output of this selection is shown in Figure 16-14.

Search Head   Configure

Master Location	Generation ID	Generation Peers
https://216.121.120.123:8089	2	C39C763E-7028-4B0A-B479-603F9A5CFBAC
		3D19C509-F845-4EFB-A2F0-E0EEEFC1C4ED
		39C31F8E-92E0-4FC6-9F34-7A5593E31F5D
		159C4E8F-B2A0-4EAF-8CB2-DC7BD283E6CF

Figure 16-14. Search head dashboard

The information presented in the search head dashboard is rather cryptic, but what you are looking for is as many lines as peer nodes under the Generation Peers title. In our example, shown in Figure 16-14, it presents four lines, which is correct. It is interesting to note that you can have more than one active cluster, and a search head, or a group of search heads, can access more than one cluster. This provides you with the required flexibility that allows you to design a solution for pretty much any situation.

With the cluster now up and running, you can review the cluster overview report as the master node, which can be found in Figure 16-15. Although this figure still presents partial information of the total report, we have added a section that shows the peer details.

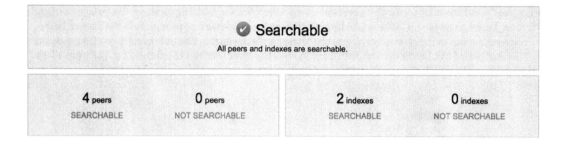

Figure 16-15. *Overview report for a complete cluster*

The steps we have done so far apply to a basic cluster, where only the main index is being used. If you use any additional indexes or applications, you will have to prepare what is called a configuration bundle on the master node, which will allow you to distribute the additional information to all of the peer nodes. Please consult the Splunk documentation of this subject for more details.

One last point to comment is on the master node. Admittedly, it can be considered a single point of failure, but it is not as weak as it seems. If the master is not available, the peers continue talking in between each other replicating data as needed, because they have all the necessary information. The lack of a master node becomes an issue when a peer becomes unavailable, as there is no master node that can coordinate the remediation activities and provide data completeness for the searches. This also applies in the case when a peer node is added to a cluster, either because it restarted or because it is a new peer, and it does not know which peers to use for replication. The Splunk documentation explains a method to set up a standby master node.

Summary

In this chapter, we have gone through the scalability of Splunk, reviewing how to increase the number of indexers, as well as the number of search heads, and we also checked other topologies that use multiple indexers and search heads. You learned the basics of clustering to achieve high availability and set up an example cluster to illustrate how this can be done. The objective is to give you a high-level idea of the clustering functionality, and we strongly suggest that you read in detail all of the documentation and practice setting up a few clusters before you create a Splunk cluster in production.

APPENDIX A

■ ■ ■

The Performance of Splunk

This appendix contains an overview of Splunk's performance as it relates to indexing, search, distributed search, and the effect of disk speed on searches. We also suggest some ways of gaining a better understanding of your Splunk environment, including search heads.

Types of Searches

We must start this appendix with a stern warning that performance cannot be generalized or extrapolated from other similar systems. No matter how much you think your application looks like another one, or the same one running in another place, you will not get close to extrapolating the performance. As a matter of fact, your application will exhibit different performance characteristics depending on the patterns of usage. Splunk is a complex piece of software and there are no formulas to calculate its performance. In this appendix we will share the performance of Splunk under various circumstances so that you can understand its behavior and set the right expectations.

The performance data presented here is only for illustrative purposes and should not be considered an accurate representation of Splunk's performance. If you were to run the same performance tests your resulting numbers will be different. This is because there are too many configuration files, not only in Splunk but also in the underlying operating system, in which a minor difference can have a huge effect.

Splunk has several moving parts, which do not allow us to provide a generic performance measurement. On a high level, these are the things that can affect the performance of Splunk:

- Types of searches
- Number of indexers
- Number of search heads
- The type of data being indexed
- The underlying hardware and operating system

The engineering team at Splunk categorizes the types of searches based on the number of events that are returned as part of the results of a search. Your expectation would be that of categorizing based on the complexity of the search, but this cannot really be measured, whereas the number of events generated by a search can and have a very big impact in the response time of the searches. The following are the categories or types of searches:

Dense: When there is one result for each one to one thousand events indexed (1 to 1,000)

Sparse: One result for each one thousand to one million events indexed (1,000 to 1,000,000)

Rare: One result for each one million to one billion events indexed (1,000,000 to 1,000,000,000)

Needle in the Haystack: One result for each one billion events or more indexed (1,000,000,000+)

One way to look at this is that a dense search, which produces a large number of events, is a panoramic view of the indexed data. A sparse search zooms in a little bit, as if you had binoculars. A rare search starts getting into the details, as if it were a microscope; and a Needle in the Haystack search really dives down to the molecular level, if you will, like an electron microscope. From a view of the forest, to the branches, to the leaves, to seeing the photosynthesis in action. The more you zoom in, the better the performance of Splunk, as we will see in this appendix.

■ **Note**   We must warn you that the performance tests presented in this appendix are done in the vacuum of a laboratory, which cannot be expected to reflect real-world situations.

Splunk does not cache or use buffers for results of searches. All the performance tests done for this appendix clear the Unix caches before executing, in order to provide correct comparisons.

Search Performance Profiles

To show how performance depends on the type of search, we ran a simple experiment. We created a typical web access log file following the combined log format, but at the end we placed some keywords such as *every2*, *every100*, and so on, and a sequence number, which is used for verification purposes. Each keyword appears in each corresponding event, so *every2* will appear on every second event, and *every100* appears in every 100th event. Multiple keywords appear as corresponds, for example, every 100th event will also include the *every2* keyword, as 100 is a multiple of 2; and every 10,000th event will include not only the *every10K* keyword but also *every2* and *every100*. The typical web access log entry looks like this:

```
150.135.220.120 - - [04/Jan/2012:23:59:59 -0700] "GET
/themes/splunk_com/img/skins/white/logos/downloads/logo_freebsd.gif HTTP/1.1" 200 -
"http://www.splunk.com/index.php/landing?ac=ostg-cpd" "Mozilla/4.0 (compatible; MSIE 7.0;
Windows NT 5.1; .NET CLR 1.0.3705; .NET CLR 1.1.4322; Media Center PC 4.0)"
"150.135.220.120.0000044400204798" every2 every100 num=000000100051200
```

We generated a little more than 3.6 billion events, about 1 Terabyte of raw data, which were indexed on one commodity server. We used another server as a search head. The test is not very realistic as nothing else is happening, no data is being indexed, and nobody else is doing other searches, but it still provides us with a pretty good idea of response times based on the categorization of searches. The tests were done using the CLI with a simple search, for example:

```
search every2
| stats count
```

The results can be seen in Figure A-1. We had to use a logarithmic scale since there are four orders of magnitude between the results of a dense search using the *every2* keyword and the speed of light test, which just does a search for the keyword *SpeedOfLight*, which we know does not exist in the data we generated. When you do a search, Splunk looks in its buckets of data to see if the string in there. The speed of light search is just to measure how long it takes to go through the Splunk infrastructure looking for a string that does not exist; in this example it is just two seconds. When it finds the string in a bucket it looks in the raw data for the actual event that contains the searched string, but to do this it has to decompress it first, as the raw data is stored compressed. The more events match the search, the more decompression it has to do, so a dense search tends to be CPU bound. As can be seen in the results chart, the very dense search for the *every2* keyword takes almost four hours as Splunk has to handle over 1.8 billion web access events. As we move to a less dense search as the *every100*, the response time drops to 45 minutes because Splunk only has to handle about 36 million events. As the results include fewer and fewer events the performance increases dramatically, as can be seen with the *every200million* keyword, which takes only 26 seconds to return the 18 events that match the search.

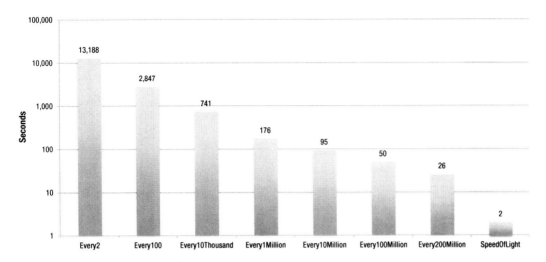

Figure A-1. *Search performance by type*

As searches go from dense to sparse, to rare, and to Needle in the Haystack, they move from being CPU-bound to I/O-bound because there is less raw data to decompress and more reads into the buckets. Figure A-2 shows the transition happening somewhere between one result for every 10,000 and 100,000 events indexed, which puts it in the middle of the sparse category of searches.

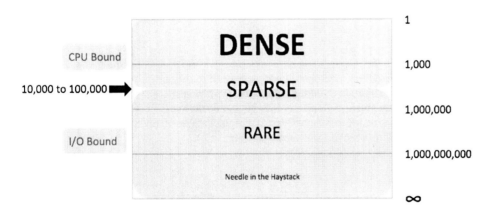

Figure A-2. *Search resource consumption*

To know if your search is CPU- or I/O-bound, just click on the button for the job inspector, which can be found on the search screen of the user interface, under the time picker as a blue square with an "i" inside it. This will start another browser window with information about the execution costs of that specific search. In Figure A-3 you can see a partial output for a simple search we used in Chapter 10, looking for the United flight with 31 hours of scheduled flight time:

```
UniqueCarrier=UA CRSElapsedTIme=1865
```

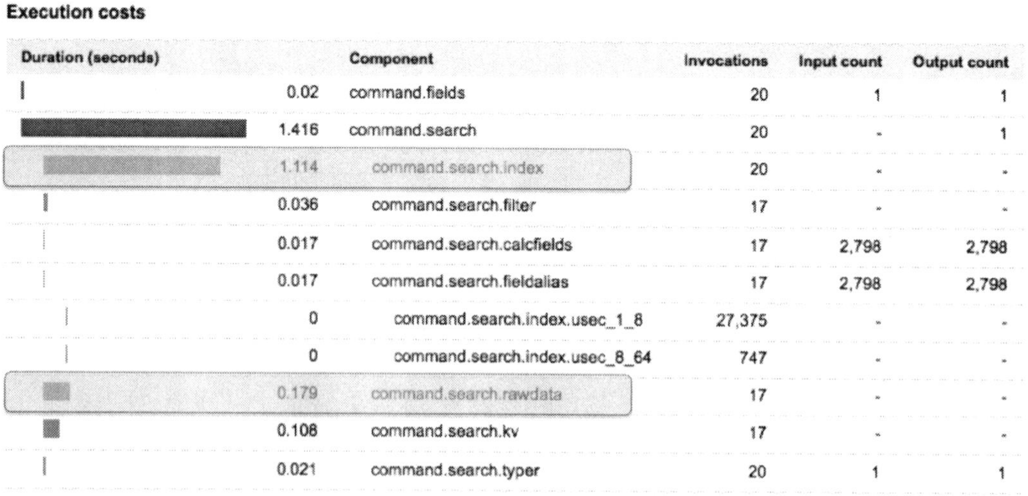

Execution costs

Duration (seconds)		Component	Invocations	Input count	Output count
	0.02	command.fields	20	1	1
	1.416	command.search	20	-	1
	1.114	command.search.index	20	-	-
	0.036	command.search.filter	17	-	-
	0.017	command.search.calcfields	17	2,798	2,798
	0.017	command.search.fieldalias	17	2,798	2,798
	0	command.search.index.usec_1_8	27,375	-	-
	0	command.search.index.usec_8_64	747	-	-
	0.179	command.search.rawdata	17	-	-
	0.108	command.search.kv	17	-	-
	0.021	command.search.typer	20	1	1

Figure A-3. Search execution costs

Looking carefully at the output of the job inspector, you can see that the amount of time spent by *command.search.index* is 1.114 seconds, whereas *command.search.rawdata* is 0.179 seconds; thus this search is I/O-bound. This was expected as this search falls under the Rare category since there was only one result out of 147 million indexed events. A final note of caution: this does not apply to searches based on the *tsidx* mechanism.

Needle in the Haystack

Setting up a performance test for a Needle in the Haystack search is very complex and time-consuming; that is why this search type was not included in the previous experiments. However, we had the opportunity to run a few separate performance tests specifically to measure this case. In a similar fashion to the previous tests, we created a set of web access logs using the access combined format and appended a unique sequence number to each event, no keywords this time. We generated 56 billion events and loaded them evenly on 10 Splunk indexers. This cluster also had a single search head. The search command we used from the CLI was the following:

```
search 0000028000000123
```

This looks for that particular string, which is unique within the 56 billion events and corresponds to event 28 billion one hundred twenty three, which can be found about halfway. The results of this search can be found in Table A-1.

Table A-1. Needle in the Haystack Performance

Bucket Size	Bloom Filter	Response (Secs)
1GB	No	186.636
1GB	Yes	31.788
10GB	No	71.459
10GB	Yes	9.727

Of course, the main focus of attention is that finding the unique event out of 56 billion took less than 10 seconds, but there are a few things we need to explain first. Let's start with the buckets. This is where Splunk stores the raw data; the default size is automatically calculated by Splunk when it is installed. It is set based on the number of cores on the server where Splunk is running and other configuration information. When the buckets are associated with the main index, the default size is most likely to be 10GB per bucket, whereas for buckets associated with a custom index, that is, an index you defined, it is most likely to default to 1GB. You can configure the bucket sizes, but in general we suggest you use the predifined *auto_high_volume* setting for high-volume indexes, such as the main index, otherwise use the *auto* setting. A high-volume index would typically be considered one that gets over 10GB of data a day.

The next item that has a huge impact on the search performance is the Bloom Filters. These are data structures that can categorically state that an element is not in a data set. Splunk uses the Bloom Filter to find if a string is not in a bucket. This way Splunk does not have to open and process buckets that do not have the string we are looking for, which improves the performance dramatically. When you combine a bigger bucket size with Bloom Filters, you will have fewer buckets, which means fewer Bloom Filters to process, thereby decreasing the processing time for a search. Bloom Filters are created at indexing time and are enabled by default in Splunk. There is a relatively small overhead to create them, but it is well worth its price when you have Rare or Needle in the Haystack searches. The overhead is broken down into the additional processing time to create the Bloom Filter during indexing and the additional amount of disk space used to store the filter. We suggest that if the majority of your searches are Rare or Needle in the Haystack you can use 10GB bucket size on your custom indexes.

Distributed Searches

One of the most frequently asked questions is when an indexer should be added to a cluster of Splunk servers. Because there is no exact answer to this question, we have prepared a set of three performance tests that will give you a pretty decent idea on the behavior of Splunk as it relates to increasing the number of indexers in a deployment.

Performance Test 1

In this set of tests we keep a fixed number of events, 100 million, evenly distributed on all indexers, while we increase the number of indexers. Our first data point is using one indexer, which has all the 100 million events. Next we increase the number of indexers to ten, but as we have a fixed number of events, each indexer will now have only 10 million events. The next data point is with 50 indexers, each with two million events and, finally, 100 indexers, each with one million events. The first set of results, for dense searches, can be seen in Table A-2.

Table A-2. *Performance Test 1—Dense Searches*

Search Heads	Indexers	Events per Indexer	Response Time (Secs)
1	1	100 Million	1,113
1	10	10 Million	116
1	50	2 Million	32
1	100	1 Million	36

As we add indexers the performance increases. There is a hint of linear scalability between 1 and 10 indexers, but there are not enough data points to be sure. However, we can see that going from 50 to 100 indexers there is an inflection point, where the response time increases. This is a point of diminishing returns, where it is no longer worth adding indexers. We attribute the degradation of performance to the communication overhead between the search head and the indexers.

In Table A-3 you can see the response times for sparse searches. Although not so obvious, we also have a hint of linear scalability between one and 10 indexers. We can also see the same inflection point as for the dense searches, which can be attributed to the same reasons as before.

Table A-3. *Peformance Test 1—Sparse Searches*

Search Heads	Indexers	Events per Indexer	Response Time (Secs)
1	1	100 Million	9
1	10	10 Million	1
1	50	2 Million	1
1	100	1 Million	2

Table A-4 shows how having too many indexers will actually hurt the performance of Rare searches. The inflection point happens earlier then with the other search types. This is because of the nature of the Rare searches, the responses of every indexer are very fast, but the communication overhead of all the additional indexers has a direct impact on the response times.

Table A-4. *Peformance Test 1—Rare Searches*

Search Heads	Indexers	Events per Indexer	Response Time (Secs)
1	1	100 Million	0.4
1	10	10 Million	0.4
1	50	2 Million	1
1	100	1 Million	2

From this performance test we can conclude:

- Increasing the number of indexers tends to improve performance
- There is a point of diminishing returns where too many indexers will have a negative impact on the performance of a search, especially a Rare search.

Performance Test 2

In this performance test we will keep a fixed number of events per indexer. What this means is that every time we increase the number of indexers we increase the total amount of events. We will only have two data points, the first one with 10 indexers, each with 10 million events for a total of 100 million events. The second one has 200 indexers, each with 10 million events for a total of 2 billion events.

In Table A-5 we can see the results of this test for Dense searches. In this case the scalability is much better than linear, but we do not have enough data points to make such an assertion. Just remember that in this set of tests we are increasing the number of indexers, each with additional data. In the previous set of tests we increased the number of indexers but diluted the number of events per indexer.

Table A-5. *Peformance Test 2—Dense Searches*

Search Heads	Indexers	Total Events	Response Time (Secs)
1	10	100 Million	116
1	200	2 Billion	1,024

Once again we can see in Table A-6 that there is a hint of the scalability being much better than linear for Sparse searches. Extrapolating, our response time with two billion events should be 20 seconds, but it is 9.

Table A-6. *Peformance Test 2—Sparse Searches*

Search Heads	Indexers	Total Events	Response Time (Secs)
1	10	100 Million	1
1	200	2 Billion	9

Finally, as shown in Table A-7, the Rare searches present a similar pattern, where there is a hint of much better than linear scalability. We can then conclude that under these conditions it makes sense to add indexers as the number of events increases and that the scalability is linear or better.

Table A-7. *Peformance Test 2—Rare Searches*

Search Heads	Indexers	Total Events	Response Time (Secs)
1	10	100 Million	0.4
1	200	2 Billion	5

Performance Test 3

We will call this one the realistic performance test. The scenarios presented in the previous two performance tests just do not happen in reality. The new events sent over to be indexed will be evenly distributed to all the indexers, so the new indexer will only receive a fraction of the total number of events. As the new indexer does not have any events when it is first included in the mix, the total number of events will not be evenly distributed across all the indexers.

For this set of tests we will start using one search head with one indexer. Our first data point is with 100 million events. Then we increase to 200 and 300 million events. At this point we add another indexer and increase the number of events to 400 and 500 million events. What happens here is that when we add the new indexer and increase the number of events by 100 million, 50 million will go to the new indexer for a total of 50 million events, and the other 50 million will go to the old indexer for a total of 350 million events. When we add 100 million events more, they will be evenly distributed between both indexers. Now the new one will have a total of 100 million events and the old one will have 400 million.

In Figure A-4 we have plotted the actual response times for the five data points, but we also added a trend line representing linear scalability. As you can see the actual response time is better than the projected linear scalability. In reality, the results would probably be better than what is displayed in this chart. This is because the searches are accessing all the data, in real life these searches tend to be done over a recent period of time that keeps rolling, thus the overall search time will improve more than what we are showing here.

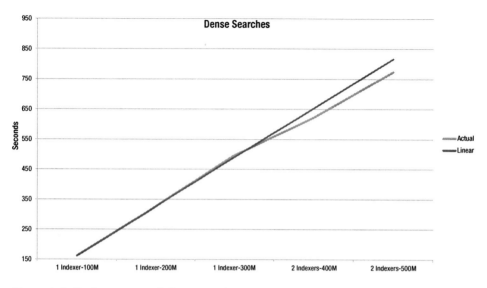

Figure A-4. *Performance test 3, Dense searches*

Another way of interpreting the chart in Figure A-5 is by looking at the linear scalability trend line. Here you can see that it reaches 80 seconds response time at 400 million events on a single indexer. When we add a second indexer it reaches 80 seconds at 500 million events, therefore contributing substantially to the performance and scalability.

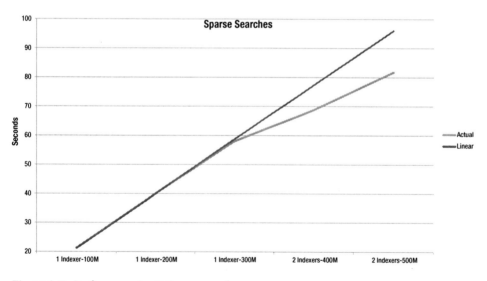

Figure A-5. *Performance test 3, Sparse searches*

In Figure A-6 we can see that adding an indexer produces a small improvement, but it seems that when the response times are so fast it might not be worth adding another indexer.

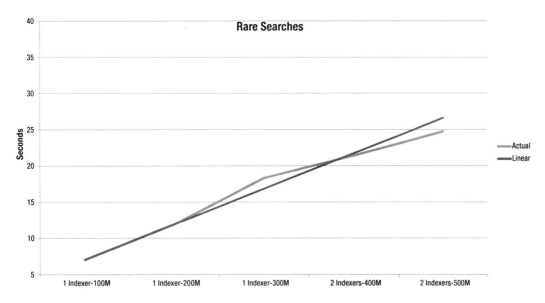

Figure A-6. *Performance test 3, Rare searches*

Based on the results of this set of performance tests we can conclude that adding another indexer is useful, especially for Sparse searches, but there do not seem to be any advantages for Rare searches. It is clear that by scaling using additional indexers we defer a degradation of the response times for a while, which can justify the increase of indexers.

Indexing Performance

Another example of why performance cannot be generalized is when we talk about indexing performance. This case is highly dependent on the type of data that is being indexed. Additionally, the transformations that are done on the data as it is being loaded into Splunk will also have an impact. This was amply discussed in Chapter 9, as we reviewed the different ways in which the timestamp of the airline on-time data could be handled. Finally, the underlying operating system and hardware will also impact the performance.

Depending on the type of data, the size of the events will vary a lot. For example, the average size of a web access log event, using the combined log format, is about 300 characters. The average event size for a syslog file varies too much because of the different types of sources. We have seen it go from 300 to 1,200 characters. Another popular one is the JBoss log file; when it contains something it is usually Java stack traces, which tend to be long, on the average about 5,000 characters. Because of all these variable event sizes, it is preferable to measure indexing performance in kilobytes per second (KBps) instead of just events.

In Figure A-7 you can see the results of measuring the indexing throughput with the three types of log files discussed earlier. The results with Linux are pretty consistent around the 14,000 KBps. The same can be said for Solaris, although with the throughput dropping to around 11,000 KBps. What is surprising is the inconsistency of indexing throughput for the Windows operating systems. Just to give a wider picture, we chose to compress the syslog file with gzip. We can attribute the slightly slower throughput on Linux and Solaris to the fact that the syslog file has to be decompressed. However, it seems that Windows thrives on compressed files.

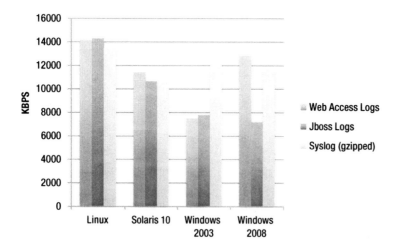

Figure A-7. *Indexing performance*

This chart is not meant as an endorsement of any specific operating system. As you know there are too many things that can affect the performance of an operating system. The tests, however, were conducted on the same commodity servers, thus Solaris 10 is on x86 instead of a SPARC based server. The purpose of this chart is to illustrate that there are differences between data types, which can also be influenced by the underlying operating system and hardware.

Disk Speed and Searches

As the slowest hardware component when compared with the other ones that make up a computer, disk drives do have an impact on the performance, especially with an application like Splunk, which stores the data on disk. In Splunk, data is stored in buckets. These can be hot, warm, cold or frozen. We will not go into the details of how they work and how data rolls from one to the other, but we can say that indexing performance is directly linked to the write performance of the device(s) where the hot and warm buckets are located. Search performance is partially linked to the read performance of the device(s) where the hot, warm and cold buckets are placed. You can specify the location of the buckets in the configuration files.

The test that we set up is focused on measuring the search performance using disk drives with different rotational speeds. To get a better idea of this behavior, we will also take into account the number of concurrent searches happening.

The chart in Figure A-8 shows the Solid State Disk (SSD) giving the best search performance, which improves as the number of concurrent searches increases. The improvement over traditional mechanical disk drives is so big that we had to use a logarithmic scale. The difference in search performance of an SSD might be enough to justify the price difference.

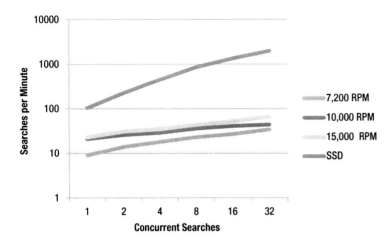

Figure A-8. *Disk speed and search performance*

Figure A-9 zooms into the performance of the mechanical disks, so we can better understand it without the distortion created by the SSD's different storage technology. The gap between a 7,200 RPM disk and a 10,000 RPM disk is substantial, averaging about 50 percent. However, the difference between 10,000 RPM and 15,000 RPM is a lot smaller. It does improve as the number of concurrent searches increases, reaching 34 percent at 32 concurrent searches.

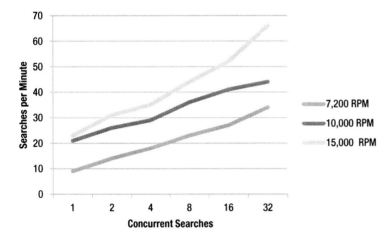

Figure A-9. *Mechanical disk speed and search performance*

Although not used in these performance tests, when working with RAID configurations, the recommended setup is RAID 10, also known as RAID 0+1. RAID 10 provides fast writes and reads with the greatest redundancy. Do not use RAID 5, because the duplicate writes make for slow indexing.

The results of this test might help you choose disk drives for your installation, but remember that you have to test with your own data, searches and patterns of usage before you make a decision. Another pointer is that concurrent searches can be originated by multiple users or by dashboards, or a combination of both. We say this because we have seen people forgetting dashboards when they calculate concurrent searches.

Understanding your Splunk Environment

SplunkWeb, the user interface of Splunk, provides a comprehensive set of reports that will give you a very good idea of the behavior of your Splunk environment. When you are in the search app, the Summary tab shows a recap of all the indexed data, by sources, source types and hosts. When you choose the Status tab, a pull down menu presents a number of options as can be seen in Figure A-10. Each one of these options provides detailed information of the inner workings of Splunk, your data, and the searches. We strongly suggest that you become familiar with these reports.

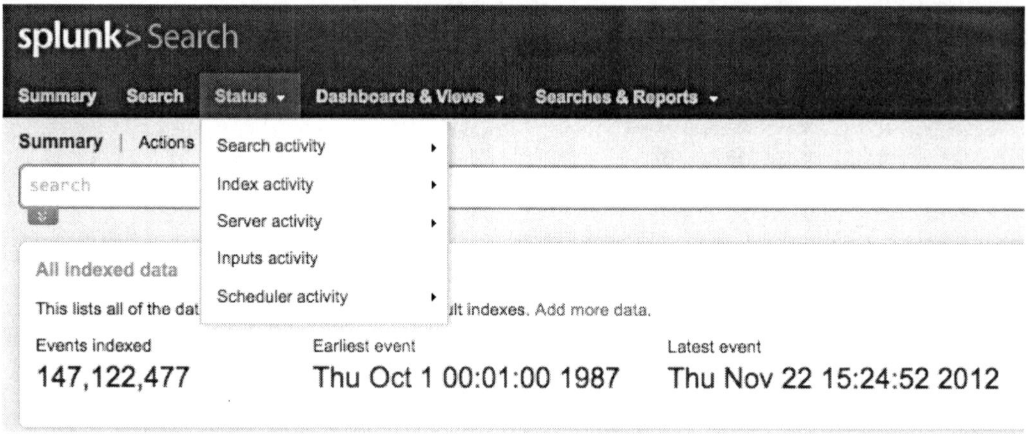

Figure A-10. *Status options*

There are a few other tips we want to share with you so that you can better understand your Splunk environment. We have discussed the categorization of searches that the engineering team of Splunk uses, so we will show you how to calculate this for your searches. But first we will tell you how to find which are your most expensive searches and what have been their responses over time. Finally, we will have a discussion regarding the performance of search heads.

Most Expensive Searches

You will probably want to start by knowing which are your most expensive searches, so that you can see if they can be improved or optimized. When you select the "Search activity overview" option of the Status pull down menu, you will get a dashboard that presents you with various panels. Scroll to the end of the page, where you will find a panel with the title "Recent searches by expense (last hour)." This contains the essence of the information we want, but the time range is only the last hour and it does not show the name of the saved search, just the actual search.

We have modified this search in such a way that it will provide the name of the saved search, and if it is not saved then it identifies the search as Ad Hoc. If you click on the bottom left of the panel, where it says "View results" it will take you to a search screen, which includes the actual search and the results. Here you can modify the search by adding a line that uses the *eval* command to create a field named *saved_name*. Also add a line to the *stats* command, so that the search name shows up in the table. Then modify the time range to something more representative of your needs, probably one month, and save it as Most Expensive Searches. The final search looks like the following, in which we have **highlighted** the changes:

```
`audit_searchlocal`
|   convert num(total_run_time)
|  eval user = if(user="n/a", null(), user)
|  `audit_rexsearch`
|  eval is_scheduled = if(search_id LIKE "scheduler%", "yes", "no")
```

```
| eval saved_name = if(savedsearch_name LIKE "", "Ad Hoc", savedsearch_name)
| stats min(_time) as _time
        first(user) as user
        first(total_run_time) as total_run_time
        first(saved_name) as saved_search_name
        first(search) as search
        first(is_scheduled) as is_scheduled by search_id
| search user=*
| sort - total_run_time
```

The output of this search can be seen in Figure A-11. By adding the column with the name of the saved search, it is really easy to identify expensive searches. The output will only show the most expensive occurrence of each search. The next step is to gain a better understanding of the questionable searches.

total_run_time ‡	saved_search_name ‡	search ‡
2971.22	Ad Hoc	search index=twitter | tscollect namespace=summary_tweets
1516.16	love	search index=twitter lang=en | where like(text, "%love%") | sentiment twitter text | stats avg(sentiment)
1178.83	love	search index=twitter lang=en | where like(text, "%love%") | sentiment twitter text | stats avg(sentiment)
586.24	love	search index=twitter lang=en | where like(text, "%love%") | sentiment twitter text | stats avg(sentiment)
532.24	Ad Hoc	search index=twitter lang=en | where like(text, "%love%") | sentiment twitter text | stats avg(sentiment)
438.49	love	search index=twitter lang=en | where like(text, "%love%") | sentiment twitter text | stats avg(sentiment)
385.49	Ad Hoc	search index=twitter lang=en | top user.screen_name
365.22	beliebers	search index=twitter lang=en| rename entities.hashtags{}.text as hashtags | fields text, hashtags | mvexpand hashtags | where like(hashtags,"
363.91	Ad Hoc	search index=twitter lang=en| rename entities.hashtags{}.text as hashtags | fields text, hashtags | mvexpand hashtags | where like(hashtags,"
342.12	Ad Hoc	search index=twitter lang=en | language text | stats values(language)

Figure A-11. *Most expensive searches*

Historic Run Times

Once you have identified the most expensive searches, we suggest that you review the behavior over time of those that concern you most. The previous report, Most Expensive Searches, only shows the most expensive for every search. Looking at the historic run times, or response times, will give us a better idea. The search for this is very simple. If it is a saved search use this:

```
index=_audit info=completed savedsearch_name="saved search name"
| chart first(total_run_time) as RunTime over _time
```

For an Ad Hoc search, you will have to type the actual search instead of the saved search name. Do not forget to enclose the name of the saved search or the actual search with double quotes (") and note that there is a space between the *over* keyword and the special field *_time*:

```
index=_audit info=completed savedsearch_name="actual search"
| chart first(total_run_time) as RunTime over _time
```

The output of either of these searches will be a chart. You will have to play with the charting options to get a clear picture. For us it usually involves adjusting the minimum value of the *y*-axis of a line chart to obtain better resolution. From our experience, variations of as much plus/minus 10 percent can be considered acceptable. If the search that you are analyzing has many data points, it will be easier to see a trend, if there is any. A trend that shows the response of a search degrading over time can be caused by the following reasons:

- You are indexing more data than before

- The search is returning more events in proportion than before, that is, it is becoming more dense

- The search is covering more time than before

Any of these reasons can cause a degradation over time for a search. You will have to analyze your setup in detail to make sure that if any of the previous reasons are occurring, they are for a known cause. If you cannot explain the degradation based on any of the exposed reasons, you have a problem with your search, and you should examine it in more detail.

Sometimes we will use a variation of the previous searches to present the number of events returned by a search. You could try to combine both, but we have found that sometimes the difference in scale between both sets of values will mess up the chart. The alternative search is:

```
index=_audit info=completed savedsearch_name="saved search name"
| chart first(event_count) as ReturnedEvents over _time
```

Categorizing Searches

Knowing which type of searches we have helps us to set the right expectations in terms of performance. To figure out into which category a search fits, you first need to calculate the ratio of events in a response to the total number of events in an index:

```
Ratio = total number of events in an index / number of events in the results of a search
```

It might seem that finding the number of events in the results of a search is not always easy to figure out, especially with searches that have a lot of clauses, that is, use lots of pipes. Searches usually start with streamable commands, which are the ones that do something with an event, maybe a modification, or just select events from an index, and pass them to the next section of the search. We want to focus on the very first command of a search, which is usually a straight search. Based on this one is that we calculate the ratio, because it is the one that is looking into the buckets for the events that match the requested condition. When dealing with searches that include subsearches, each one of them should be considered an individual search for the purposes of categorizing them.

Finding out the number of events in an index is as simple as using the *evencount* command or going to the summary page in the search app. If the resulting ratio is between:

- One and one thousand, the search is Dense

- One thousand and one million, it is a Sparse search

- One million and one billion, we have a Rare search

- More than one billion, it is a Needle in the Haystack search

Granted, doing this calculation is a moving target as the indexes are typically growing, especially in cases where the searches are based on a rolling time frame. Unfortunately, Splunk does not keep a history of the number of events stored in the indexes, so we cannot find out into which category a search falls over time.

Search Head Performance

Most of the research we have done on the performance of Splunk has focused on searches, indexing, and topologies. As we are seeing that dashboards with multiple panels and sometimes complex searches are growing in popularity, we realize that we have not put much effort into understanding the load imposed on the actual search heads themselves. Most searches that make up dashboards are based on a rolling time period, for example, the last minute or the last five minutes. Even if these searches are dealing with a relatively small amount of events, rolling summaries are dense searches, and using report acceleration or tsidx acceleration does not always make a difference, as the summarization has to be done at some point in time before the next refresh of the panel, which might be in such a short period of time that it cannot be done.

Dashboards have a tendency to proliferate like rabbits and you can very quickly find yourself in a situation in which you have a large set of dashboards with a dozen or so panels each for a range of audiences that go from executives, to management, to the operations people that actually use them. As part of this we are seeing more people deploying search heads, which are dedicated exclusively to process the searches for dashboards.

The question then becomes, when should I add another search head? Before we attempt to answer that question, we need to explain how dashboards work. All the dashboards are made up of scheduled searches. A scheduled search is supposed to run at a specific interval, for example, every five minutes or every hour on the hour. If there is a lag in the start time of a search, it is quite possible that it is because the system is overloaded. Minor irregular lags are acceptable, as are average lags in the range of single seconds, but when the lags are consistently big you should seriously consider adding another search head.

Another metric is the number of skipped searches. A search is skipped when the previous run took so long that it is time to run the one that follows the skipped one. For example, say we have a search that is scheduled to run every minute, but it takes two minutes to run, so it will skip the next execution, which is now in the past and run in the following interval. If this search takes, say 45 or 50 seconds and now it is taking longer, then it is very likely that you have a load problem, which can be identified due to a lag or a skip. Of course, if your searches take longer than the intervals you have assigned them, this is not a load problem, just a logistics issue.

So, back to our question. The answer has two parts and we must say that it is based more on experience and gut feeling than formal experiments like those we have done in this appendix. The first part is based on using the comprehensive information that Splunk offers under the Status tab in the user interface, in particular the "Scheduler activity overview" page. There you will find two important panels, one that provides us with the number of started and skipped searches over the last 24 hours, and the other one that presents the average execution lag, also for the last 24 hours. With the previous explanation you should be able to quickly assess the load on your search head(s).

The second part of the answer focuses on metrics of the Unix/Linux operating system. The general statement is that you should add another search head when any resource, such as CPU or memory is reaching capacity. That is too generic, so let us have a look at something a bit more specific. Our commodity servers are running CentOS 6.2, where the *top* command provides us with 2 metrics that we consider critical, the Unix load, and the percentage of I/O wait. There are other commands that provide these metrics also. In Figure A-12 you can see where to find these metrics in the output of the *top* command.

```
top - 10:17:32 up 31 days,  8:35,  2 users,  load average: 2.14, 1.72, 1.02
Tasks: 161 total,   1 running, 160 sleeping,   0 stopped,   0 zombie
Cpu(s): 15.3%us,  0.7%sy,  0.0%ni, 83.9%id,  0.0%wa  0.0%hi,  0.0%si,  0.0%st
Mem:   8028504k total,  7950164k used,    78340k free,   167708k buffers
Swap:  4240116k total,   123976k used,  4116140k free,  5421992k cached

  PID USER      PR  NI  VIRT  RES  SHR S %CPU %MEM    TIME+  COMMAND
18978 root      20   0  174m  41m 9732 S 210.3  0.5   8:51.34 splunkd
28755 root      20   0  692m  88m 9.9m S  4.7  1.1  89:28.78 splunkd
28891 root      20   0 2080m 278m 4136 S  3.0  3.6  18:31.20 python
19692 root      20   0 15012 1344  988 S  0.3  0.0   0:00.52 top
    1 root      20   0 19204  844  700 S  0.0  0.0   0:01.11 init
    2 root      20   0     0    0    0 S  0.0  0.0   0:00.00 kthreadd
```

Figure A-12. Output of the Unix top command

The load average can be found on the top right corner. It presents three numbers, the load average for the last minute, the last 5 minutes, and the last 15 minutes. If these numbers are consistently higher than the number of CPU cores that your server has, then it is time to seriously consider adding another search head. The second metric is the I/O wait, which can be found a couple of lines right below the load average title. In Figure A-12 it shows as 0.0%wa. This metric presents the amount of time the CPU cores are waiting for I/O. Starting at 0.2 percent, this metric tends to increase exponentially very quickly, and you will reach chaos in a rather short time. Now you know as much as we do about search heads and their performance.

Summary

In this appendix you learned about the types of searches and their performance. We also reviewed when we should consider adding another indexer to our environment. A rule of thumb is that adding another one will almost always improve the overall search time. We saw the importance of disk I/O for search performance. We shared some tips on getting to know your Splunk setup, specifically looking for the most expensive searches; and we suggested that you also look at the history of their search time and try to understand the reason for degradation if any. Finally, we reviewed how to monitor the search heads trying to avoid unacceptable performance.

APPENDIX B

■ ■ ■

Useful Splunk Apps

This chapter contains an overview of few useful Splunk Apps and Add-Ons that fall in the categories of monitoring, security, databases, applications, application servers, and networking.

Splunk Apps

In the previous chapters of this book, we have explored and used a few Splunk Apps and Add-Ons such as Splunk App for Windows, *Nix, Google Maps, Globe, Web Intelligence, Sentimental Analysis, Splunk DB Connect, Splunk Deployment Monitor, and so on. Before we dive into this chapter, let's do a quick recap on Splunk Apps.

Splunk is designed as a platform that serves as an infrastructure where third-party developers or Independent Software Vendors (ISVs) can build specialized applications that provide extensions to Splunk. There are two ways to build these extensions, Apps, and Technology Add-Ons. Splunk Apps package the extended functionality together with standard features such as saved searches, dashboards, and defined inputs. Additionally, they bundle their own user interface layered on top of Splunk's user interface. By contrast, Add-Ons, or simply TAs, are smaller components as compared to Apps, which include the additional functionality without their own user interface. We will have to use the standard Splunk Search application against the indexed data configured through Add-Ons. Apps and Add-Ons can be written by anybody with a decent programming knowledge. Splunk has a vibrant community that constantly creates and shares Apps and Add-Ons. It is hosted at http://splunk-base.splunk.com/apps/.

At the time of writing, there were at least a few hundred Splunk Apps and Add-Ons available on Splunkbase. We have categorized some of the popular and interesting Apps into broad categories of monitoring, networking, security, databases, applications, and application servers. The idea of this chapter is get you familiar with Splunkbase and give a short list of popular Apps and Add-Ons that you can leverage with your Splunk deployments. As these Apps and Add-Ons are revised frequently, we suggest that you refer to the specific Apps pages in Splunkbase to find the latest news. The details provided for each of the Apps or Add-Ons in this chapter are only basic with the intent of providing a jumpstart on what the App does, the sources of data it captures, and reports or dashboard it provides. The chapter does not go into the details of installation, deployment, configuration, and usage of the Apps or Add-Ons listed. You can leverage the learnings and understanding of the Apps and Add-Ons used in previous chapters and explore some of the listed Apps that you might find useful for your environment. The Apps and Add-Ons we will look at are:

- Applications, Databases, and Application Servers

    - Splunk for Oracle WebLogic Server

    - Splunk Hadoop Connect

    - Splunk App for Microsoft Exchange

    - Technology Add-On for Microsoft SQLServer 2012 and 2008r2

    - Splunk App for AWS usage tracking

- Security
    - Splunk for FISMA
    - Splunk App for PCI Compliance
    - Splunk App for Enterprise Security
    - Splunk for RSA SecurID Appliances
- Monitoring
    - Splunk App for VMware
    - Splunk App for HadoopOps
    - Splunk App for Active Directory
    - Splunk on Splunk (S.o.S.)
    - Splunk for Nagios
    - Splunk App for Snort
- Networking
    - Cisco Security Suite
    - Splunk for Palo Alto Networks
    - Splunk for Juniper Firewalls
    - Splunk for Barracuda Networks Web Application Firewall

Splunk for Oracle WebLogic Server

Field	Description
Category	Applications, Databases, and Application Servers
App Name	Splunk for Oracle WebLogic Server
Author/Vendor	Function1_Inc
License	Creative Commons BY 3.0
Price	Free
Splunk version computability	5.x, 4.3, 4.2
Support from Splunk	No
App URL	http://splunk-base.splunk.com/apps/72283/splunk-for-oracle-weblogic-server
Description	Splunk for Oracle WebLogic Server is an App that monitors WebLogic server components by collecting different logs.
	Splunk for Oracle WebLogic Server Technology Add-On (TA) is packaged with this app. The packaged App polls WLS AdminServers at different intervals and loads the data into Splunk.
	The App comes with a topology overview of all domains and applications. It provides a set of reports for users to a drill-down to a WLS instance level, a specific application or applications within a domain, servers, JMS instances, JDBC data sources, and JVM heap size. This App gathers data for entire WLS deployments and provides simple and easy to understand reports and charts.

Splunk Hadoop Connect

Field	Description
Category	Applications, Databases, and Application Servers
App Name	Splunk Hadoop Connect
Author/Vendor	Splunk
License	Splunk Software License Agreement
Price	Free
Splunk version computability	5.x, 4.3
Support from Splunk	Yes
App URL	`http://splunk-base.splunk.com/apps/57216/splunk-hadoop-connect`
Description	Splunk Hadoop Connect App helps to easily move data between Splunk and Hadoop.
	Three key use cases where you can make use of Hadoop connect are:
	Export data collected in Splunk to HDFS for further batch processing. You can export a subset of collected events from Splunk.
	Navigate HDFS directories and files using the app user interface.
	Import data from HDFS to Splunk and make use SPL and visualization capabilities to build reports, charts, and dashboards.

Splunk App for Microsoft Exchange

Field	Description
Category	Applications, Databases, and Application Servers
App Name	Splunk App for Microsoft Exchange
Author/Vendor	Ahall_splunk
License	Creative Commons BY 3.0
Price	Free
Splunk version computability	5.x, 4.3
Support from Splunk	Yes
App URL	http://splunk-base.splunk.com/apps/28976/splunk-app-for-microsoft-exchange
Description	The Splunk App for Microsoft Exchange consumes logs from Microsoft Exchange environment and provides monitoring capability for Exchange deployments.
	The Splunk App for Microsoft Exchange collects the following data:
	Internet Information Server (IIS) logs for the Exchange servers, Blackberry Enterprise Server (BES) v5.03 logs, Windows Event logs, Security Logs, Exchange audit logs, Application logs, Performance monitoring data, and Senderbase/reputation data.
	Splunk App for Microsoft Exchange provides dashboards that give overview of topology, message tracking, client behavior, operations, capacity planning, identify infrastructure problems, monitor the performance of all servers throughout in messaging environment, track messages throughout in messaging environment monitor client usage, monitor security events, such as virus outbreaks and anomalous logons, track administrative changes to the environment, and analyze long-term mail operations trends.

Technology Add-on for Microsoft SQLServer 2012 and 2008r2

Field	Description
Category	Applications, Databases, and Application Servers
App Name	Technology Add-On for Microsoft SQLServer 2012 and 2008r2
Author/Vendor	Ahall_splunk
License	Creative Commons BY 3.0
Price	Free
Splunk version compatibility	5.x
Support from Splunk	No
App URL	http://splunk-base.splunk.com/apps/63361/technology-add-on-for-microsoft-sql-server-2008r2
	http://splunk-base.splunk.com/apps/62876/technology-add-on-for-microsoft-sql-server-2012
Description	The Add-On for Microsoft SQL Server 2012 and 2008R2 collects database logs related to configuration and audit information.
	These Add-Ons can be used along with Splunk TA for Windows that we have installed and configured in Chapter 2 of this book. When used with Windows TA you can collect data from SQL Server logs and get insights about Microsoft SQL Server 2012 and 2008R2 usage.
	These Add-Ons provide information about the SQL Server configuration and database utilization. You can make use of standard Splunk search to process and analyze information out of security logs and build custom dashboards.
	At the time of writing this chapter, both add-ons are available as a beta release only.

Splunk App for AWS usage tracking

Field	Description
Category	Applications, Databases, and Application Servers
App Name	Splunk App for AWS usage tracking
Author/Vendor	Nkhetia
License	Creative Commons BY 3.0
Price	Free
Splunk version computability	5.x, 4.3, 4.2, 4.1, 4.x
Support from Splunk	No
App URL	http://splunk-base.splunk.com/apps/65926/splunk-app-for-aws-usage-tracking
Description	Splunk App for AWS collects data related to used compute and storage on AWS along with billing data. Collected data through this App helps to analyze AWS usage and also optimize usage of Amazon EC2 environment. AWS App provides several prepackaged reports and dashboards that show billing analysis, spending trends by subaccounts, billing projections, month-over-month billing comparison, spending alerts based on user-specified spend-limit, instance usage, on-demand instances usage, reserved instances usage, unused reserved instances at any hour, instance usage analysis by instance-type, availability zones, tags and subaccounts, baseline calculations, and recommendations.

Splunk for FISMA

Field	Description
Category	Security
App Name	Splunk for FISMA
Author/Vendor	Mike Wilson
License	Creative Commons BY 3.0
Price	Free
Splunk version computability	5.x, 4.3, 4.2, 4.1, 4.x
Support from Splunk	No
App URL	http://splunk-base.splunk.com/apps/44883/splunk-for-fisma
Description	The FISMA App helps enterprises to audit based on NIST 800-53 guidelines.
	Unlike most of the Splunk Apps that capture data, FISMA App is a framework and relies on other technology Add-Ons such as Splunk TA for Windows and the Splunk for Unix and Linux technology add-on (*Nix) to get the data. Other sources of data are proxy servers, firewall, wireless security, vulnerability scanners, network scanners, anti-virus systems, and so on.
	Once the data from external Add-Ons are mapped, FISMA App provides searches that are part of the dashboard where you can view data related to access control, audit and accountability, security assessment and authorization, configuration management, contingency planning, identification and authentication, incident response, personnel security, risk assessment, system and communications protection, and system and information integrity.

Splunk App for PCI Compliance

Field	Description
Category	Security
App Name	Splunk App for PCI Compliance
Author/Vendor	Splunk
License	Splunk Software License Agreement
Price	Contact Splunk
Splunk version computability	5.x, 4.3
Support from Splunk	Yes
App URL	http://splunk-base.splunk.com/apps/55494/splunk-app-for-pci-compliance
Description	The Splunk App for PCI Compliance collects data and provides monitoring capabilities based on the requirements of PCI Data Security Standard (PCI DSS).
	This App provides integration with different technology Add-Ons that can be used to collect data from wireless devices, proxies, firewalls, intrusion detection systems, networking devices, antivirus software, vulnerability management systems, operating systems, and databases such as Oracle.
	The dashboards in the Splunk App for PCI Compliance provide a high-level overview of cardholder data environment and the ability to investigate particular events or compliance issues. Additional dashboards that are included with the App are:
	<ul><li>The PCI compliance posture dashboard, which provides a current compliance status, and can be used to monitor PCI compliance status daily.</li><li>The incident review dashboard helps to identify threats and respond to those threats quickly.</li><li>The audit dashboards validate continuous monitoring of the environment, which helps view audit changes in the incident review dashboard.</li><li>The asset center dashboard helps to identify assets included in cardholder data environment.</li></ul>

Splunk App for Enterprise Security

Field	Description
Category	Security
App Name	Splunk App for Enterprise Security
Author/Vendor	Splunk
License	Commercial
Price	Contact Splunk
Splunk version computability	5.x
Support from Splunk	Yes
App URL	http://splunk-base.splunk.com/apps/22297/splunk-app-for-enterprise-security
Description	Splunk App for Enterprise Security is an App to analyze security data collected in Splunk.
	This App helps in collecting and correlating system performance information with endpoint, network, and access security data. Technology Add-Ons can be used on conjunction with his App to collect security-related data coming from sources such as firewalls, intrusion detection systems, and so on.
	The Splunk App for Enterprise Security is focused on security events data collection, correlation, and provides a monitoring solution with alerts. Users can easily drill down from aggregated data to raw system data, and visualize data using dashboards. Workflow capabilities in the App allows the user to investigate events across data sources. Dashboards help to review incidents, endpoints, network, identity, and audits.

Splunk for RSA SecurID Appliances

Field	Description
Category	Security
App Name	Splunk for RSA SecurID Appliances
Author/Vendor	Joshd
License	Creative Commons BY 3.0
Price	Free
Splunk version computability	5.x, 4.3, 4.2, 4.1, 4.x
Support from Splunk	No
App URL	http://splunk-base.splunk.com/apps/33495/splunk-for-rsa-securid-appliances
Description	This Splunk App provides useful data that relates to the activity happening with RSA SecurID appliances. The App is designed to work with RSA SecurID Appliance 130 and 230 models.
	This App makes use of SNMP to capture data from RSA. The Splunk server indexes the incoming data from SNMP traps and makes the events available for search and analysis.
	This App provides several reports that provide summary data for All Users Accessing the Device(s), Count of Events, Total Failed/Successful Logins, Top Ten Connecting Hosts, and Top Ten Actions.
	In User Activity category reports include:
	Successful Actions, Failed Actions Successful Action Reasons, Failed Action Reasons, Login Failures by User, After Hours events, System Level Actions, Runtime Level Actions, and Admin Level Actions. In Network Activity category reports include Received KBytes by Interface, Transferred KBytes by Interface Total Inbound Packets by Interface, Total Outbound Packets by Interface, Total TCP In/Out Segments, Total UDP In/Out Segments, Total TCP Active/Passive Connections Opened, Total TCP and UDP Error Counts, and so on.

Splunk App for VMWare

Field	Description
Category	Monitoring
App Name	Splunk App for VMware
Author/Vendor	Splunk
License	Splunk Software License Agreement for Splunk for VMWare solution
Price	Free
Splunk version computability	5.x, 4.3
Support from Splunk	Yes
App URL	http://splunk-base.splunk.com/apps/28423/splunk-app-for-vmware
Description	Splunk App for VMWare comes with dashboards and saved searches that provide insights into virtualization layer.
	This App has four main components, The Splunk App for VMWare, TA for vCenter (used to forward vCenter data to Splunk), Splunk forwarder virtual appliance for VMWare also known as Splunk FA VM (used to collect data from VMWare environment) and TA for VMWare (it is packaged with the Splunk FA VM). Making use of these components you can collect logs and performance metrics directly from the vSphere hosts.
	Splunk App for VMWare comes with reports, charts, searches, and indexing definitions for VMWare data, which is collected from the virtualization layer. The App provides operational intelligence by collecting the data from VMWare environment. You can use standard Splunk search to analyze the data and create reports and alerts on the data in the VMWare environment.

Splunk App for HadoopOps

Field	Description
Category	Monitoring
App Name	Splunk App for HadoopOps
Author/Vendor	Splunk
License	Splunk Software License Agreement
Price	Free
Splunk version computability	5.x, 4.3
Support from Splunk	Yes
App URL	http://splunk-base.splunk.com/apps/57004/splunk-app-for-hadoopops
Description	Apache Hadoop typical deployments are run as clusters, which makes it challenging to manage and monitor Hadoop clusters. This is where Splunk App for HadoopOps comes into play, and supports Hadoop releases from Cloudera and Hortonworks.
	This App provides a holistic view of the cluster along with the ability to index individual log events, alerts, notifications to provide monitoring, analysis, and visualization of Hadoop operations.
	This App comes with a set of packaged reports in the dashboard that help to view the current health of the cluster and status of individual nodes and services running within individual nodes. The default dashboard of the App provides a view of cluster health and status. The nodes dashboard provides charts related to nodes within the cluster, the activities dashboard provides current job activity, and the job history dashboard provides the historical job activity.

Splunk App for Active Directory

Field	Description
Category	Monitoring
App Name	Splunk App for Active Directory
Author/Vendor	Ahall_splunk
License	Creative Commons By 3.0
Price	Free
Splunk version computability	5.x, 4.3
Support from Splunk	Yes
App URL	http://splunk-base.splunk.com/apps/51338/splunk-app-for-active-directory
Description	Splunk App for Active Directory provides monitoring capabilities for Windows Server Active Directory (AD) deployment.
	The App collects Windows event logs that include security, application, system, Distributed File System Replication, NT File Replication Services, DNS server AD schema changes, AD forestwide health, information and replication statistics, domain controller health and performance metrics, and DNS server health information.
	The App provides different dashboards. The Operations dashboard includes information about AD topology, domain, and DNS status.
	The Security dashboard provides information about the AD security profile and operations. On top of this, the dashboard also provides information on logon failures, attempts to controvert user security settings, and user utilization, as well as display audits and reports on all AD objects in environment. Prebuilt audit reports provide detail of the current status of users, computers, groups, and group policy objects. The change management dashboard provides details of the recent changes made to the AD environment.

Splunk on Splunk (S.o.S)

Field	Description
Category	Monitoring
App Name	Splunk on Splunk (S.o.S.)
Author/Vendor	Splunk
License	Splunk Software License Agreement
Price	Free
Splunk version computability	5.x, 4.3, 4.2, 4.1, 4.x
Support from Splunk	Yes
App URL	`http://splunk-base.splunk.com/apps/29008/sos-splunk-on-splunk`
Description	Splunk on Splunk (S.o.S.) is an App that makes use of Splunk's monitoring capabilities to analyze and troubleshoot problems in Splunk environment.
	This App gathers data from Splunk logs and other configuration files along with memory and CPU usage for Splunk Web, splunkd, and search processes as well as other system resource information.
	This App provides different reports and dashboards that show information about Splunk, CPU/Memory usage for different Splunk processes.

Splunk for Nagios

Field	Description
Category	Monitoring
App Name	Splunk for Nagios
Author/Vendor	Luke Harris
License	Lesser GPL
Price	Free
Splunk version computability	5.x, 4.3, 4.2, 4.1, 4.x
Support from Splunk	Yes
App URL	http://splunk-base.splunk.com/apps/22374/splunk-for-nagios
Description	Nagios is a monitoring system that enables organizations to identify and resolve IT infrastructure problems.
	Splunk for Nagios App integrates with Nagios and makes it easy to load log files from Nagios server to Splunk.
	This App provides several reports, dashboards, and scheduled saved searches in Splunk to send alerts to Nagios. Some of the dashboards include:
	Status dashboard that shows recent alerts and notifications. Alerts dashboard to display alert history. Host dashboard that has charts to display different metrics such as CPU, memory, swap, load, disk usage, network interface utilization, processes gathered from Nagios. NAS dashboard provides graphs of storage and quota usage. Cisco Network dashboard comes with graphs for network utilization.

Splunk App for Snort

Field	Description
Category	Monitoring
App Name	Splunk App for Snort
Author/Vendor	Ayn
License	Creative Commons BY – NC-SA 2.5
Price	Free
Splunk version computability	5.x, 4.3, 4.2, 4.1, 4.x
Support from Splunk	No
App URL	`http://splunk-base.splunk.com/apps/22369/splunk-for-snort-splunk-4x`
Description	Snort is an open source network intrusion prevention and detection system (IDS/IPS) developed by Sourcefire. Splunk for Snort captures Snort logs and provides dashboards to view the captured data as different reports and charts.
	The most basic feature provided by this App is to extract fields from Snort logs. The following fields are extracted:
	Source IP address, Destination IP address, Source port, Destination port, Network protocol, ID values, Network interface, name, Signature name, Signature category, Signature classification, Signature priority, Type of Service, Packet IP header length, Packet total length. The App includes a custom search interface.
	Map view Splunk for Snort provides a dashboard for viewing geographical location of source IPs that have triggered alerts.

Cisco Security Suite

Field	Description
Category	Networking
App Name	Cisco Security Suite
Author/Vendor	Will Hayes
License	Splunk Software License Agreement
Price	Free
Splunk version computability	5.x, 4.3, 4.2,4.1, 4.x
Support from Splunk	No
App URL	`http://splunk-base.splunk.com/apps/22300/cisco-security-suite`
Description	The Cisco Security Suite App provides a common interface for all the data collected by other Apps and Add-Ons that are part of the Cisco Security Suite.
	The Apps and Add-Ons that are integrated with Cisco security Suite are:
	Splunk for Cisco Client Security Agent (CSA)
	Splunk for Cisco IronPort Email Security Appliance (ESA)
	Splunk for Cisco IronPort Web Security Appliance (WSA)
	Splunk for Cisco Firewalls (PIX, FWSM, ASA)
	Splunk for Cisco IPS
	Splunk for Cisco MARS
	This suite App makes use of one or more Apps or Add-Ons to collect the data from different Cisco appliances. At least one of the other Apps or Add-Ons along with this main App has to be configured to see data.
	This suite App comes with reports, views, and searches that make use of the data captured from various Cisco appliances and indexed into Splunk.

Splunk for Palo Alto Networks

Field	Description
Category	Networking
App Name	Splunk for Palo Alto Networks
Author/Vendor	Monzy
License	Creative Commons BY 3.0
Price	Free
Splunk version computability	5.x
Support from Splunk	No
App URL	http://splunk-base.splunk.com/apps/22327/splunk-for-palo-alto-networks
Description	Splunk for Palo Alto Networks app makes use of data provided by Palo Alto Networks's firewalls and provides a simple to use reporting and analysis tool for Palo Alto Network appliances.
	The Palo Alto devices come with different logs and this Splunk App is designed to work with the default log configuration.
	This App helps users to correlate data across different sources such as application and user activity in different network domains.
	This App provides comprehensive set of reports and dashboards that cover threat details, ability to add and remove particular host, IP addresses, userID, and so on. Splunk standard search mechanism can also be used to create custom reports on top of data captured from Palo Alto Network appliance logs.

Splunk for Juniper Firewalls

Field	Description
Category	Networking
App Name	Splunk for Juniper Firewalls
Author/Vendor	Defensive-ISS
License	Creative Commons BY 3.0
Price	Free
Splunk version computability	5.x, 4.3, 4.2, 4.1, 4.x
Support from Splunk	No
App URL	`http://splunk-base.splunk.com/apps/57437/splunk-for-juniper-firewalls`
Description	The Splunk for Juniper Firewalls App is used for collecting data from Juniper firewall devices ISG and SSG platforms.
	The Splunk for Juniper Firewall App provides a number of different dashboards including:
	Threat summary dashboard, deep inspection summary dashboard, firewall summary dashboard, geographical summary dashboard, and change management dashboard.

Splunk for Barracuda Networks Web Application Firewall

Field	Description
Category	Networking
App Name	Splunk for Barracuda Networks Web Application Firewall
Author/Vendor	Barracuda_networks
License	Creative Commons BY 3.0
Price	Free
Splunk version computability	5.x, 4.3, 4.2, 4.1, 4.x
Support from Splunk	No
App URL	`http://splunk-base.splunk.com/apps/67085/splunk-for-barracuda-networks-web-application-firewall`
Description	Splunk App for Barracuda Networks Web Application Firewall collects logs provided by firewall along with reports and dashboards.
	This App works with version 7.7 or higher versions of the Barracuda Web Application Firewall firmware and collects data using TCP/UDP.
	The real-time dashboards, reports, and search fields can be used to investigate collected data. Specific reports are available to see security, traffic, and audit data. Reports also make use geolocation mapping to show the location of attacking clients.

Index

▓ D

CPSIA information can be obtained at www.ICGtesting.com
Printed in the USA
LVOW091955230513

335257LV00004B/12/P

9 781430 257615